Mathematical
Programming
Applications

MATHEMATICAL PROGRAMMING APPLICATIONS

George Hayhurst

Macmillan Publishing Company
New York
Collier Macmillan Publishers
London

Macmillan Publishing Company
866 Third Avenue, New York, New York 10022

Collier Macmillan Canada, Inc.

Hayhurst, George.
  Mathematical programming applications.

  Includes index.
  1. Programming (Mathematics)   I. Title
T57.7.H388   1987   519.7   87-21571
SBN: 0-02-352740-4

Printing: 1 2 3 4 5 6 7 8   Year: 6 7 8 9 0 1 2 3 4

ISBN 0-02-352740-4

Contents

Dedicated to

CHARLES and CATHRYN, our children

*Who quite rightly observe that book writing interferes
with the proper objective of life, which is to have fun*

Introduction

What is the best way of using our scarce resources?

This is a basic problem for individuals and organizations engaged in industry, business, commerce, economics, and life in general. Mathematical programming provides a partial solution to this problem.

Mathematical programming has proven to be a very powerful tool in the disciplines encompassing business, commerce, and economics. Many successful practitioners within these fields are able mathematicians; but a great many more find the standard mathematical texts too complex to be particularly helpful. At the same time, some texts which at first sight seem readable are later discovered to be simplistic and useless for the real problems encountered.

The same thing is often true of computer software. With the recent explosive growth in the microcomputer industry, many software packages have emerged that claim to cover the routines required by mathematical programming procedures. However, what was acceptable as a program to illustrate a procedure to students, written in BASIC and implemented with an interpreter on, say, an 8K Pet computer in the late 1970s, is totally inappropriate for the 1980s, and was never suitable for real business applications.

Because mathematical programming is a methodology for large-scale problems, it makes sense to use proper commercial packages that have been tested over many applications. It also makes sense to run these

1

programs on large, commercial computers where they can be implemented efficiently. A number of personal computers are capable of interacting with these large computers, allowing the data entry to be carried out on the small computer, instructions to be sent to the host mainframe via the phone system for the "number crunching," with the results sent back to the personal computer for analysis at leisure.

This book uses the LINDO package, which is widely available on many mainframe and minicomputers. It is also available as a restricted package called LINDO/PC for the IBM PC 256K, and as Super LINDO/PC for larger installations. Very large problems are run on mainframes using packages that require the IBM MPS format for input files, so the procedure to convert LINDO files to MPS format is covered in Chapter 15. A brief discussion of other mainframe alternatives is found in Appendix 3.

LINDO interfaces particularly well with personal computers because of the file structure of its input and output. Because students and practitioners will wish to access this package from different sites, both office and home, the method of access shown illustrates the use of an IBM PC (or compatible microcomputer) as representative of a professional machine. For a representative home computer, a Commodore 64 is illustrated. Perhaps the ultimate upgrade for any PC, professional or home, is the addition of a Cyber or other mainframe via a communications package, together with well-tested professional software, at minimal cost.

Because this text is computer-file oriented, the representation of the mathematical variables has been made consistent with computer keyboard inputs. This means that a variable that might normally be represented in conventional mathematical notation as x^{12} will be represented as X12, as this is in line with the input typed on a computer keyboard and handled by the computer package used. Mathematical constants and functions that are conventionally represented by Greek letters will be translated into nonambiguous Roman letters. Italics and boldface will be used only in the event that their absence would be confusing.

COMPUTER HARDWARE REQUIREMENTS

The prices that appear in this section are only for guidance. The hardware market is changing rapidly, and many bargains can be obtained by using "last year's" technology.

Minimal Base (Around $100)

At the most basic level, the LINDO computer package is accessible by using a terminal or a minimal home microcomputer, such as a Timex or Commodore VIC, together with a Vicmodem, emulating a "dumb" terminal. At this level, the problems discussed in the text will need to be typed directly into the computer and read from the screen. With a more sophisticated communications package, small files can be loaded from a cassette tape back-up and output might be saved in the computer memory and on cassette tape. For practical purposes, however, the minimal hardware system will act simply as a dumb terminal without any file-handling capability.

Home Computer (Around $500)

A viable system capable of file preparation, handling, and storage can be obtained from a home computer equipped with a disk drive, modem, and telecommunications package.

To illustrate the approach, I used the family Commodore 64 ($150), 1541 disk drive ($200), Vicmodem, (purchased used for $20), and VIP communications software (currently $44). This price can be improved by substituting for the used Vicmodem and VIP the "Total Communications" package (at a cost of around $50), which includes an autodial autoanswer modem together with file transfer software.

Personal Computer ($2,000 and Up)

To access the LINDO package from an office environment, generate files for the mainframe, and receive from the mainframe output files that can be incorporated into memoranda, reports, and integrated models such as Symphony from Lotus or Ashton Tate's Framework, an IBM PC compatible machine can be used. I used my personal Columbia VP Portable linked to an ancient 300-baud acoustically coupled modem, utilizing the Crosstalk XVI telecommunications package (Microstuf, Inc.). Office environments using the IBM AT with hard disks, in contrast, will probably link to the mainframe using the Crosstalk or Smartcom (by Hayes) package via a Hayes "Smartmodem" operating at 1200 or 2400 baud. Integrated software packages like Framework contain their own communications options. The Mite communications package, which is available within Framework, can be used on its own.

In order to illustrate the minimal machine dependence of the approach used in this text, I prepared some files, some of general text, and some text files of characters required for the input to the LINDO package using the proprietary Commodore word-processing software package "Easyscript." These files were then stored onto the Commodore 64 disk. The VIP program was then loaded from the disk. The California State University, Fullerton, Computer Center dial-in number was called and the Commodore logged on the UNIX system. There, the general text files were downloaded into my UNIX account and my UNIX mailbox files downloaded onto the Commodore disk for later consideration. (UNIX is an operating system developed by Bell Labs and widely used by universities.)

I then logged out of the Fullerton UNIX system and, still holding the connection to Fullerton, linked through to the California State University Central Cyber System in downtown Los Angeles. There, I opened my account, downloaded the LINDO files previously prepared using Easyscript into local files in my workspace on the Central Cyber. I then ran the LINDO package, using these local files as input, and diverted the output from LINDO into other local files. Having completed my tasks on LINDO, I downloaded these local files directly to the Commodore disk for later analysis. Also at this stage, I put the more interesting input and output files into my permanent filestore on the Central Cyber system.

At a later date, I logged on the UNIX system from my hard-wired office terminal and formatted the general text files originally prepared with Easyscript, using NROFF. In this session, I also ran more LINDO models, saving the output in further permanent files on the Central Cyber.

The system was also accessed from my home using my Olivetti portable computer with an internal 1200-baud Hayes modem, linked to a 35-megabyte hard disk and 60-megabyte tape backup, to modify files and download student assignments to the Cyber.

Back home again in Irvine, I downloaded the NROFF ed file from the Fullerton UNIX system, together with the LINDO output files from the Central Cyber, into my Columbia VP portable, converted the resulting DOS files into Easywriter format, and placed the disk containing these files into an IBM PC portable for incorporation into a document. The resulting document was printed from an IBM XT running an expensive printer.

Although this appears to be a long-winded way to prepare a document, it illustrates that the approach of this text is in no way dependent upon your own computer hardware. A student can use a number of different means to access the computer solutions and maintain the same instructions and structure from many diverse locations.

The method illustrated in the text uses the California State University Central Cyber system, which can be accessed by dial-up line to any of the California State University campus sites. Thus mathematical programming optimizations can be solved on the Central Cyber while at home or in the office, with a cellular telephone in the car, and even with a duplex marine radio channel on your yacht! A student can thus obtain the skills needed for successful course completion using very little home computer equipment at minimal entry cost, and will find that these skills need no modification when applied with front-line technology to solve real problems.

The approach used in this book requires only character files; it does not use any graphics. It is therefore suitable for virtually any computer. Those computers used for interaction with databases such as Videotex (with the NAPLPS character sets) can also be used, but only the ASCII characters are needed.

The utility of this approach depends heavily upon the speed of the telecommunications available to the user. At 300 baud, the transmission of small files is of little inconvenience. However, one of the problems solved in Chapter 4 takes a session of some 40 minutes to transmit and receive the files at 300 baud. A 300-baud modem can be obtained very cheaply (about $25); a used 1200-baud modem might be obtained for the same cost, but is likely to cost about four times as much as a used 300-baud modem. A 2400-baud modem is two or three times the cost of a 1200-baud modem at the time of writing, and can only be used with software that can handle 2400 baud.

Current developments in the field of communications will make the approach taken in this book even more attractive. Pacific Bell is evaluating a new technology, code-named "Victoria," which provides a telephone subscriber with two voice channels and five data channels over a single telephone wire. One of these data channels will be capable of transmitting computer data at 9600 baud without a modem. Thus the problem of Chapter 4 that took 40 minutes with a cheap 300-baud acoustically coupled modem could be handled in a minute, without a modem, by a subscriber using this system.

This digression has been included because of the very varied computer-skill levels of prospective users of mathematical programming. Readers with computer experience will find nothing exciting in moving files between different machines. But by far the majority of readers will have little experience with computers, and most of that gained from former stereo salesmen now selling personal computers, who know little of the technicalities of their products. To these readers, the capabilities of such interactions are little short of miraculous. After reading this text, supported by a course of instruction, the only "Gee whiz" effect remaining is "Gosh, is it as easy as that?"

METHOD OF ACCESS BY PERSONAL COMPUTER

(For readers who do not have access to a personal computer, or who wish to use the text by accessing the LINDO package by terminal, this section may be omitted.)

1. Prepare files for input using a word processor or editor.

	Commodore 64	IBM PC
Wordprocessor	Easyscript	Easywriter
		Wordstar
Editor	Screen Editor	Edlin

2. Save the files as system-readable text files on disk. On the Commodore 64 this is automatically achieved by Easyscript. However, in order to aviod problems in the file transfer stage, do not save the text file with tabs. (It is all right to use tabs in preparing the file, and the file will remain tabulated, but reject the option that Easyscript offers of saving the file with a record of tabs.) Using Easywriter on an IBM PC, the saved file will need to be converted to a DOS file using the housekeeping routines of Easywriter. Edlin works directly on DOS files.

3. Load the communications software. Using the Commodore, insert the program disk, type **LOAD"*",8,1**; then when the VIP logo appears (indicating the communications option), enter a carriage return. After loading the Crosstalk files into the PC, the startup command is XTALK. Both systems are well documented, with the VIP terminal program for the Commodore being well supported by on-system help pages. The VIP program is particularly well presented using Apple-MacIntosh-like icons. An icon of a hand with a pointing finger can be maneuvered to an appropriate section using a joystick and zapped using the fire button to emulate mouse control. More conservative operators might prefer to use the cursor position keys. The VIP software will accommodate several modem types and speeds, including the Hayes "Smartmodem." Crosstalk XVI takes a more conservative approach of presenting a table of parameters that can be toggled on and off. In either case, after setting the characteristics of your modem, a file can be created inside the program that will automatically initiate the connection procedures in future calls. Different procedures can be stored for different host computers. Help calls are available in both systems, the VIP program having basic help and icon indicators held in memory, with more detailed explanations held on disk. (However, the VIP can be tedious in paging its responses from the single slow Commodore disk.) Phone calls can be made manually or automatically. Both programs allow autodial and autoanswer options.

4. Dial the host computer system. When the connection is established, the screen of the PC appears exactly the same as an ordinary terminal using an 80-column display. The Commodore can also be set to display 80 columns, but a good-quality monitor is needed to read the resulting display. Using an ordinary television screen, the attempt at 80 columns is unreadable due to the lack of adequate bandwidth available on the TV set. Screens with 60 or 40 columns are available; probably the most convenient is the 40-column set, because the line wraps around, presenting two lines for each normal 80-column line. At this stage, the keyboard of the personal computer is used in exactly the same way as an ordinary terminal of the mainframe, with the screen reflecting the same information.

If you are dialing into one of the California State University computer centers, you will receive an acknowledgment message such as

```
CSU Fullerton Computer Center
Which System ?
```

At this stage, you can request access to any system for which you have authorization—the UNIX system on a machine at Fullerton, or RSTS or NOS systems on other machines at Fullerton. However, the programs you will need to access for LINDO may not be available on the local Cyber, and must be accessed at the Central Cyber Site. To link into the Central Cyber the response is CCS. Thus

```
Which System ? CCS<CR>
CCS -enter 1 <CR> after GO
GO<CR>
CSU X 25 Network, Pad 508102/1
Pad>
com
WELCOME TO THE NOS SOFTWARE SYSTEM.
COPYRIGHT CONTROL DATA 1978, 1984.
[date] [time] A2A0906
CAL STATE CYBER 170-730.              NOS
2.2 - 596C_19.
FAMILY:<CR>
```

(Note: Within the chapters, use of the carriage return after entering commands and responses will be assumed and not printed.)

The information now displayed on your screen shows that you have accessed the X 25 packet switch system to the central Cyber system. The date and time are presented, the host computer shown, and the version of the operating system (NOS) presented. The host computer is awaiting your response to the question "which family?" At this stage you respond with a carriage return. Next you receive the prompts

```
USER NAME: XXXXXX
PASSWORD: XXXXXX
```

Following valid responses in the XXXXXX fields, the host computer continues with two sequences of characters, for example:

```
JSN: AZHL, NAMIAF
```

These are the sequence characters that your job has been allocated. If you lose the connection to the host for any reason while you are still working there, then reconnecting within about 15 minutes will give you the option of continuing your processing at the point where the line was lost. You now have the NOS command prompt "/" on your screen awaiting your next instruction.

If you wish to know what files are currently held in your permanent filestore, respond with the NOS command CATLIST:

```
/CATLIST<CR>
```

and when the names of all your permanent files will be listed.

Suppose that you have a file in there called FRED and you would like to transfer that file to your microcomputer. First, you must get a copy of the file FRED from your permanent filestore and put the copy into your local work area.

```
/GET,FRED<CR>
```

will cause a copy of file FRED to be placed in your local workspace. This local workspace is actually still at the remote host computer site. Suppose that you wish to look at the contents of file FRED prior to saving it in your microcomputer.

```
/COPY,FRED<CR>
```

will cause the contents of file FRED to be copied to the default output, the terminal screen. Because your microcomputer is still behaving as a dumb terminal, the contents will be displayed on your personal computer screen.

If you are satisfied that this is the correct file you want to work on in your microcomputer, you must do two things. First, because you have operated on file FRED, you are now at the end of file FRED, not the beginning. To get back to the start you must "rewind" the file, thus:

```
/REWIND,FRED<CR>
```

(/REWIND,* will reset all your local files.) Next, you must instruct the communications package running on your personal computer that you wish to capture the next set of data transferred. This may be captured into memory or directly to disk. On the Commodore, the icons at the bottom of the screen remind you which of the function keys toggle the workspace or disk open or closed. Using Crosstalk on the PC, keys can be customized, but F5 is the default toggle on and off for capture. If capture directly to disk is desired, then a filename for the new file to be created on the disk is needed. This is achieved by toggling into the communication command mode. The toggle is called the switch key in Crosstalk, and its default setting is the "Home" key (Shift 7) on the PC. Using MITE on the PC, you toggle into the command mode by typing the "CONTROL" and "J" keys together. This "CONTROL" key is depicted as "^" in the instructions.

```
Command?
```

appears on the status line (using Crosstalk), with the communications link held open.

```
Command? CA LOCFRED<CR>
```

will cause the next data set sent to be captured directly to a disk file named LOCFRED on the PC. If a file LOCFRED already exists on the disk, then following the command with / E will erase the old file or with / A will append to the old file.

   Using the Commodore, the procedure is very similar. The Commodore screen has informative icons for the various options available, each option being associated with a function key. F7 is the function key used for file capture.

   Toggling back to talk to the host mainframe by pressing the "Home" key again on the PC, the NOS command

```
/COPY,FRED<CR>
```

will cause the file to be sent to the personal computer once more, but this time, in addition to the characters appearing on the screen, they are also written onto the disk in the file LOCFRED.

Other available options are detailed in the documentation that accompanies the software packages.

Now that files have been successfully captured, the dual procedure of sending files can be considered. There is a little more to consider in this direction. Because files prepared in different ways will contain different control characters such as carriage returns, their handling will be slightly different. Files prepared using Easyscript 64 for the Commodore are the easiest to handle, for they are stored as one long string. The only difficulty encountered with the Commodore will be if you saved the file to disk and also saved the tab information. Remember, you must reject the option of saving the tab-setting information with the file for successful file transfer. When using Crosstalk, read the manual for the option you should use to handle the files you have prepared.

Assuming that you have a file you wish to send to the host computer, and you have decided on the command to use with the PC, then toggle the "Home" key to set up the host computer to recieve the file.

When using the California State University Central Cyber, operating under the NOS operating system, decide on the name you want to give to the file when it is in your workspace at the host computer. Suppose that you decide that JOHN is a good name. You need to tell the host that you want to open a file called JOHN and write data into it. One way to achieve this is to use an editor program on the NOS system. We will illustrate the use of XEDIT to receive the data into file JOHN.

```
/XEDIT<CR>
```

calls the editor.

```
XEDIT 3.1.00
EMPTY FILE / CREATION MODE ASSUMED ??
```

The "??" prompt appears shows that you are no longer talking to the NOS operating system (which uses the "/" prompt). This "??" prompt is the command level prompt for XEDIT, indicating that XEDIT is awaiting your instructions. You wish to indicate that you want XEDIT to enter the input mode; this is achieved by simply pressing the carriage return key. Thus

```
??<CR>
```

will change the prompt to

```
INPUT
?
```

       The new prompt "?" indicates that everything the host computer now receives will be input into the newly opened file. It is now time to set up the personal computer to send the file to the host.
       Toggle the PC using the "Home" key to get into the command mode and issue the command you chose to send the file, together with the filename under which the file is stored in the personal computer. A similar procedure is used by the Commodore, and the user is guided through the needed steps by on-screen help. With MITE, J followed by U will provide on-screen menus to guide the user through the upload procedures.
       The file has now been sent to the host computer and captured into the file we previously opened there. At the end of the data transfer you will see the screen showing the "?" prompt on a single line. You are still in the file-creation mode of XEDIT at the host computer. You must exit this input mode and store the file into your filestore. This is achieved by entering a single carriage return on a line by itself. (For this reason, your input file should not contain blank lines.)

```
?<CR>
```

will cause the input mode to be terminated and a return to the XEDIT command mode.

```
EDIT
??
```

is the response from the host, indicating that XEDIT is now awaiting a further command. You wish to quit XEDIT, saving the file under the label JOHN. This is achieved by

```
??Q,JOHN<CR>
```

On receipt of this command, the host computer returns to NOS and replies with

```
/JOHN IS A LOCAL FILE
```

This indicates that you have a file in your workspace that is labeled JOHN, but as it is only a local file, it will be erased should you log off. (If you are cut off at this stage because of a line fault, you can resume processing by referencing the job numbers you are currently working on if you reestablish the connection within 15 minutes.) In order to place the file into your permanent filestore, you should instruct the system with

```
/SAVE,JOHN
```

which produces the response

```
/JOHN IS A PERMANENT FILE
```

To terminate the session at the host computer, a normal logoff procedure is followed, exactly as would be used for a remote terminal session.

```
/BYE
```

causes the logoff messages to be displayed in your PC.

```
[user name]
UN=XXXXXX   LOG OFF     13.50.03
JSN=APLN           SRU-S    12.595
IAF       CONNECT TIME  00.32.53.

HOST DISCONNECTED   CONTROL
CHARACTER=(C/Z)
ENTER INPUT TO CONNECT TO HOST

cir pad 0

cir conf

channel 05 dest 01010200
time 1326 1359 charge 0
pkt 00100 00387 seg 00100 00944
char 00007 00253 reset 000

CSU X 25 Network, Pad 758101/5
```

You have now successfully logged off the host computer and can hang up the phone line and leave the communications program.

It may seem quite complicated to remember all these different commands, each for a different system, but it is much easier in practice than it appears in describing it. Essentially, there are three elements involved:

1. Your personal computer. At first, this appears to many people to be a complex piece of apparatus to operate. However, with practice and familiarity, it can become a very helpful tool for you. Much advice can be obtained from local computer clubs and from students who have experience with the same machine.

2. The host computer. In universities, students have to become familiar with the operating system of the university computers in order to complete assignments. This is true whether you use a school terminal or your own microcomputer.

3. The communications package. This is simply a device that allows you the option of communicating with host computers, just like a school terminal, or switching into a mode where you issue instructions to the microcomputer to recieve or send files. With practice, this becomes just like using any other program, and is probably easier to use than many popular programs. Many communications packages are available, such as PC TALK. Some are in the public domain of software, and may be freely downloaded from electronic bulletin boards. Crosstalk XVI (for the IBM PC) is capable of transferring files in a more sophisticated manner than the simple method illustrated here. For large files, more error checking might be appropriate, so protocols such as XMODEM are available when using this software. Many schools also support other protocols that allow error-checked file transfers. KERMIT might be available on your host system. The support staff of your host computer center will be able to provide guidance on the packages supported there.

THE PLAN OF THE TEXT

The following workplan is suggested for a 15-week semester course.

Week	Chapter	Topics
1	Introduction 1	Introduction to the school computer system and its operating system
		Graphical solutions to linear programming
2	1	Simplex equations, the LINDO package
3	1	Simplex tableaux, LINDO solutions
4	2	Alternative constraints and duality
5	3	Parametric programming
6		Review and midterm assessment
7	4 and 5	Transportation assignment
8	6	Decomposition
9	7	Games theory
10	8	CPM/PERT
11	9	Integer techniques
12	10	Nonlinear techniques
	11	Quadratic programming
	12	General optimization
13	13	Dynamic programming
14	14	Goal and stochastic programming
15		Review and project submission Final examination

1

Linear Programming

Linear programming is a type of mathematical programming in which the mathematical functions describing a constrained optimization problem are all linear. This linearity is best described with an example.

Suppose that a food processing firm makes only two products, cooking oil and margarine. In order to produce these two products, raw materials, in the form of sunflower oil, are converted using labor in the firm's plant. Since there are limited amounts of these resources available, they therefore act as constraints on the options of the firm's product mix.

Consider the sunflower oil resource. This is in short supply and the firm has only 6 tons available for the time period under consideration. This is sufficient to make up to 6 tons of cooking oil or 6 tons of margarine, or amounts of both that add up to that sum. This is defined as a linear relationship, because if all the alternatives were plotted out on a graph with cooking oil on the vertical axis and margarine on the horizontal axis, they would fall on a straight line (see Figure 1.1).

The capacity of the plant, another resource, also acts as a constraint on how management decides on the best product mix. In the case under consideration, the capacity of the plant is such that it can produce up to

Figure 1.1

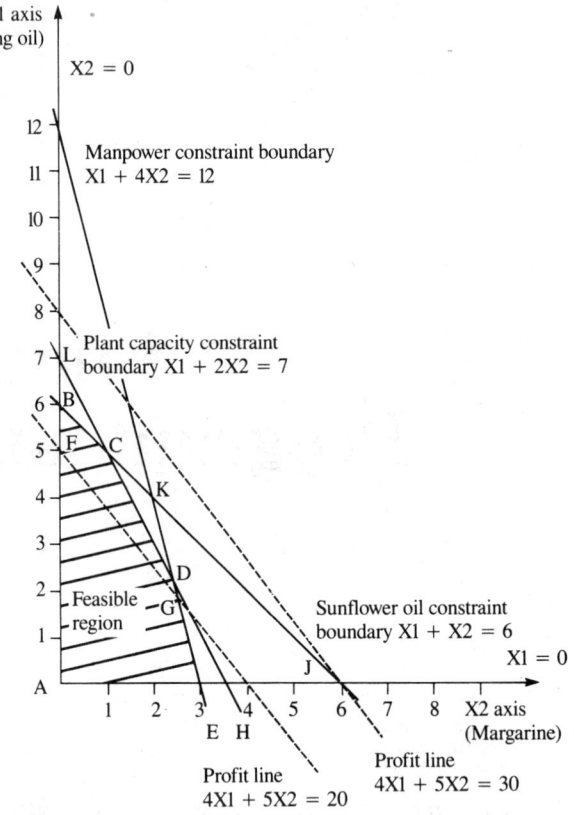

X1 axis
(Cooking oil)

X2 = 0

Manpower constraint boundary
X1 + 4X2 = 12

Plant capacity constraint
boundary X1 + 2X2 = 7

Feasible
region

Sunflower oil constraint
boundary X1 + X2 = 6

X1 = 0

X2 axis
(Margarine)

Profit line
4X1 + 5X2 = 20

Profit line
4X1 + 5X2 = 30

7 tons of cooking oil, or up to 3½ tons of margarine, or amounts prorated in between, another linear relationship shown in Figure 1.1.

The labor needed to process margarine is 4 times that needed for cooking oil, and there are 12 man-hours available, sufficient to make 12 tons of oil.

These are the constraints on the solution. All are linear. Of the options the firm might choose in order to convert their limited resources into products, some are more attractive than others. In this case, the contribution to profit achieved by the conversion of 1 ton of margarine is $5; for cooking oil it is $4. How can profit be maximized?

Let the amounts made of the two products be designated as X1 tons of cooking oil and X2 tons of margarine. These are the variables X1 and X2. (You may be familiar with the conventional notation for variables, which displays them as italicized lower-case letters with subscripted numbers. In this text we are using a much simpler form of notation that more accurately reflects the computer orientation of contemporary linear programming.) Considering the sunflower oil constraint, X1 plus X2 may add up to 6, that is, be less than or equal to 6. This may be written as the linear expression

$$X1 + X2 \leq 6$$

because if only cooking oil were produced, then X1 could take a value up to 6, and vice versa for X2. Similarly, considering all the other constraints, the expressions may be written as

$X1 + X2 \leq 6$	Sunflower oil constraint
$X1 + 2\,X2 \leq 7$	Plant capacity constraint
$X1 + 4\,X2 \leq 12$	Labor constraint

There are two further constraints on the solution, namely that negative amounts of product may not be manufactured. That is, both X1 and X2 must be greater than or equal to zero, never below it:

$$X1 \geq 0 \text{ and } X2 \geq 0$$

Note: The sign $\leq$ signifies "less than or equal to"; $\geq$ signifies "greater than or equal to."

These constraints set the parameters for the solutions that are feasible for the problem. This feasible region is bounded by the straight lines

$$
\begin{aligned}
X1 + X2 &= 6 \\
X1 + 2\,X2 &= 7 \\
X1 + 4\,X2 &= 12 \\
X1 \phantom{{}+ 4\,X2} &= 0 \\
X2 &= 0
\end{aligned}
$$

This region is depicted in Figure 1.1.

The graphical representation of the constraints shows that any point, defined by values of X1 and X2, that lies within the feasible region satisfies the constraints of the problem. As there are many such points, the aim is now to decide which is the best point for the problem. For this case, the profit on the two products is considered. The total profits Z produced by making X1 tons of cooking oil and X2 tons of margarine is

$$Z = 4\,X1 + 5\,X2$$

Hence, we want a point inside the feasible region that maximizes this function of Z and that is also of the same linear form as the constraints.

Thus, for two different levels of profit, Z will take different values, say Za and Zb. Therefore, the two lines

$$Za = 4\,X1 + 5\,X2 \text{ and}$$
$$Zb = 4\,X1 + 5\,X2$$

will be parallel. Let us set Z at $20 and $30. The two corresponding profit lines,

$$4\,X1 + 5\,X2 = 20 \text{ and}$$
$$4\,X1 + 5\,X2 = 30$$

are shown in Figure 1.1. Observe that all the points (X1,X2) lying along the portion of $4\,X1 + 5\,X2 = 20$ between the points F and G will give a profit of $20. It is not possible to obtain a profit of $30 because no part of the line $4\,X1 + 5\,X2 = 30$ enters the feasible region.

In order to maximize profit, we must choose the line with maximum Z that has a gradient equal to $-\frac{5}{4}$ (the ratio of the coefficients of X1 and X2) and yet enters the feasible region. This will occur at point C when $X1 = 5$ and $X2 = 1$, corresponding to a policy of manufacturing 5 tons of cooking oil and 1 ton of margarine. A maximum profit of $(4 \times 5) + (5 \times 1)$ = $25 results.

It will be noted that the solution is on the boundary of the feasible region and also coincides with at least one intersection of two constraints. (More intersections are obtained when the gradient of the profit line is equal to the gradient of the constraint and they coincide at the optimum solution.)

This simple representation is all very well for a two-variable problem where a geometrical picture of the problem may be drawn with the two variables represented by the two dimensions of the graph. With three, four, or more variables it becomes impossible to draw multidimensional graphs (a representation of a three-dimensional case is shown in Figure 1.5), and this approach cannot be used for these larger and more common problems. An alternative, algebraic method must be used.

Returning to the constraint representations:

$X1 + X2 \leq 6$	(i)	**Equation Set 1.1**
$X1 + 2\,X2 \leq 7$	(ii)	
$X1 + 4\,X2 \leq 12$	(iii)	
$X1 \geq 0$	(iv)	
$X2 \geq 0$	(v)	

As inequalities are cumbersome to handle, these representations can be re-written as equalities by the addition of "slack" variables, which take whatever value is necessary to preserve equality:

$X1 + X2 + X3 = 6$	(i)	**Equation Set 1.2**
$X1 + 2\,X2 + X4 = 7$	(ii)	
$X1 + 4\,X2 + X5 = 12$	(iii)	

If only positive or zero values of any of the variables are allowed, then Equation Sets 1.1 and 1.2 are identical. In this representation, X3 is the amount of unused resource (i), sunflower oil, for any feasible choice of product amounts X1 and X2, X4 is the amount of unused plant capacity, and X5 is the amount of unused labor for any solution.

It was previously noted that the optimum solution in a totally linear case would occur at an intersection of two constraint boundaries. This is equivalent to noting that the optimum solution will use up at least two of the slack variables when the optimum solution produces positive values of all products.

It will be observed that if any two of the five variables take the value zero, then three simultaneous equations are formed in three unknowns. The solution of the simultaneous equations then corresponds to an intersection of two of the constraints—in particular, those whose variables were set to zero. For example, $X3 = 0$, $X4 = 0$ produces the intersection of

$$X1 + X2 = 6 \text{ with}$$
$$X1 + 2\,X2 = 7$$

$X1 = 0$, $X2 = 0$ produces the origin—that is, the initial condition before anything is produced. To elaborate on this point, Equation Set 1.2 is composed of three equations and five variables; thus there are many solutions satisfying them. Rewriting this set as

Equation Set 1.3

$$X3 = 6 - X1 - X2 \qquad \text{(i)}$$
$$X4 = 7 - X1 - 2\,X2 \qquad \text{(ii)}$$
$$X5 = 12 - X1 - 4\,X2 \qquad \text{(iii)}$$

and then adopting the convention that the variables to the right of the equality sign should be set to zero, we obtain three equations in three unknowns that have the solution

$$X3 = 6$$
$$X4 = 7$$
$$X5 = 12$$
$$X1 = 0$$
$$X2 = 0$$

This corresponds to point A in Figure 1.1 on the boundary of the feasible region, at an intersection, in this case the origin.

Exchanging one variable from the left-hand side with one variable from the right-hand side of the equality sign will produce another point on the boundary of the feasible region at another intersection. However, there is the problem of selecting which two variables to exchange in order to optimize the solution. The profit function, or optimization function, was written as

$$Z = 4\,X1 + 5\,X2$$

Rewriting this as

$$X0 = 0 - 4\,X1 - 5\,X2$$

$X0$ will be equal to $-Z$, or negative profit. The purpose of this apparent clumsiness is to enable the set of equations to be written consistently in the format

Equation Set 1.4

$$X0 = 0 - 4\,X1 - 5\,X2 \qquad \text{(i)}$$

$$X3 = 6 - X1 - X2 \qquad \text{(ii)}$$
$$X4 = 7 - X1 - 2\,X2 \qquad \text{(iii)}$$
$$X5 = 12 - X1 - 4\,X2 \qquad \text{(iv)}$$

Remembering the convention that variables to the right-hand side of the equality take the value zero, Equation Set 1.4 may be interpreted as

$$X3 = 6, X4 = 7, X5 = 12, X1 = 0, X2 = 0, X0 = 0$$

that is, make no product, have all resources unused, and make no profit. This is a representation of the initial state, before any decisions are made. How may this position be improved?

Consideration of Equation 1.4(i) shows that for every extra unit (ton) of X1 manufactured, a profit of $4 is recovered, while every ton of X2 produces an extra $5. Hence it is initially more profitable to produce X2, that is, proceed along the X2 axis in Figure 1.1. The question now is, how many units, or how far along the axis?

The answer, from looking at the graph, is to point E, the intersection of X1 = 0 and the labor constraint boundary, X1 + 4 X2 = 12. This is the solution of the simultaneous equations

$$X1 \qquad = \quad 0 \text{ with}$$
$$X1 + 4 X2 = 12$$

that is, the point X1 = 0, X2 = 3.

If we were to proceed farther than X2 = 3, say to point H at X1 = 0, X2 = 3.5, or to point J at X1 = 0, X2 = 6, we would be at the intersection of X1 = 0 with other constraint boundaries: H is the solution of the simultaneous equations

$$X1 \qquad = \quad 0 \text{ with}$$
$$X1 + 2 X2 = 7 \ (X1 = 0; X2 = 3.5)$$

and J is the solution of

$$X1 \qquad = \quad 0 \text{ with}$$
$$X1 + X2 = 6 \ (X1 = 0; X2 = 6)$$

Clearly, from our figure, E is the correct choice, as this constraint is reached first. Returning to Equation Set 1.4, points E, H, and J are produced from the equations by substituting X2 for X5, X4, and X3, respectively. Without the prior knowledge gained by inspecting the graph, how can we make the correct substitution? The answer is that we must use the equation that indicates the smallest amount of the variable being introduced into the solution.

A wrong substitution will result in one or more of the variables taking a negative value, which is not permitted by the structure. That would be an infeasible solution. If one of the product variables, also referred to as a *structural variable,* became negative, the solution would call for a negative amount of product, and if a slack variable became negative, the solution would require a negative amount of that spare resource and call for more of the raw material than is available.

A useful pointer to the correct choice of equation for substitution is to take the one that produces the smallest positive value when -1 times the coefficient of the variable to be exchanged is divided into the constant term of the equation. This is much simpler than it sounds. For instance, let this ratio be designated Q. (The conventional notation for this ratio is θ or theta; again, we are using a notation more compatible with computer-assisted linear programming.) For Equation Set 1.4 we have

Equation Set 1.5

$$X0 = 0 - 4\,X1 - 5\,X2 \tag{i}$$

$$
\begin{aligned}
X3 &= 6 - X1 - X2 & Q &= (6/1) = 6 & \text{(ii)}\\
X4 &= 7 - X1 - 2\,X2 & Q &= (7/2) = 3.5 & \text{(iii)}\\
X5 &= 12 - X1 - 4\,X2 & Q &= (12/4) = 3 & \text{(iv)}
\end{aligned}
$$

To summarize, the variable chosen to be exchanged is the one with the largest negative coefficient in Equation 1.5(i). This is known as the pivot column (here X2). The variable chosen to be exchanged for this variable, X2, is that with the smallest Q, in this case X5. This is known as the pivot row, Equation 1.5(iv). The coefficient of the variable where both the pivot row and the pivot column intersect is known as the pivot element, in this case -4, but more of this later.

We have decided to rewrite Equation Set 1.5 so that X2 appears on the left-hand side in place of X5. Thus we have, from Equation 1.5(iv),

$$X5 = 12 - X1 - 4\,X2$$

which becomes

$$4\,X2 = 12 - X1 - X5$$

or

$$X2 = 3 - 1/4\,X1 - 1/4\,X5$$

On substituting this expression for X2 in all the other equations, we arrive at

$$X0 = 0 - \quad 4\,X1 - 5(3 - 1/4\,X1 - 1/4\,X5)$$

$$X3 = 6 - \quad X1 - (3 - 1/4\,X1 - 1/4\,X5)$$
$$X4 = 7 - \quad X1 - 2(3 - 1/4\,X1 - 1/4\,X5)$$
$$X2 = 3 - 1/4\,X1 - \qquad\qquad\qquad 1/4\,X5$$

which results in

$$X0 = -15 - 2.75\,X1 + 1.25\,X5 \qquad\qquad \text{(i)} \qquad \textbf{Equation Set 1.6}$$

$$X3 = \quad 3 - 0.75\,X1 + 0.25\,X5 \qquad\qquad \text{(ii)}$$
$$X4 = \quad 1 - 0.5\ \ X1 + 0.5\ \ X5 \qquad\qquad \text{(iii)}$$
$$X2 = \quad 3 - 0.25\,X1 - 0.25\,X5 \qquad\qquad \text{(iv)}$$

The presentation of these equations may be further improved for convenience in writing and interpretation by using the semitabular presentation

		X1	X5	
X0 =	−15	−2.75	+1.25	(i)
X3 =	3	−0.75	+0.25	(ii)
X4 =	1	−0.5	+0.5	(iii)
X2 =	3	−0.25	−0.25	(iv)

Equation Set 1.7

This solution corresponds to point E in Figure 1.1, where X3 = 3, X4 = 1, X2 = 3, X1 = 0, and X5 = 0, with a profit of $15.

We may now proceed to examine this Equation Set in exactly the same manner as the Equation Set 1.4 was examined, which is to consider Equation 1.7(i) and search for negative coefficients. The most negative (here the only one) in this case is −2.75 for X1. This indicates that X1 should be brought into the solution to further improve the value of the optimization function, and that each extra unit of X1 produced will improve the profit by $2.75. (Each unit of X1 does contribute $4 to the profit, but to make X1 requires that some of the other product being produced be displaced from the current solution, thus the $2.75 is the *net* increase in profit resulting from such a rearrangement.)

X1 is our new pivot column for this iteration. The calculation for Q indicates Equation 1.7(iii) as the pivot row, so X1 should be exchanged for X4. The iteration becomes

$$0.5\,X1 = 1 - X4 + 0.5\,X5$$
$$X1 = 2 - 2\,X4 + X5$$

that is

$$X0 = -15 - 2.75(2 - 2\,X4 + X5) + 1.25\,X5$$

$$
\begin{aligned}
X3 &= 3 - 0.75(2 - 2\,X4 + X5) + 0.25\,X5 \\
X1 &= 2 - 2\,X4 + X5 \\
X2 &= 3 - 0.25(2 - 2\,X4 + X5) - 0.25\,X5
\end{aligned}
$$

which becomes

Equation Set 1.8

		X4	X5	
X0 =	−20.5	+5.5	−1.5	(i)
X3 =	1.5	1.5	−0.5	(ii)
X1 =	2	−2	1	(iii)
X2 =	2.5	0.5	−0.5	(iv)

This is equivalent to point D in Figure 1.1, with the solution X1 = 2, X2 = 2.5, X3 = 1.5, X4 = 0, X5 = 0, and profit = \$20.50.

At this solution, all the plant capacity and labor are fully used up, but 1.5 tons of sunflower oil are unused. Equation 1.8(i) shows that this solution can be further improved because X5 has a negative coefficient, indicating that a better utilization of resources is possible and indicating a new pivot column. Iteration of the equations in exactly the same manner as before produces

Equation Set 1.9

		X4	X3	
X0 =	−25	1	3	(i)
X5 =	3	3	−2	(ii)
X1 =	5	1	−2	(iii)
X2 =	1	−1	1	(iv)

As there are no negative coefficients in the X0 row, Equation 1.9(i), this is the optimum solution, which maximizes the profit function for this problem. It corresponds to point C in Figure 1.1 and is interpreted as X1 = 5, X2 = 1, X5 = 3, X4 = 0, X3 = 0, with a profit of $25.

CONVEXITY

At this point it is useful to consider the question of convexity. For the above procedure to discover the true optimum solution to the constrained optimization problem, it is important that the constraints bounding the feasible region form a convex set. This is defined as follows: It must be possible to take any two points within the convex set (the feasible region) and join them with a straight line segment that never leaves the feasible region. This is illustrated in two dimensions by Figure 1.2.

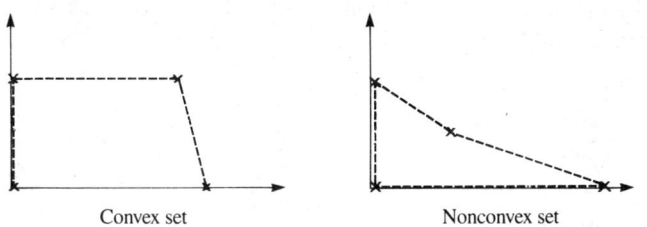

Convex set   Nonconvex set

Figure 1.2

It is important that linear programming be applied only to convex feasible regions, or this method of linear programming might fail to determine the optimum point. For example, see Figure 1.3. The procedure just described, the basis of the simplex method of linear programming, may be likened for illustration to an attempt to discover the highest point of an irregularly shaped field on a planar hillside, bounded by straight walls. As the

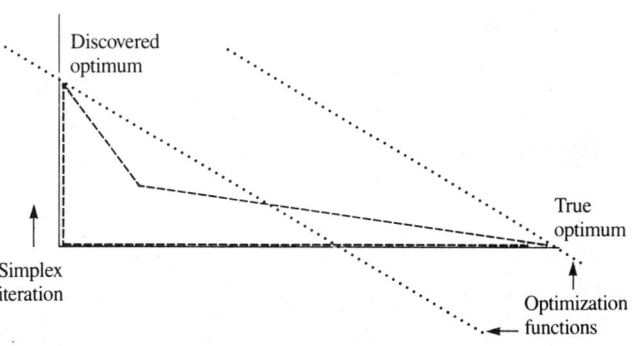

Figure 1.3

hillside is shrouded in mist, the only information available to a person starting at the junction of the two walls at the bottom of the hill is the steepness of the slopes in the direction of each boundary wall. By keeping to the wall boundaries, the person will eventually find the wall intersection at the highest point of the field. The best way to proceed is to choose the wall ascending most steeply. When the wall from the next intersection starts to descend, one can be sure the top point has been reached given a convex set of constraints, but possibly not given a nonconvex set.

With constraints of the form that we are discussing—half-spaces cut by a linear constraint, where the line, plane, or hyperplane (a plane in more than three dimensions) cuts the space into two parts—the feasible region will be bounded by a set of points that form a convex set. It is important to note that one or more of these points might be at infinity. When this is coupled with an optimization function that attempts to select such a point, the problem is called *unbounded*. Alternatively, the constraints might restrict the solutions in such a way that no feasible region can be defined. In this case, the problem is said to be *inconsistent*. These situations are illustrated in two dimensions by Figure 1.4.

Figure 1.4

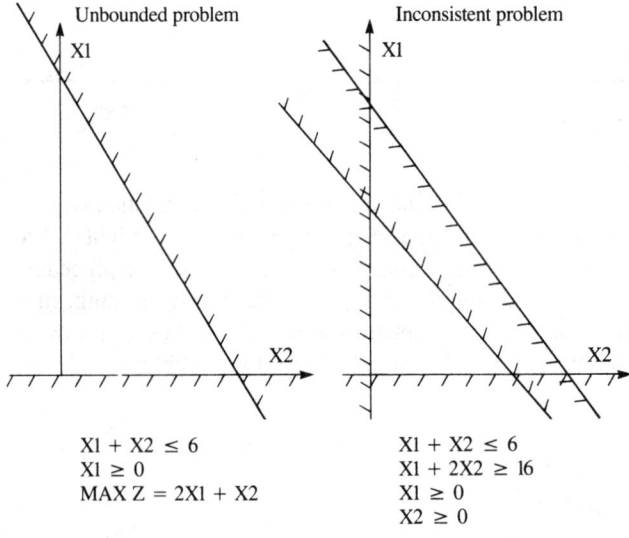

Unbounded problem

$X1 + X2 \leq 6$
$X1 \geq 0$
MAX $Z = 2X1 + X2$

Inconsistent problem

$X1 + X2 \leq 6$
$X1 + 2X2 \geq 16$
$X1 \geq 0$
$X2 \geq 0$

NOTE: As X2 is unrestricted in sign Z may be increased to ∞.

MORE THAN TWO VARIABLES

When the problem is expanded in number of variables, the graphical problem increases in dimensionality. With three variables, a three-dimensional graph is necessary. A four-variable problem neccessitates a four-dimensional graph, which is somewhat difficult to handle. For illustration, we will expand the problem with cooking oil and margarine by adding an additional product; baking fat, at a profit of $3 per ton, made from the same raw materials, so that the relationships defining the problem now become, for example,

Maximize	$4 X_1 + 5 X_2 + 3 X_3$	(i)	**Equation Set 1.10**
subject to	$X_1 + X_2 + 4 X_3 \leqslant 6$	(ii)	
	$X_1 + 2 X_2 + X_3 \leqslant 7$	(iii)	
	$X_1 + 4 X_2 + 2 X_3 \leqslant 12$	(iv)	
with	X_1, X_2, X_3 all $\geqslant 0$		

The graphical representation of this problem is shown in Figure 1.5. The boundaries of the constraints are now six planes:

$X_1 + X_2 + 4 X_3 = 6$	(i)	**Equation Set 1.11**
$X_1 + 2 X_2 + X_3 = 7$	(ii)	
$X_1 + 4 X_2 + 2 X_3 = 12$	(iii)	
$X_1 = 0$	(iv)	
$X_2 = 0$	(v)	
$X_3 = 0$	(vi)	

The points bounding the feasible region are now given by the points of intersection of these planes with each other. The intersection of the first three planes,

$X_1 + X_2 + 4 X_3 = 6$	(i)	**Equation Set 1.12**
$X_1 + 2 X_2 + X_3 = 7$	(ii)	
$X_1 + 4 X_2 + 2 X_3 = 12$	(iii)	

lies at the solution of these three simultaneous equations (because the point must be on all these three planes simultaneously). There will be a unique solution to these equations provided that none of the planes is parallel to any other. This is defined as the point $X_1 = 2$; $X_2 = 2.2857$; $X_3 = 0.4286$, which is conveniently written as (2,2.2857,0.4286).

Figure 1.5

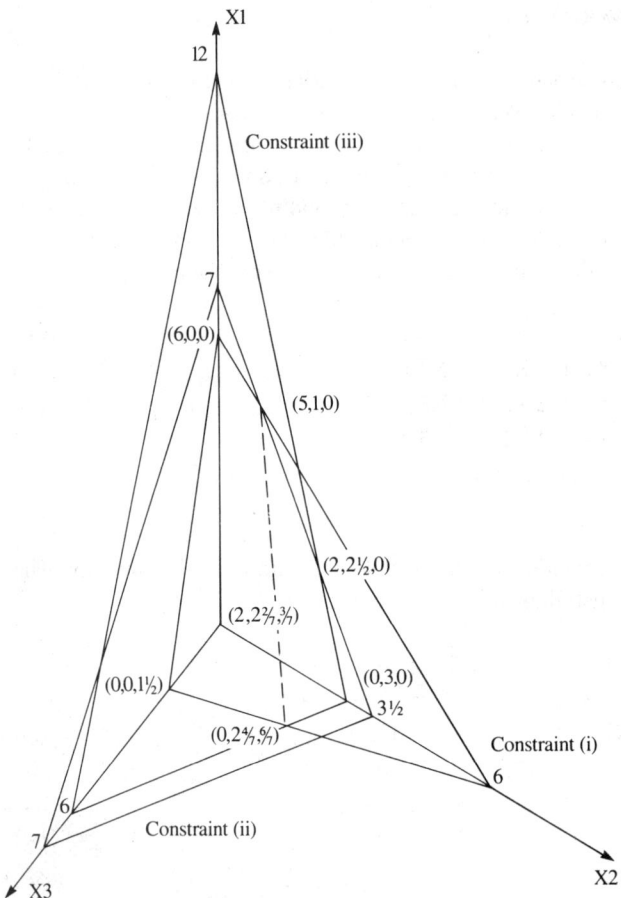

The intersection of the three planes defined by the first, third, and fourth constraints,

Equation Set 1.13

$$X1 + X2 + 4X3 = 6 \qquad \text{(i)}$$
$$X1 + 4X2 + 2X3 = 12 \qquad \text{(ii)}$$
$$X1 = 0 \qquad \text{(iii)}$$

is the point (0,2.5714,0.8571); other points are similarly determined. As in the two-dimensional case, not all the intersections define the boundaries of the feasible region.

SIMPLEX EQUATION SOLUTION OF THE THREE-VARIABLE PROBLEM

As in the two-variable case, the three-variable problem may be conveniently presented as the task of maximizing an objective (optimization) function:

Maximize	$4 X1 + 5 X2 + 3 X3$	(i)	**Equation Set 1.14**
subject to	$X1 + X2 + 4 X3 \leqslant 6$	(ii)	
	$X1 + 2 X2 + X3 \leqslant 7$	(iii)	
	$X1 + 4 X2 + 2 X3 \leqslant 12$	(iv)	

Introducing slack variables X4, X5, and X6, the solution proceeds in exactly the same fashion as previously. (Caution: In the two-dimensional problem, X3 was a slack variable; here it is a structural variable representing a product.) The set of equations

$X0 = 0 - 4 X1 - 5 X2 - 3 X3$	(i)	**Equation Set 1.15**
$X4 = 6 - X1 - X2 - 4 X3$	(ii)	
$X5 = 7 - X1 - 2 X2 - X3$	(iii)	
$X6 = 12 - X1 - 4 X2 - 2 X3$	(iv)	

represents the origin point (0,0,0) with a profit of $0. This may be iterated as before to

	X1	X6	X3		**Equation Set 1.16**
$X0 =$	-15	-2.75	$+1.25$	-0.5	(i)
$X4 =$	3	-0.75	0.25	-3.5	(ii)
$X5 =$	1	-0.5	0.5	0	(iii)
$X2 =$	3	-0.25	-0.25	-0.5	(iv)

representing the point (0,3,0) and a profit of $15 (compare with Equation Set 1.7).

Developing further,

	X5	X6	X3		**Equation Set 1.16**
$X0 =$	-20.5	$+5.5$	-1.5	-0.5	(i)
$X4 =$	1.5	1.5	-0.5	-3.5	(ii)
$X1 =$	2	-2	1	0	(iii)
$X2 =$	2.5	0.5	-0.5	-0.5	(iv)

we arrive at the point (2,2.5,0) with a profit of $20.50 (compare with Equation Set 1.8). Finally,

Equation Set 1.18

		X5	X4	X3	
X0 =	− 25	1	3	10	(i)
X6 =	3	3	−2	−7	(ii)
X1 =	5	1	−2	−7	(iii)
X2 =	1	−1	1	3	(iv)

is interpreted as the point (5,1,0) in the three dimensions of Figure 1.5. It is equivalent to the solution:

Maximum of objective function (profit) = $25

Structural Variables (product values)	Slack Variables (unused resources)
X1 (tons of cooking oil) = 5	X4 (tons sunflower oil) = 0
X2 (tons of margarine) = 1	X5 (plant capacity) = 0
X3 (tons of baking fat) = 0	X6 (labor man-hours) = 3

SIMPLEX TABLEAUX

To perform linear programming by rewriting equations in terms of different variables is quite time consuming. Accordingly, several variations of coding methods have been devised. Many textbook authors have their own pet formulations, and their texts have slight but irritating differences, much to the annoyance of students. In general, the coding methods fall into two groups: those using an extended simplex tableau and those using a condensed form.

The Extended Simplex Tableau

Simplex iterations are performed using a tableau or graphical form such as that shown in Figure 1.6. The simplest way of describing its completion is to discuss the entries alongside the example, that is,

Equation Set 1.19 Maximize $4 X1 + 5 X2 + 3 X3$ (i)

subject to $X1 + X2 + 4 X3 \le 6$ (ii)

$X1 + 2 X2 + X3 \le 7$ (iii)

$X1 + 4 X2 + 2 X3 \le 12$ (iv)

Figure 1.6

Insertion of slack variables gives us

Maximize	$4 X1 + 5 X2 + 3 X3 + 0 X4 + 0 X5 + 0 X6$	(i)	**Equation Set 1.20**
subject to	$X1 + X2 + 4 X3 + X4 = 6$	(ii)	
	$X1 + 2 X2 + X3 + X5 = 7$	(iii)	
	$X1 + 4 X2 + 2 X3 + X6 = 12$	(iv)	

The optimization function has the zero coefficients of the slack variables included. The six variables are included in the extended simplex tableau thus:

Cj			4	5	3	0	0	0		**Tableau 1.1**
Ci	Basis	B	X1	X2	X3	X4	X5	X6	Q	
0	X4	6	1	1	4	1	0	0		
0	X5	7	1	2	1	0	1	0		
0	X6	12	1	4	2	0	0	1		
	Zj									
	Zj − Cj									

The Cj row is composed of the coefficients of the optimization function. The basis column shows the variables that form the basis of the solution at this point. The other variables, not included in the basis column, are referred to as *nonbasic variables* and take the value zero at this point. The B column shows the value of the basic variables for this solution. These entries are equivalent to the right-hand side values appearing in Equation Set 1.20. The Ci column contains the coefficient of the corresponding basic variable in the optimization function (the same as the Cj appearing above the variable's column entry). The remainder of the tableau entries correspond to the left-hand side coefficients appearing with the variable column heading. It now remains to calculate the remaining entries for the initial tableau.

The Z_j row entries evaluate the amount of decrease in the objective function caused by disturbing the current solution enough to reallocate sufficient resources to produce one unit of X_j (where X_j is X_1 or X_2 etc.). Each entry in the Z_j row is evaluated by multiplying each X_i entry in the body of the matrix by its corresponding value in row C_i, and then summing these for the column. Hence

$$
\begin{array}{llll}
Z_1 = & 0 \times 1 & Z_2 = & 0 \times 1 & Z_3 = & 0 \times 4 & Z_4 = & 0 \times 1 \\
 & + 0 \times 1 & & + 0 \times 2 & & + 0 \times 1 & & + 0 \times 0 \\
 & + 0 \times 1 & & + 0 \times 4 & & + 0 \times 2 & & + 0 \times 0 \\
 = & 0 & = & 0 & = & 0 & = & 0
\end{array}
$$

and so forth. In this initial tableau all entries in the Z_j row will be zero, since each C_i entry is zero, corresponding to a slack variable.

Since each Z_j value is the amount by which the optimization function is reduced to allow one unit of nonbasic variable to enter the basis, the value of $Z_j - C_j$ is the net reduction when the contribution to the objective function of one unit of the nonbasic variable is added. It is therefore the net overall reduction of the objective function caused by the introduction of one unit of X_j.

In order to improve the solution, we want to introduce the variable that gives us the minimum net decrease in the optimization function, that is, the variable with the largest negative coefficient in the $Z_j - C_j$ row. This is identified as the pivot column. Completing the tableau gives

Tableau 1.2

Cj			4	5	3	0	0	0	Q
Ci	Basis	B	X1	X2	X3	X4	X5	X6	
0	X4	6	1	1	4	1	0	0	6
0	X5	7	1	2	1	0	1	0	3.5
0	X6	12	1	4	2	0	0	1	3 *
	Zj	0	0	0	0	0	0	0	
	Zj − Cj		−4	−5	−3	0	0	0	
				**					

Note that the asterisks indicate the pivot row and pivot column.

The Q column is evaluated, as in the case of the equations, to discover which row will determine the smallest amount of the variable defined by the pivot column to enter the basis. It is evaluated by dividing each element of the pivot column into each corresponding element of the B column. The row with the smallest value is the pivot row. The element at the intersection of the pivot column ($X2$) and the pivot row ($X6$) is known as the *pivot element* (4).

Compare this tableau with Equation Set 1.15 depicting the solution at the point (0,0,0) of Figure 1.5. Note that, without column $X3$, the tableau depicts Equation Set 1.4 at the solution point (0,0,) shown in Figure 1.1.

The entry in the Z_j row under the B column is a convenient place to record the value of the objective function for this solution point. It is evaluated by multiplying the B entries by the corresponding C_i entries and summing them; in this case the objective function value is zero.

The current tableau shows the solution

$$X1 = 0, X2 = 0, X3 = 0, X4 = 6, X5 = 7, X6 = 12.$$

To iterate the tableau, we must replace the variable ($X6$) in the basis column of the pivot row with the variable in the pivot column ($X2$). We must transfer its corresponding C_j coefficient into the C_i column and divide every element of the pivot row by the pivot element (4) so that the row is transformed from

Ci	Basis	B	X1	X2	X3	X4	X5	X6
0	X6	12	1	4	2	0	0	0

to

Ci	Basis	B	X1	X2	X3	X4	X5	X6
5	X2	3	0.25	1	0.5	0	0	0.25

Note the correlation with Equation Set 1.16(iv).

The other elements within the body of the tableau are next manipulated so that the coefficients of $X2$ become zero. Consider the line corresponding to the first constraint:

Basis	B	X1	X2	X3	X4	X5	X6
X4	6	1	1	4	1	0	0

A zero in the $X2$ coefficient will be produced if each coefficient of the modified pivot row is subtracted from the corresponding coefficient in the first constraint row:

Basis	B	X1	X2	X3	X4	X5	X6
X4	6	1	1	4	1	0	0
X2	3	0.25	1	0.5	0	0	0.25
	3	0.75	0	3.5	1	0	-0.25

The row corresponding to the second constraint is rather different; it is

Basis	B	X1	X2	X3	X4	X5	X6
X5	7	1	2	1	0	1	0

To produce a zero for the X2 coefficient, we must double each coefficient of the modified pivot row and subtract them:

Basis	B	X1	X2	X3	X4	X5	X6
X5	7	1	2	1	0	1	0
X2 ($\times 2$)	6	0.5	2	1	0	0	0.5
	1	0.5	0	0	0	1	-0.5

The new tableau then becomes

Tableau 1.3

Ci	Basis	B	Cj 4 X1	5 X2	3 X3	0 X4	0 X5	0 X6	Q
0	X4	3	0.75	0	3.5	1	0	-0.25	4
0	X5	1	0.5	0	0	0	1	-0.5	2*
5	X2	3	0.25	1	0.5	0	0	0.25	12
	Zj	15	1.25	5	2.5	0	0	1.25	
	Zj − Cj		-2.75	0	-0.5	0	0	1.25	
			**						

At this stage, a limited check may be made on the calculations. In the above tableau, the $Zj - Cj$ row was prepared by first calculating the Zj entries, then subtracting the corresponding Cj elements. If, however, the Zj − Cj row is updated by the same procedure as the other rows of the tableau, the same entries should result:

Old Zj − Cj	0	−4	−5	−3	0	0	0
3 × old pivot	15	1.25	0	2.5	0	0	1.25
	15	−2.75	0	−0.5	0	0	1.25

Work on the tableau proceeds by identifying the pivot column (**), calculating the Q column for its smallest entry to identify the pivot row (*), and finally discovering the new pivot element (0.5). Iterating again, we arrive at

	Cj		4	5	3	0	0	0		
Ci	Basis	B	X1	X2	X3	X4	X5	X6	Q	
0	X4	1.5	0	0	3.5	1	−1.5	0.5	3 *	
4	X1	2	1	0	0	0	2	−1		
5	X2	2.5	0	1	0.5	0	−0.5	0.5	5	
	Zj	20.5	4	5	2.5	0	5.5	−1.5		
	Zj − Cj		0	0	−0.5	0	5.5	−1.5		
								**		

Tableau 1.4

This solution, at X1 = 2, X2 = 2.5, X3 = 0, X4 = 1.5, X5 = 0, X6 = 0, with its optimization function, profit = \$20.5, compares with Equation Set 1.17. Also compare Equation Set 1.8 of the original two-dimensional problem with this tableau and the X3 column. Iterating this tableau now results in

	Cj		4	5	3	0	0	0	
Ci	Basis	B	X1	X2	X3	X4	X5	X6	Q
0	X6	3	0	0	7	2	−3	1	
4	X1	5	1	0	7	2	−1	0	
5	X2	1	0	1	−3	−1	1	0	
	Zj	25	4	5	13	3	1	0	
	Zj − Cj		0	0	10	3	1	0	

Tableau 1.5

This tableau is optimal, since there are no negative entries in the Zj − Cj row. Notice that in iterating the previous tableau to achieve this final tableau, the X1 row was not considered as a possible pivot row. Q was not calculated since there was a negative element in that position of the pivot column. To use a negative element would result in a negative value appearing in the B column of the next tableau. This would be interpreted as a solution that calls for a negative amount of structural variable (product) to be manufactured or a negative amount of slack variable (resource) to be used. The final tableau indicates a solution at $X1 = 5$, $X2 = 1$, $X3 = 0$, $X4 = 0$, $X5 = 0$, $X6 = 3$, with the optimization function (profit) $= \$25$. Compare this with Equation Sets 1.9 and 1.18.

The entries in the $Zj - Cj$ row are of further interest. The entry for X3 takes the value 10. X3 is not in the current basis. If it were to enter the basis, as its $Zj - Cj$ value is positive, the optimization function (profit) would be reduced by $10 for each unit of X3 produced. X4 and X5 also have entries in the $Zj - Cj$ row, but these variables are not structural (product) variables. The value for X4 is 3; thus if X4 were to be brought into the basis, the optimization function would be reduced by $3 for each unit of X4 (ton of sunflower oil) "produced," that is, left unmanufactured. The same effect comes about when the amount of resource, here 6 tons, is reduced by one unit. As the functions are linear, the converse argument holds that if the amount of resource is increased by one unit, (from 6 tons to 7 tons), then the marginal increase in the profit function will be $3. These positive values in the $Zj - Cj$ row of the nonbasic variables are known as *shadow prices*. They will be discussed further in Chapter 3.

The Condensed Simplex Tableau

Inspection of the extended simplex tableaux will show that the columns of the tableaux associated with the variables contained in the current basis are all 1 or 0. In fact, these columns make up an identity matrix. It is therefore convenient to devise a coding method that removes the chore of updating this section of the matrix. This is known as the condensed tableau method.

Consider again the same problem:

Maximize $4 X1 + 5 X2 + 3 X3$
subject to $X1 + X2 + 4 X3 \leq 6$
                $X1 + 2 X2 + X3 \leq 7$
                $X1 + 4 X2 + 2 X3 \leq 12$

Place this in the following tableau:

	X1	X2	X3		
X4	1	1	4	6	
X5	1	2	1	7	
X6	1	4	2	12	
	-4	-5	-3	0	D = 1

The bottom row of this tableau is equivalent to the $Z_j - C_j$ row of the extended tableau. To interpret, read the top row as the nonbasic variables and their associated shadow prices. The left-hand column variables are the variables forming the current basis of the solution, with their values presented in the right-hand column of the tableau. The value of the objective function for this solution point is depicted in the bottom right element of the tableau. The D value is important; it is the current "multiplier." In this method, if we start with integer values, the values arising from our computations will remain integer, avoiding time-consuming divisions. In order to interpret the value of a variable at any stage, the value presented in the matrix must be divided by D.

To iterate the tableau, proceed as before to identify the pivot column in order to bring a nonbasic variable into the basis. The nonbasic variable is identified by choosing the largest negative entry on the bottom row of the tableau (-5), identifying X2 as the pivot column. The pivot row is identified, as previously, by dividing the elements of the pivot column into the elements of the current basis (Q), and then choosing the smallest Q value to identify the pivot row. Iteration then proceeds as follows:

1. Exchange the pivot column and pivot row labels.

	Old				New		
	X1	X2	X3		X1	X6	X3
X4				X4			
X5				X5			
X6				X2			

2. Copy the pivot row from the old matrix into the new, *without* the pivot element.

	X1	X6	X3	
X4				
X5				
X2	1		2	12

3. Multiply each element of the old pivot column by -1 and copy it into the new matrix, *without* the pivot element.

	X1	X6	X3	
X4		-1		
X5		-2		
X2	1		2	12

4. All other elements of the tableau are calculated from the old tableau in the following manner:

P = Pivot element
N_{ij} = Tableau entry to be updated
K_i = Element of the pivot column in the same row as N_{ij}
R_j = Element of the pivot row in the same column as N_{ij}

Thus

Figure 1.7

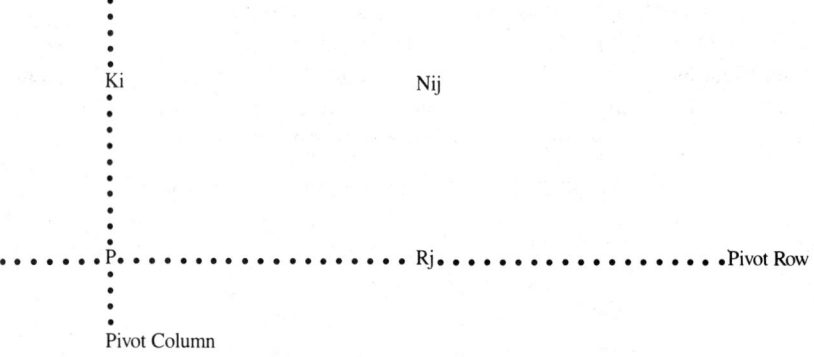

The value that replaces N_{ij} is calculated from

$$\frac{(P \times N_{ij}) - (R_j \times K_i)}{D}$$

For example,

$$4 \text{ is replaced by } \frac{(4 \times 4) - (2 \times 1)}{1} = 14$$

$$6 \text{ is replaced by } \frac{(4 \times 6) - (12 \times 1)}{1} = 12$$

$$0 \text{ is replaced by } \frac{(4 \times 0) - (12 \times -5)}{1} = 60$$

and so forth. This generates

	X1	X6	X3	
X4	3	-1	14	12
X5	2	-2	0	4
X2	1		2	12
	-11	5	-2	60

5. The final step is to exchange the old pivot element with the D value:

	X1	X6	X3		
X4	3	-1	14	12	
X5	2	-2	0	4	
X2	1	1	2	12	
	-11	5	-2	60	D = 4

Compare this with Equation Set 1.16 and the corresponding extended simplex tableau. Its interpretation is the solution $X1 = 0$, $X2 = 3$, $X3 = 0$, $X4 = 3$, $X5 = 1$, $X6 = 0$, with optimization function (profit) = \$15. In other words, each tableau entry is divided by D for interpretation. The next iteration proceeds thus:

	X5	X6	X3		
X4	-3	1	7	3	
X1	4	-2	0	4	
X2	-1	1	1	5	
	11	-3	-1	41	D = 2

and the next,

	X5	X4	X3		
X6	−3	2	7	3	
X1	−1	2	7	5	
X2	1	−1	−3	1	
	1	3	10	25	D = 1

which is optimal and corresponds to the entries in the final extended simplex tableau.

MATRIX METHODS

The simplex method, applied by equation manipulation or the coding method of tableaux, is essentially a paper-and-pencil manipulation system. With cheap and easily accessible computation power available, why bother? For most applications we don't have to bother, but it is important to understand what is going on inside the machine. Furthermore, there are some instances, for example, where a quick evaluation of a decision is required, when access to a machine is limited or inappropriate. The algorithms used by computer programs, while they follow the concepts described, arrange their computations in a rather different manner for efficiency. There are basically two implementations of the simplex algorithm used in commercial computer routines. These are the revised simplex form using the explicit form of the inverse and the revised simplex using the product form of the inverse.

Revised Simplex with Explicit Inverse

Consider again our standard problem:

Maximize $\quad 4 X1 + 5 X2$

subject to
$$X1 + X2 + X3 \qquad\qquad = 6$$
$$X1 + 2 X2 \qquad + X4 \qquad = 7$$
$$X1 + 4 X2 \qquad\qquad + X5 = 12$$

The following vectors and matrices are defined:

$$A = \begin{bmatrix} 1 & 1 & 1 & 0 & 0 \\ 1 & 2 & 0 & 1 & 0 \\ 1 & 4 & 0 & 0 & 1 \end{bmatrix} \quad b = \begin{bmatrix} 6 \\ 7 \\ 12 \end{bmatrix}$$

$$c' = (4\ 5\ 0\ 0\ 0)$$

$$B = \begin{bmatrix} 1 & 0 & 0 \\ 0 & 1 & 0 \\ 0 & 0 & 1 \end{bmatrix} \quad xB = \begin{bmatrix} X3 \\ X4 \\ X5 \end{bmatrix} \quad cB = \begin{bmatrix} 0 \\ 0 \\ 0 \end{bmatrix}$$

$$B^{-1} = \begin{bmatrix} 1 & 0 & 0 \\ 0 & 1 & 0 \\ 0 & 0 & 1 \end{bmatrix}$$

Note that in these and the following matrices and vectors, letters representing specific values are shown as boldface.

Step 1. Compute $P' = c'B\ B^{-1}$:

$$(0\ 0\ 0)\begin{bmatrix} 1 & 0 & 0 \\ 0 & 1 & 0 \\ 0 & 0 & 1 \end{bmatrix} = (0\ 0\ 0)$$

It is necessary only to evaluate nonbasic variables, thus for X1: $Z1 - C1 = P'a1 - c1$:

$$(0\ 0\ 0)\begin{bmatrix} 1 \\ 1 \\ 1 \end{bmatrix} - 4 = -4$$

For X2: $Z2 - C2 = P'a2 - c2$:

$$(0\ 0\ 0)\begin{bmatrix} 1 \\ 2 \\ 4 \end{bmatrix} - 5 = -5$$

As X2 is min($P'aj - cj$), or -5, this is chosen to enter the basis.

Step 2. Now compute $\mathbf{y} = \mathbf{B}^{-1} \mathbf{a}2$:

$$\begin{bmatrix} 1 & 0 & 0 \\ 0 & 1 & 0 \\ 0 & 0 & 1 \end{bmatrix} \begin{bmatrix} 1 \\ 2 \\ 4 \end{bmatrix} = \begin{bmatrix} 1 \\ 2 \\ 4 \end{bmatrix}$$

Currently $\mathbf{b} =$

$$\begin{bmatrix} 6 \\ 7 \\ 12 \end{bmatrix}$$

For a positive $\mathbf{y}i$, compute $Q = \mathbf{b}i/\mathbf{y}i$, and choose the minimum:

$$i = 1, \quad Q = 6/1 \ = 6$$
$$i = 2, \quad Q = 7/2 \ = 3.5$$
$$i = 3, \quad Q = 12/4 = 3$$

Therefore choose the third row.

Step 3. Determine new inverse and new $\mathbf{b}$:

cB	xB	b	Old Inverse			y
0	X3	6	1	0	0	1
0	X4	7	0	1	0	2
0	X5	12	0	0	1	4

Step 4. The entries for the new $\mathbf{b}$ and the new inverse are found in precisely the same way as in the extended simplex tableau, by using row operations.

cB	xB	b	New Inverse			y	y for next iteration
0	X3	3	1	0	−0.25	0	0.75
0	X4	1	0	1	−0.5	0	0.5
5	X2	3	0	0	0.25	1	0.25

Step 5. The final column of the above table is calculated first by computing $\mathbf{P}'$ as done previously (equivalent to calculating $Zj - Cj$, then choosing the pivot column in the extended simplex tableau), then, as in step 2 above, computing $\mathbf{y}$ for the new pivot column

$$(X1): \quad \mathbf{y} = \mathbf{B}^{-1} \, \mathbf{a1} = \begin{bmatrix} 1 & 0 & -0.25 \\ 0 & 1 & -0.5 \\ 0 & 0 & 0.25 \end{bmatrix} \begin{bmatrix} 1 \\ 1 \\ 1 \end{bmatrix} = \begin{bmatrix} 0.75 \\ 0.5 \\ 0.25 \end{bmatrix}$$

Step 6. Continue until the solution is found.

cB	xB	b	New Inverse			y	yn
0	X3	1.5	1	−1.5	0.5	0	0.5
4	X1	2	0	2	−1	1	−1
5	X2	2.5	0	−0.5	0.5	0	0.5

The final solution is

cB	xB	b	New Inverse			y
0	X5	3	2	−3	1	1
4	X1	5	2	−1	0	0
5	X2	1	−1	1	0	0

Compare with the extended simplex tableaux.

Revised Simplex with Product Form of the Inverse

The revised simplex method using the product form of the inverse is more economical than the method using the explicit form in its consumption of computer space. In both forms, the revised simplex method offers computational advantages when the number of nonbasic variables exceeds the number of basic variables. The product-form method, widely adopted for commercial packages, differs from the explicit-form method only in the way that the inverse is handled. In essence, much of the activity in computing the successive iterations of the simplex method is performed because a nonbasic variable might eventually enter the basis. If this does not happen, then the computation carried out on that variable is wasted. This procedure, suitable for computers if not for individuals, remembers the steps completed and only iterates that part of the simplex which is relevant. Specifically, an identity matrix is iterated instead of making use of the old inverse. For the same problem, using the previously defined matrices:

cB	xB	b	Identity Matrix			Incoming Variable (X2)
0	X3	6	1	0	0	1
0	X4	7	0	1	0	2
0	X5	12	0	0	1	4

Using the same calculations as before to identify the pivot row, and iterating as in the simplex method, this identity matrix is transformed into the first elementary matrix:

cb	xB	b	New Identity Matrix	First Elementary Matrix (E1)	New Incoming Variable (X1)
0	X3	3	$\begin{bmatrix} 1 & 0 & 0 \\ 0 & 1 & 0 \\ 0 & 0 & 1 \end{bmatrix}$	$\begin{bmatrix} 1 & 0 & -0.25 \\ 0 & 1 & -0.5 \\ 0 & 0 & 0.25 \end{bmatrix}$	$\begin{bmatrix} 0.75 \\ 0.5 \\ 0.25 \end{bmatrix}$
0	X4	1			
5	X2	3			

After one iteration, the first elementary matrix is equivalent to the new inverse previously computed. The iterations proceed as before, but instead of performing row operations on the new inverse (first elementary matrix), a new identity matrix is used in the iteration process. This generates the second elementary matrix:

cb	xB	b	Identity Matrix	Second Elementary Matrix (E2)	New Incoming Variable (X5)
0	X3	1.5	$\begin{bmatrix} 1 & 0 & 0 \\ 0 & 1 & 0 \\ 0 & 0 & 1 \end{bmatrix}$	$\begin{bmatrix} 1 & -1.5 & 0 \\ 0 & 2 & 0 \\ 0 & -0.5 & 1 \end{bmatrix}$	$\begin{bmatrix} 0.5 \\ -1 \\ 0.5 \end{bmatrix}$
4	X1	2			
5	X2	2.5			

The new inverse of the explicit form is calculated from the two elementary matrices E1 and E2 by evaluating their product $E2 \cdot E1$:

$$\begin{bmatrix} 1 & -1.5 & 0 \\ 0 & 2 & 0 \\ 0 & -0.5 & 1 \end{bmatrix} \begin{bmatrix} 1 & 0 & -0.25 \\ 0 & 1 & -0.5 \\ 0 & 0 & 0.25 \end{bmatrix} = \begin{bmatrix} 1 & -1.5 & 0.5 \\ 0 & 2 & -1 \\ 0 & -0.5 & 0.5 \end{bmatrix}$$

Calculation may then be made using either E2 E1 or the current inverse B^{-1}:

$$P' = cB \ B^{-1} = cB \ E2 \ E1$$

$$= (0 \quad 4 \quad 5 \quad) \begin{bmatrix} 1 & -1.5 & 0 \\ 0 & 2 & 0 \\ 0 & -0.5 & 1 \end{bmatrix} \begin{bmatrix} 1 & 0 & -0.25 \\ 0 & 1 & 0.5 \\ 0 & 0 & 1 \end{bmatrix}$$

$$= (0 \quad 5.5 \quad -1.5 \ )$$

Calculation proceeds as previously to the final solution:

cb	xB	b	Third Elementary Matrix
0	X5	3	
4	X1	5	
5	X2	1	

$$\begin{bmatrix} 2 & 0 & 0 \\ 2 & 1 & 0 \\ -1 & 0 & 1 \end{bmatrix}$$

The new inverse is $E3 \cdot E2 \cdot E1$

Evaluating $(\ 0 \ \ 4 \ \ 5 \ ) \ E3 \ \cdot \ E2 \ \cdot \ E1$
produces $(\ \ 3 \ \ 1 \ \ 0 \ \ )$

As these shadow values are all positive, the solution is achieved.

USE OF LINEAR PROGRAMMING COMPUTER PACKAGES

Because linear programming is computationally intensive, it makes sense to have the iterations performed electronically. There are many packages commercially available, some better than others, and many, particularly those for small systems, little better than useless. The mainframe package described here, LINDO, is representative of good packages and may be used in either batch or interactive modes. It is supported on many systems from DEC to Cyber and is widely available.

First log on your computer system. (California State University students will find LINDO installed on the Central Cyber System.) After logging on successfully with your user name and password, you will be in the operating system level. For the Cyber this is the NOS operating system. The prompt character at this level is "/". To access LINDO you should now type

```
FIND,LINDO
```

followed by a carriage return.

On most computers, LINDO has two prompt characters:

: indicates that LINDO is awaiting a command.
> indicates that LINDO is awaiting input data.

You will initially be faced with a command prompt, for example ":". Typing HELP will generate some basic information. Typing QUIT will return you to the initial prompt. Suppose that you merely wish to try the package by using data input from your terminal keyboard. Using the same example as previously, you then type

```
: MAX 4X1 + 5X2
>ST
>X1 + X2 < 6
>X1 + 2X2 < 7
>X1 + 4X2 < 12
>END
```

After each line of input you should enter a carriage return; this will send the line, and position your cursor at the beginning of the next line, giving you a ">" prompt until you input a further command, in this case END. The system will respond by giving you a ":" prompt; instruct the computer to begin processing your program by typing GO and a carriage return. The system will then respond with

```
LP OPTIMUM FOUND AT STEP 3
OBJECTIVE FUNCTION VALUE
1)  25.0000000
VARIABLE          VALUE          REDUCED COST
X1               5.000000            .000000
X2               1.000000            .000000

ROW          SLACK OR SURPLUS   DUAL PRICES
2)                 .000000         3.000000
3)                 .000000         1.000000
4)                3.000000          .000000
NO. ITERATIONS= 3
DO RANGE (SENSITIVITY) ANALYSIS? >
```

At this point, we are not interested in proceeding further. In order to leave the program, respond to the > prompt with the character N and a carriage return. The computer will respond with the command prompt, ":". Now type QUIT and a carriage return. The computer will respond

```
NORMAL END OF EXECUTION
$REVERT
```

This will bring you to the system prompt, which in a Cyber installation is a slash (/). To log off the computer entirely, give the appropriate command to your operating system (or type HELP if you have forgotten it).

There are some interesting observations about this computer solution.

1. You can use real names for the variable names.

2. The "$<$" character is interpreted as the soft "$\leq$" character, because most terminal keyboards are not provided with the true combined mathematical symbol.

USING FILES FOR LINEAR PROGRAM PROBLEMS

Inputting the linear programming formulation directly into the LP program via the keyboard of the terminal works well for small problems, but when problems increase in size it is an advantage to prepare the problem first, using an editor to correct any mistakes, and then run the program on the corrected input. Using this approach, you can also store problem formulations to use at a later time.

All operating systems have some form of editor available. The newer and better systems allow you to edit your files using a screen editor, which means that anything displayed on your terminal screen can be altered simply by moving the cursor to the desired position and replacing characters that you wish to alter. Students familiar with word processors or microcomputers will be used to these facilities. Older, batch-oriented operating systems have editors that work in a line mode; each line to be modified must be identified specifically prior to modification. Because this method is more involved, and a line editor is used on the California State University Central Cyber System, its use will be illustrated.

The editor used by the CCS System 50 is XEDIT. To access XEDIT, first access the operating system and receive the appropriate prompt

```
/
```

In response to the prompt type XEDIT and a carriage return. The system will access XEDIT and return the message

```
XEDIT 3.1.00
EMPTY FILE / CREATION MODE ASSUMED
??
```

The "??" is the command-level prompt for XEDIT. Type a carriage return and the prompt will change:

```
INPUT
?
```

You are now in the input mode and will remain in that mode as long as the "?" is returned. Type your file exactly as before:

```
? MAX 4X1 + 5X2
? ST
? X1 + X2 < 6
? X1 + 2X2 < 7
? X1 + 4X2 < 12
? END
? GO
?
```

Notice that we are still in the input mode at this stage. Notice also that we have placed in this file the character string END, which was used in the interactive mode to indicate the end of constraint input, and also the GO command, which was used to instruct LINDO to solve the problem.

We now wish to terminate this editing session, and save the file. We terminate the editing session by entering a carriage return. The computer responds

```
EDIT
??
```

The "??" is XEDIT's prompt for a name under which to store the information just entered. Normally we can choose almost any name for our files in the computer, but LINDO has the idiosyncracy that any file submitted to it must be called TAPE##, where ## is any number between 20 and 99. We need therefore to exit the editor and make sure that the file is called TAPE##.

There are various commands available (in this system and others) to achieve the next few effects; but on leaving the editor, it is always necessary to save the file you have prepared, or you will lose it and be very frustrated. One command to achieve this effect is

```
?? Q,TAPE20
```

On receiving this command XEDIT knows that you wish to quit the editing mode and save your file in a box labeled TAPE20. This will be passed to the system, which will respond with

```
/TAPE20 IS A LOCAL FILE
```

This means that your file is now available for use as long as you remain logged on the system. A good analogy is that the file is now on your desk and available for use. When you log off the system all files are cleared, the analogy being that the cleaning staff throws into the trash all papers not put back in the file drawers. Should you wish to save this file for a future date, you should put it into your permanent filestore with the command

```
/SAVE,TAPE20
```

The response,

```
/TAPE20 IS A PERMANENT FILE
```

confirms that it is now safe for future use.

When you log on the system at a future date, in order to use previously prepared files you must get them out of your filestore with the command

```
/GET,TAPE20,TAPE32
```

This will copy the two permanent files into your local work area. Put them into the local work area or you will receive frustrating complaints from the computer that it doesn't have the files.

If you have used the files (for example, if you wish to remember what you put into them), you can copy the output of the file onto an output device such as the terminal screen. In fact, if you tell the computer to COPY a file, without defining where to put the output, it will default to your terminal screen:

```
/COPY,TAPE20
```

will achieve this. However, after copying out the contents, the computer file pointers will be positioned at the end of the file. Should you wish to work on the file contents with LINDO, or even look at the file again, the commands that previously worked will produce the complaint

```
EOI ENCOUNTERED,
```

In order to get the pointers at the beginning of the file again, the command

```
/REWIND,TAPE20
```

is needed. In fact, it is good practice to use the command

```
/REWIND,*
```

occasionally to reset all local files. Furthermore, if you have too many files out in your local area, it will fill up, so when you have finished with some files they should be removed with

```
/CLEAR,TAPE20,TAPE24
```

If these files were previously made permanent, their copies will remain in the permanent filestore. To throw away unwanted files from your permanent store,

```
/PURGE,TAPE20,TAPE45
```

will do it.

In order to work on a file prepared by XEDIT, access LINDO. Once you have the LINDO command prompt, respond with the command

```
: TAKE
```

The computer will reply

```
FILE NUMBER ?
>
```

Entering "34" will cause TAPE34 to be loaded into the LINDO work area for analysis. If you want to put the output from your LINDO analysis into a file, rather than having it displayed immediately on your screen, you can use the DIVERT command in LINDO. Having accessed LINDO, but prior to running the optimization, input

```
: DIVERT
```

The system will respond with

```
FILE NUMBER ?
>
```

to which you should respond with the tape number you would like to use. The output from your optimization run (1) may then be placed into a local file that can be listed at a later time before you log off the system, **or** (2) may be made permanent for analysis at some future session, **or** (3) if you are using a microcomputer to access the mainframe, may be transferred to the microcomputer store for later examination.

The method of using a personal computer to access the host mainframe is discussed in the Introduction.

CHAPTER 1 PROBLEMS

The solutions to questions 1 through 10 will be found at the end of the chapter.

1. The Ding-he Boat Company markets two types of fiberglass boats from a single molding. The more expensive boat is a sailboat, which sells for $1,000. The other is a rowboat, which can be used with an outboard motor, that is marketed at $750. A sailboat takes twice as long to complete as a rowboat, and if only rowboats were produced, the company would have time to complete a maximum of 1,000 per year. The molding capacity for the basic hulls is 800 per year (sailboats and rowboats combined). The supply of sail fittings amounts to 400 sets per year, and rowboat fittings can be supplied at a maximum rate of 700 sets per year. Decide the production schedule that will maximize the Ding-he Boat Company cash flow next year.

2. Clone Computers has just received a government order for its "Psi" Clone computers. It can sell all its production as a result of this order, and production lines for "Psi" Clone government model and "Psi" Clone civilian model have been set up. All the ICs (chips) used in government computers must be type-approved. Commercial computers can use some of these ICs, but all government computers must have special ICs that provide electromagnetic pulse protection; and these must not be allowed to reach the civilian market. Fortunately, the type of commercial IC used by Clone Computers has now been type-approved by the government for nonsensitive parts of the circuits.

The marketing department has insisted that computers for the commercial market must have a cosmetic finish, which takes 10 minutes per computer. Government computers are finished in a satin-finish titanium box, supplied by another contractor, which takes 3 minutes to install. The factory finishing department works a standard 7-hour day, and there are 50 standard working days allocated to the next planning period.

The numbers of ICs required are

	Government	Civilian	Available
Government-approved ICs	450	250	2,500,000
Secret ICs	100	0	100,000
Commercial ICs (nonsensitive)	175	350	1,050,000

Clone makes 90 percent more profit on government contracts than it makes on civilian work. What product plan should Clone adopt?

3. A fraternity has decided to provide liquid refreshment to a visiting sorority group. They have two recipes for punch. These have the names "Friendship" and "Loveboat." On their assessment scale, Friendship ranks an average 7.6 and Loveboat 9.6. The two punch recipes are based on two different beverages. Of the first of these they have enough to make 19 bottles of Friendship or 34 bottles of Loveboat. The second base beverage in stock is sufficient to produce 33 bottles of Friendship or 13 bottles of Loveboat. These beverages may be apportioned pro rata for varying amounts of punch. Loveboat must be served with passion fruit topping, and they have a supply of passion fruit sufficient for 10 bottles of Loveboat. How should the fraternity apportion its beverages?

4. Mill Key Industries, a dairy products manufacturer, has a shipment of 2,000 tons of milk due for delivery. This can be turned into 2,000 tons of yogurt or 2,000 tons of cheese. It takes 50 percent more plant capacity to produce cheese than yogurt, and there is sufficient plant capacity to process 2,700 tons of yogurt if all the plant were devoted to that process. All plant facilities are interchangeable between yogurt and cheese. It takes twice as much labor to produce yogurt than cheese, and there is sufficient labor to process 3,000 tons of cheese. Yogurt has a contribution to profit of $190 per ton, and cheese has a contribution to profit of $250 per ton. How should the milk be processed in order to maximize contribution to profit?

5. New C Foods has invested heavily in offshore seaweed cultivation. It has decided to specialize in two nutritious sea plants, California kelp and Oregon sea ragwort. When processed these sell for $1.10 and 55¢ per pound, respectively. New C has leased a tract of the sea bed from the government and has determined that this tract would supply up to 20 cubic hectares of sea ragwort per year if it were all devoted to this crop. California kelp requires four times the space of sea ragwort. It takes 13 man-years to process a cubic hectare of sea ragwort but only 6 man-years to process the same quantity of kelp. The operation has 78 experienced operatives and will be unable to train any more in time to handle the next year's crops. California kelp will require 4 vessels per cubic hectare of cultivation, but sea ragwort can be managed with 3 vessels per cubic hectare. These vessels are highly specialized and take one year to order and construct. New C has 24 of these vessels, which have been newly commissioned. When the two crops are processed, the weight of processed crop is directly proportional to the raw volume. What proportions of kelp and ragwort should New C plan to cultivate next year?

6. The warehouse manager of Fred's Superstores has a storage facility of 2,000 cubic feet. He currently stocks three major lines, Video dewiginators, Sterio confabulators, and Mobile entropisers. Each dewiginator costs $100, occupies 1 cubic foot of volume, and contributes $2 to the profit of the company. Confabulators have a price tag of $120, take up 2.5 cubic feet of space, and contribute $3 to company profits. Entropisers cost $180 each, occupy 5 cubic feet, and contribute $4 to company profits. Fred's current weekly turnover is $122,000. What is the best policy for the warehouse manager to adopt?

7. Eileen Dover owns a franchise chain specializing in fragrance merchandising. Her current creation, called "Cliff Hangar," is due for media launch. She wishes to distribute this media launch among TV, radio, and newspapers. The data gathered by her marketing advisers is detailed here:

	Media		
	TV	Radio	Newspapers
Cost per exposure	$2,000	$250	$500
Audience reached	250,000	25,000	100,000

Her promotional budget is set at $25,000, and she wishes to limit the maximum exposures in any one medium to 25. What marketing strategy reaches the most potential customers?

8. Robin Banks is the executive vice president of First Natural Savings and Loan. He is currently considering the organization's loan policy in the light of difficulties in other institutions. The categories of loans made by First Natural are:

Type of Loan	Annual Percentage Rate
Home loans	12.00
Second trust deeds	14.00
Automobile	16.50
Business loans	15.25

In addition First Natural invests in base securities, which are risk free but only return 9 percent per annum.

There are however, some constraints imposed by regulatory commissions and state and federal laws. These are:

1. Risk-free loans may not exceed one-third of total deposits.

2. Home loans and second trust deeds may not exceed the amount invested in risk-free loans.

3. Business loans may not exceed 49 percent of total loans.

4. Automobile loans may not exceed 50 percent of home loans and second trust deeds together.

First Natural is currently scheduling for $9 million in deposits in the next period. How should the loan policy be optimized?

9. Rich Hustle is an investment counselor for Creative Financial Arts, a firm considering its investment opportunities. It is considering a two-year period, and Rich has identified three possibilities. These are a particular fund, an interesting bond, and a useful stock. The details are as listed.

Investment	Available	Return (%)	Amount ($)
Fund	Now	10	No limit
Bond (1 yr)	Now	13	100,000
Stock	in 1 yr.	18	50,000

Rich wishes to invest $150,000, which C. F. Arts has just received as "greenmail" from its latest unsuccessful takeover bid. How should Rich distribute his optimal investment?

10. B. Quick & Associates has had considerable success in manufacturing and distributing half-length computer cards. It has two particularly interesting new designs, one of which makes an IBM PC emulate an IBM AT for the price of $150. The other card enables an IBM PC to read and write computer floppy disks automatically in any unknown format. This product sells for $65. Due to very heavy demand, the firm has decided to open up a new manufacturing operation. The current operations have enabled B. Quick to set aside $5,000 to start up the new operation. The manager of the local bank has agreed to make available a line of credit not to exceed $15,000 at a rate of interest of 18 percent per annum on the average amount borrowed. Following good banking practice, the manager has stipulated that the outstanding debt on this arrangement must be covered one and a half times by the liquid assets in the company.
Currently the AT card generates a profit margin of $25 and the floppy disk (DK) card generates a profit margin of $12. B. Quick has discovered that the average rate of turnover of accounts receivable is 90 days in this industry. How should B. Quick plan the next three months of its start-up operation?

HOMEWORK PROBLEMS

The solutions to these problems do not appear in this text. You're on your own!

11. Pumpkin Patch Dolls is planning production of two major new dolls, Cilly and Jimmy. Cilly will sell for $50 and Jimmy for $15. Shipments must be organized according to materials and labor availability. Cilly requires 5 times as much stuffing as Jimmy, and there is sufficient stuffing material on hand to produce 40 shipments of Jimmy dolls. A shipment of Cilly dolls requires 5 man-weeks of manufacturing capacity while a similar shipment of Jimmy dolls needs 3 man-weeks to make. The manufacturing capacity available to this project is 60 man-weeks. Finally, packaging and preparation is required, with Jimmy dolls needing twice as much packaging as Cilly dolls. There is sufficient resource in the packaging department to cope with 32 shipments of Cilly dolls. How should Pumpkin Patch Dolls plan its production?

12. Cal. St. Inc. is a major fertilizer producer whose products are spread throughout the state. The production manager is currently scheduling two products, "Yellow Rain" and "Acid Drops." Yellow Rain wholesales for $22 per ton and Acid Drops wholesales at $15 per ton. Each ton of

Yellow Rain needs 42 lb of steer manure and each ton of Acid Drops uses 12 lb of steer manure. There is currently available 504 lb of fresh steer manure in the organization. Yellow Rain's formula calls for 16 kg of nitrates and 14 kg of phosphates; the Acid Drops formula requires 18 kg of nitrates and 30 kg of phosphates. There are 288 kg of nitrates and 420 kg of phosphates available for allocation. How many tons of each fertilizer should Cal. St. Inc. produce?

13. All the beer drunk in the state must be approved by the S tate Water Commissioner, who has stipulated four criteria of quality. In a standard vat of beer, there must be less than 6.45 mg of suspended iodine, less than 4.59 mg of borates, less than 403 mg of chlorates, and less than 1,000 mg of inert suspended solids. The Baja Beer Company owns two major sources of water at the Mojave River Junction and Tijuana Springs. Analysis of the wells revealed the following:

Elements (mg/100 gal)	Mojave River	Tijuana Springs
Iodine	0.43	0.15
Borates	0.27	0.17
Chlorates	13.00	31.00
Suspended Solids	10.00	100.00

How many gallons of water from each source can the Baja Beer Company include in each standard vat and still meet the government purity standard? (The state water supply system contains none of the listed impurities.)

14. General Caward E. Custard is planning an assault on two fronts, Juno and Vixen. He considers it 30 percent more desirable to advance along front 2 (Vixen) than to advance an equivalent distance along front 1 (Juno). The forces he has at his disposal are:

Troops	798 companies under his command
Tanks	528 available
Air strikes	522 may be called
Naval bombardments	561 are possible

Each mile advanced along Juno will need a commitment of 42 companies of ground troops, supported by 22 tanks and 18 air strikes. It will also be necessary to call for 17 naval barrages per mile of advance.

Each mile advanced along the Vixen front will require 19

companies of ground troops, 24 tanks, and 29 air strikes; 33 naval bombardments will also be needed.

In order to achieve maximum strategic advantage, what number of miles advanced along each front should General Custard plan?

15. Cecil B. DeMilo makes commercials for the "Harmless Toys" group. He is paid by the minute of final commercial produced. Currently he has two projects in hand: One commercial is for the M16 play stick and the other is advertising the Borgia cooking set. He has contracted to supply final commercial footage at $80 per second for the M16 project and $180 per second for the Borgia cooking set. Each minute of final footage produced requires 6 minutes of filming for the M16 project and 33 minutes for the Borgia project. He has 3 hours 18 minutes of shooting time booked. Each minute of shot film needs 10 minutes of editing for the M16 project and 15 minutes of editing for the Borgia project, and he has been allocated 2½ hours' editorial time. Each minute of final produced footage is assessed an overhead cost of $700 for the M16 project and $2,200 for the Borgia project. DeMilo's budget is currently standing at $15,400. Finally, the extras used in the filming are paid $42 per minute of final film production on the M16 project and $10 per minute on the Borgia project. The agency has agreed to supply $420 worth of extras under this arrangement. How should DeMilo schedule work on the two commercial projects?

16. Lee Axelrodd is managing director of a company that supplies specialized vehicles under government contract. He has received instructions to supply quantities of two-wheeled assault vehicles and a single-wheeled all-terrain vehicle, both of which are powered by a process that does not utilize the thermodynamic cycle. The company makes a profit of $3,500 on each two-wheeled vehicle and $7,500 on each single-wheeled vehicle to recover development costs. Each two-wheeled vehicle requires 9,500 standard production operations, and each one-wheeled vehicle calls for 6,750 standard production operations. The manufacturing facility is able to supply 64,125,000 standard production operations. Each two-wheeled vehicle requires 225 government-approved fasteners and the one-wheeled vehicle must have 825 of these fasteners. The company currently has 18,562,500 approved fasteners in stock.

The government contract calls for the supply of not more than 6,500 of two-wheeled assault vehicles and not more than 1,500 single-wheeled ATVs. How should Lee Colacoca maximize his profit from the contract?

17. E. Z. Pickins is considering three possible takeover bids. As a first strategy he would invest in all of the three companies. These are Costa Oil, Duocal, and Williams Petroleum. Currently he has $7 million to invest in the projects. Costa Oil is standing at $1 per share, Duocal has moved to $1.80 per share, and Williams Petroleum is currently 80¢ a share. There are controlling interests at 5 million shares of Costa, 8 million shares of Duocal, and 10 million shares of Williams. When it becomes known that he has moved into the market for these shares, Pickins believes that the managements of these companies will react by repurchasing the shares that he holds at the following premiums:

Costa Oil	15%
Duocal	20%
Williams Petroleum	9%

To maximize profit, how should Pickins proceed with his plans?

18. I. C. Winds produces freeze-dried vegetables. It sells its products in three major operations, each operation having different characteristics:

Distributor	Advertising/ Container	Salesforce Man-hours	Profit/ Container
Major chains	$20	5	$75
Specialty stores	$15	3	$95
Direct supply	$10	3	$80

The sales budget consists of 1,500 man-hours and $7,000 for this month's production of 800 containers. How should I. C. Winds allocate its marketing budget?

19. The Lone Star Company has advertised making loans to customers for next year at the following rates:

Type of Loan	Interest Rate (%)
Second trust deed	16
Auto	18
Business	20
Bail	25
Gambling	50

The family that owns the company has stipulated that loans shall be made only on the following conditions:

1. Gambling loans shall not exceed 50 percent of bail loans.

2. Bail loans and gambling loans together shall not exceed 50 percent of the family's investment of $1 million.

3. Business, bail, and gambling loans together shall not exceed 30 percent of total loans.

How should Lone Star apportion its money?

20. Sir Vival's Encampment is a recreational resort on the edge of the National Forest. It emphasizes proficiency in medieval skills to live off the land in the event of a disaster and promotes skills in medieval weapons. The managers of Sir Vival have decided to promote their activities aggressively via media exposure. The costs of alternative media advertisements are:

Media	Cost/ Advertisement	Maximum Slots Available	Number of Customers Reached
Evening TV	$5,000	10	10,000
Daytime TV	1,000	20	1,000
Daily newspaper	500	30	2,000
Sunday newspaper	1,000	4	5,000
Radio WXYZ	100	30	500

Sir Vival has set aside $50,000 to promote this campaign.

Other considerations are that not more than $25,000 should be spent with any medium and that the ratio of newspaper advertisements to radio advertisements should not exceed 2:1.

How can Sir Vival reach the maximum audience for its expenditure?

CHAPTER 1 SOLUTIONS

1. Let the number of sailboats be A and the number of rowboats be B. The constraints are

$$
\begin{aligned}
2A + B &\leq 1{,}000 \quad &\text{Time} \\
A + B &\leq 800 \quad &\text{Hull supply} \\
A &\leq 400 \quad &\text{Sailboat fittings} \\
B &\leq 700 \quad &\text{Rowboat fittings}
\end{aligned}
$$

Writing the optimization function in terms of $100 cash units, the formulation becomes

Maximize	$10A + 7.5B$

$$
\begin{aligned}
\text{subject to} \quad 2A + B + S1 &= 1{,}000 \\
A + B + S2 &= 800 \\
A + S3 &= 400 \\
B + S4 &= 700
\end{aligned}
$$

The simplex tableau iterations are

C_i	Basis	B	C_j → 10 A	7.5 B	0 S1	0 S2	0 S3	0 S4	Q
0	S1	1,000	2	1	1	0	0	0	500
0	S2	800	1	1	0	1	0	0	800
0	S3	400	1	0	0	0	1	0	400*
0	S4	700	0	1	0	0	0	1	—
	Z_j	0	0	0	0	0	0	0	
	$Z_j - C_j$		−10	−7.5	0	0	0	0	
			**						
0	S1	200	0	1	1	0	−2	0	200*
0	S2	400	0	1	0	1	−1	0	400
10	A	400	1	0	0	0	1	0	—
0	S4	700	0	1	0	0	0	1	700
	Z_j	4,000	10	0	0	0	10	0	
	$Z_j - C_j$		0	−7.5	0	0	10	0	
				**					

(continued)

Ci	Cj Basis	B	10 A	7.5 B	0 S1	0 S2	0 S3	0 S4	Q
7.5	B	200	0	1	1	0	−2	0	—
0	S2	200	0	0	−1	1	1	0	200*
10	A	400	1	0	0	0	1	0	400
0	S4	500	0	0	−1	0	2	1	250
	Zj	5,500	10	7.5	7.5	0	−5	0	
	Zj − Cj		0	0	7.5	0	−5	0	
							**		

7.5	B	600	0	1	−1	2	0	0
0	S3	200	0	0	−1	1	1	0
10	A	200	1	0	1	−1	0	0
0	S4	100	0	0	1	−2	0	1
	Zj	6,500	10	7.5	2.5	5	0	0
	Zj − Cj		0	0	2.5	5	0	0

The solution is interpreted as

Total cash flow = 6500 × 100 = $650,000
Produce   200 sailboats
                 600 rowboats
with   200 spare sailboat fittings
                 100 spare rowboat fittings

The shadow value of overtime is 2.5 × 100 = $250 per unit. The shadow value of extra hulls is 5 × 100 = $500 per unit.

If extra resource units can be obtained at costs below these prices, the company can profitably expand operations.

The schematic representation of this solution is shown in Figure 1.7. The solution by computer looks like this:

```
/FIND,LINDO
1 LINDO PROGRAM (CSU JAN-10-84)
: MAX 10A+7.5B
>ST
>2A+B<1000
>A+B<800
>A<400
>B<700
>END
: GO

LP OPTIMUM FOUND AT STEP 3
OBJECTIVE FUNCTION VALUE
1)  6500.00000

VARIABLE          VALUE          REDUCED COST
A              200.000000              .000000
B              600.000000              .000000

ROW         SLACK OR SURPLUS    DUAL PRICES
2)                   .000000       2.500000
3)                   .000000       5.000000
4)                200.000000        .000000
5)                100.000000        .000000
NO. ITERATIONS= 3

DO RANGE (SENSITIVITY) ANALYSIS? N
:QUIT
/BYE
```

The solution to this problem is illustrated graphically in Figure 1.8.

Figure 1.8

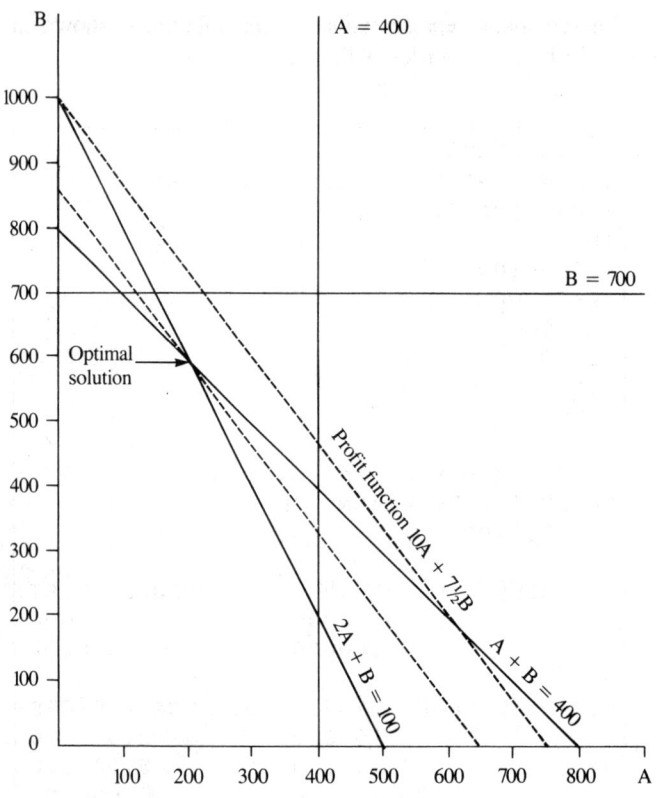

2. Let

Number of government computers = G
Number of civilian computers = C

Then

Maximize	$1.9\,G + \quad C$	
subject to	$450\,G + 250\,C \leq 2{,}500{,}000$	Govt. approved ICs
	$100\,G \qquad\qquad \leq \quad 100{,}000$	Secret ICs
	$175\,G + 350\,C \leq 1{,}050{,}000$	Commercial ICs
	$3\,G + \quad 10\,C \leq \quad 21{,}000$	Finishing minutes

or

Maximize $1.9\,G + \quad C$
subject to $9\,G + \quad 5\,C \leqslant 50{,}000$
                $G \qquad \quad \leqslant \quad 1{,}000$
                $G + \quad 2\,C \leqslant \quad 6{,}000$
          $3\,G + 10\,C \leqslant 21{,}000$
$G, L \geqslant 0$

Solution:

Manufacture 1,800 civilian computers,
            1,000 government computers.

3. Let

Bottles of Friendship produced $= F$
Bottles of Loveboat produced $= L$

Maximize $7.6\,F + 9.6\,L$
subject to $34\,F + \quad 19\,L \leqslant 646$
          $13\,F + \quad 33\,L \leqslant 429$
                    $L \leqslant \quad 10$
$F, L \geqslant 0$

Solution:

Produce 15 bottles of Friendship (15.12)
    and   7 bottles of Loveboat (7.04)

4. Let

$X1 =$ Tons of yogurt produced
$X2 =$ Tons of cheese processed

Maximize $190\,X1 + 250\,X2$
subject to $X1 + \quad X2 \leqslant 2{,}000$
          $2\,X1 + \quad 3\,X2 \leqslant 5{,}400$
          $2\,X1 + \quad\quad X2 \leqslant 3{,}000$
$X1, X2 \geqslant 0$

Solution:

Process   600 tons of yogurt
     and 1400 tons of cheese,
     giving a contribution to profit of
     $190 × 600 + $250 × 1400
= $464,000

5. Let

$X1$ = Number of cubic hectares of California kelp planned
$X2$ = Number of cubic hectares of Oregon sea ragwort planned

Then

Maximize   110 X1 + 55 X2
subject to   4 X1 +   X2 ≤ 20   Space
                  4 X1 + 3 X2 ≤ 24   Vessels
                  6 X1 + 13 X2 ≤ 78   Labor
              X1, X2 ≥ 0

Solution:

$X1$ = 4.5 cubic hectares
$X2$ = 2   cubic hectares

6. Let

$X1$ = Number of dewiginator units stocked
$X2$ = Number of confabulator units stocked
$X3$ = Number of entropiser units stocked

Then

Maximize   2 X1 +   3 X2 +   4 X3
subject to   100 X1 + 120 X2 + 180 X3 ≤ 122,000   Budget
                 1 X1 + 2.5 X2 +   5 X3 ≤   2,000   Space
              X1, X2, X3 ≥ 0

Solution:

Store 500 dewiginator units
  and 600 confabulators

Total contribution to Fred's profits
  = $ 28,000 per week.

Return on capital = 15.38%

Economic storage rate = 46¢ per cubic foot

7. Let

$X1$ = Number of TV advertisements
$X2$ = Number of radio advertisements
$X3$ = Number of newspaper advertisements

Then

Maximize	$250{,}000\ X1 + 25{,}000\ X2 + 100{,}000\ X3$
subject to	$2{,}000\ X1 + 250\ X2 + 500\ X3 \leq 25{,}000$
	$X1 \leq 25$
	$X2 \leq 25$
	$X3 \leq 25$
	$X1, X2, X3 \geq 0$

Solution:

$X1$ = 6.25 TV advertisements
$X2$ = 0 Radio advertisements
$X3$ = 25 Newspaper advertisements

Objective function value = 4,062,500 customers reached.

(Note that 6.25 TV advertisements are called for. This can be interpreted as a quarter-length time slot or as three sessions of 6 quarter-length time slots followed by one of 7, if the process is ongoing.)

8. Let

X1 = Dollars invested in home loans
X2 = Dollars invested in second trust deeds
X3 = Dollars invested in automobile loans
X4 = Dollars invested in business loans
X5 = Dollars invested in risk-free base securities

Then

Maximize $0.12 X1 + 0.14 X2 + 0.165 X3 + 0.1525 X4 + 0.09 X5$
subject to

$$X5 \leq 3,000,000$$
$$X1 + X2 \leq X5$$
$$X4 \leq 0.49(X1 + X2 + X3 + X4 + X5)$$
$$X3 \leq 0.5(X1 + X2)$$
$$X1 + X2 + X3 + X4 + X5 \leq 9,000,000$$

or

Maximize $0.12 X1 + 0.14 X2 + 0.165 X3 + 0.1525 X4 + 0.09 X5$
subject to

$$X5 \leq 3,000,000$$
$$X1 + X2 - X5 \leq 0$$
$$-0.49 X1 - 0.49 X2 - 0.49 X3 + 0.51 X4 - 0.49 X5 \leq 0$$
$$-0.50 X1 - 0.50 X2 + X3 \leq 0$$
$$X1 + X2 + X3 + X4 + X5 \leq 9,000,000$$
$$X1, X2, X3, X4, X5 \geq 0$$

Solution:

X1 = $0 invested in home loans
X2 = $1,836,000 invested in second trust deeds
X3 = $918,000 invested in auto loans
X4 = $4,410,000 invested in business loans
X5 = $1,836,000 invested in risk-free securities

Objective function value
= $1,246,275 interest received from this portfolio per year

9. Let

X11 = Dollars invested in fund during first year
X12 = Dollars invested in fund during second year
X21 = Dollars invested in bond during first year
X32 = Dollars invested in stock during second year

Then

Maximize $0.10\,X11 + 0.10\,X12 + 0.13\,X21 + 0.18\,X32$
subject to

$$X11 + X21 \leq 150{,}000$$
$$X12 + X32 \leq 150{,}000 + 0.1\,X11 + 0.13\,X21$$
$$X21 \leq 100{,}000$$
$$X32 \leq 50{,}000$$

or

Maximize $0.10\,X11 + 0.10\,X12 + 0.13\,X21 + 0.18\,X32$
subject to

$$X11 + X21 \leq 150{,}000$$
$$-0.10\,X11 + X12 - 0.13\,X21 + X32 \leq 150{,}000$$
$$X21 \leq 100{,}000$$
$$X32 \leq 50{,}000$$

$X11, X12, X21, X32 \geq 0.$

Solution:

X11 = $ 50,000 invested in fund during first year
X12 = $118,000 invested in fund during second year
X21 = $100,000 invested in bond during first year
X32 = $ 50,000 invested in stock during second year

Objective function value
   = $38,800 return from this strategy

10. Let

X1 = Units of AT cards produced with internal funds
X2 = Units of DK cards produced with internal funds
X3 = Units of AT cards produced with external funds
X4 = Units of DK cards produced with external funds

Profit margins:

$X1 = \$25$

$X2 = \$12$

$X3 = \$25 - ((150 - 25) \times 0.18 \times 0.25) = \$19.375$

$X4 = \$12 - ((65 - 12) \times 0.18 \times 0.25) = \$ 9.615$

Financial ratio stipulation of bank manager:

Cash + receivables = 1.5 (outstanding loan + interest)

$\text{Cash} = 5{,}000 - 125\,X1 - 53\,X2$

$\text{Receivables} = 150\,X1 + 65\,X2 + 150\,X3 + 65\,X4$

$\text{Loan} = 125\,X3 + 53\,X4$

$\text{Interest} = (0.18 \times 0.25) \times (125\,X3 + 53\,X4)$

$= 5.625\,X3 + 2.385\,X4$

that is,

$$5{,}000 - 125\,X1 - 53\,X2 + 150\,X1 + 65\,X2 + 150\,X3 + 65\,X4$$
$$\geq 1.5(125\,X3 + 53\,X4 + 5.625\,X3 + 2.385\,X4)$$

Then

Maximize $25\,X1 + 12\,X2 + 19.375\,X3 + 9.615\ X4$

subject to $125\,X1 + 53\,X2 \qquad\qquad\qquad\qquad \leq 5{,}000$

$\qquad\qquad\qquad\qquad 125\,X3 + \quad 53\,X4 \leq 15{,}000$

$-25\,X1 - 12\,X2 + 45.9375\,X3 + 18.0775\,X4 \leq 5{,}000$

$X1,X2,X3,X4 \geq 0$

Solution:

$X1 = 0$ AT cards produced with internal funds

$X2 = 94.34$ DK cards produced with internal funds

$X3 = 0$ AT cards produced with external funds

$X4 = 283.02$ DK cards produced with external funds

Objective function value

$= \$3{,}853.30$ return from this strategy

B. Quick should therefore plan to produce

        Zero AT computer cards and
        377.35 DK computer cards in the next 90 days.

(Noninteger values may be interpreted as work in progress.)

2

Alternative
Constraints
and Duality

In Chapter 1, the optimization problem was illustrated in the form of a maximization of profit by the manufacture of products from scarce resources. These resource constraints were expressed in the form of a mathematical relationship being less than, or equal to, the right-hand-side constraint coefficient, for example,

Maximize $\quad\quad$ $4 X1 + 5 X2$ $\quad\quad\quad\quad\quad\quad\quad\quad\quad\quad\quad\quad$ **Equation Set 2.1**
subject to $\quad\quad\quad$ $X1 + 2 X2 \leq 7$ etc.

It may have already occurred to the reader that the constraints do not always take such convenient forms. The plant capacity constraint may be fixed, so that an equality constraint results:

$$X1 + 2 X2 = 7$$

Contemplation might then produce the following solution: choose one of the variables, solve the equation for the other, and a linear optimization problem in a single variable results. For example, by solving for X1, the problem of Chapter 1 becomes

Equation Set 2.2

$$X1 = 7 - 2\,X2$$

Maximize

$$4(7 - 2\,X2) + 5\,X2$$

subject to

$$(7 - 2\,X2) + X2 \leq 6$$
$$(7 - 2\,X2) + 4\,X2 \leq 12$$

This is all very well, and a solution is obtained (with compatible relationships), but it makes postoptimal analysis difficult (see Chapter 3 on parametric programming), because information concerning the X1 variable has been removed from the presentation. It also means that we must decide which variable to remove.

Assuming that such a substitution is not made, where does the linear programming procedure start, and what initial solution is required? One method is to adopt an artificial variable (not a slack variable) in that equation so that

$$X7 = 7 - X1 - 2\,X2$$

Inserting this into the simplex equations for the constraints, we obtain

Equation Set 2.3

$$X3 = 6 - X1 - X2$$
$$X7 = 7 - X1 - 2\,X2$$
$$X5 = 12 - X1 - 4\,X2$$

The interpretation of these equations via the simplex convention is

$$X3 = 6,\ X7 = 7,\ X5 = 12,\ X1 = 0,\ X2 = 0$$

Clearly this solution is not feasible since it violates the original constraint, X1 + 2 X2 = 7.

There are two basic ways of handling this state of affairs. One way is to discover a feasible solution by inserting an objective function composed of the sums of the infeasibilities; in this case minimize X7 (or maximize $-X7$). We then must write the equation as $X7 = 7 - X1 - 2\,X2$, with either X1 or X2 on the right-hand side, as previously. With more than one infeasible constraint, a more efficient choice of variable from the left-hand side is allowed.

An alternative way to produce a feasible solution, while still optimizing the objective function, is to rewrite the objective function in terms of all the variables. For example,

Maximize

$$4 X1 + 5 X2 + 0 X3 + 0 X5 - M X7$$

subject to

$$X1 + X2 + X3 = 6$$
$$X1 + 2 X2 + X7 = 7$$
$$X1 + 4 X2 + X5 = 12$$

Equation Set 2.4

Here, a large negative coefficient $(-M)$ has been inserted in the objective function for the artificial variable. This means that as the objective function is maximized, the artificial variable will be forced from the solution and cannot remain in the basis at an optimal solution (provided that the other constraints define a feasible solution).

A further complication might arise should the problem under consideration have a constraint such that the amount of a resource used includes a minimum amount of that resource. For example, if the problem were to define that at least 6 tons of sunflower oil be used, the constraint becomes

$$X1 + X2 \geqslant 6$$

This situation calls for us to introduce a negative slack variable:

$$X1 + X2 - X3 = 6$$

or

$$X3 = -6 + X1 + X2$$

Formulation of the complete simplex equations then becomes

$$X0 = 0 - 4 X1 - 5 X2$$

Equation Set 2.5

$$X3 = -6 + X1 + X2$$
$$X4 = 7 - X1 - 2 X2$$
$$X5 = 12 - X1 - 4 X2$$
$$X3 = -6, X4 = 7, X5 = 12, X0 = 0, X2 = 0$$

Because one of the variables in the basis takes a negative

value ($X3 = -6$), the formulation appears to violate the rules that have been followed previously. Feasibility will be restored by pivoting between $X1$ or $X2$ and $X3$. This example is quite trivial, but should several of the variables take negative values, the problem of choosing the correct pivot becomes more complex. An incorrect sequence of pivots on an infeasible solution could result in a hopeless cycling outside the feasible region.

The problem of entering the feasible region is the same as the one encountered with the problem of equality constraints, and the same solution offers itself, namely to include an artificial variable in addition to the slack variable:

Equation Set 2.6

$$\text{Maximize} \quad 4\,X1 + 5\,X2 + 0\,X3 + 0\,X4 + 0\,X5 - M\,X6$$

$$\text{subject to} \quad
\begin{aligned}
X1 + X2 - X3 \phantom{{}+X4+X5} + X6 &= 6 \\
X1 + 2\,X2 \phantom{{}-X3} + X4 \phantom{{}+X5+X6} &= 7 \\
X1 + 4\,X2 \phantom{{}-X3+X4} + X5 \phantom{{}+X6} &= 12
\end{aligned}$$

The initial solution is

$$X6 = 6, \; X4 = 7, \; X5 = 12$$

For example,

Equation Set 2.7

$$\text{Minimize} \quad 6\,Y1 + 7\,Y2 + 12\,Y3$$

$$\text{subject to} \quad
\begin{aligned}
Y1 + Y2 + Y3 &\geq 4 \\
Y1 + 2\,Y2 + 4\,Y3 &\geq 5
\end{aligned}$$

Notice that this is a *minimize* optimization function. This can be dealt with by either

1. Choosing the largest positive values in the $Zj - Cj$ row of the simplex tableau, completing the solution when there are only negative entries left; or

2. Changing the sign of the coefficients in the objective function by multiplying each element by -1, and then maximizing the resultant function.

We shall use the latter approach.

Maximize $\quad -6\,Y1 - 7\,Y2 - 12\,Y3 + 0\,Y4 + 0\,Y5 - M\,Y6 - M\,Y7$ **Equation Set 2.8**

subject to

$$Y1 + \quad Y2 + \quad Y3 - \quad Y4 \qquad + \quad Y6 \qquad\qquad = 4$$

$$Y1 + 2\,Y2 + 4\,Y3 \qquad\qquad - \quad Y5 \qquad\qquad + \quad Y7 = 5$$

The simplex iterations then become

Tableau 2.1

Cj			-6	-7	-12	0	0	$-M$	$-M$	
Ci	Basis	B	Y1	Y2	Y3	Y4	Y5	Y6	Y7	
$-M$	Y6	4	1	1	1	-1	0	1	0	
$-M$	Y7	5	1	2	4	0	-1	0	1	*
	Zj	$-9M$	$-2M$	$-3M$	$-5M$	M	M	$-M$	$-M$	
	Zj$-$Cj		$(-2M+6)$	$(-3M+7)$	$(-5M+12)$	M	M	0	0	
				**						
$-M$	Y6	2.75	0.75	0.5	0	-1	0.25	1	-0.25	*
-12	Y3	1.25	0.25	0.5	1	0	-0.25	0	0.25	
	Zj	$(-2.75M-15)$	$(-.75M-3)$	$(-.5M-6)$	-12	M	$(-.25M+3)$	$-M$	$(.25M-3)$	
	Zj$-$Cj		$(-.75M+3)$	$(-.5M+1)$	0	M	$(-.25M+3)$	0	$(1.25M-3)$	
				**						
-6	Y1	3.67	1	0.67	0	-1.33	0.33	1.33	-0.33	
-12	Y3	0.33	0	0.33	1	0.33	-0.33	-0.33	0.33	*
	Zj	-26	-6	-8	-12	4	2	-4	-2	
	Zj$-$Cj		0	-1	0	4	2	$(-4+M)$	$(-2+M)$	
				**						
-6	Y1	3	1	0	-2	-2	1	2	-1	
-7	Y2	1	0	1	3	1	-1	-1	1	
	Zj	-25	-6	-7	-9	5	1	-5	-1	
	Zj$-$Cj		0	0	3	5	1	$(-5+M)$	$(-1+M)$	

Remember that * indicates pivot row and ** indicates pivot column.

This solution is optimal. Note that once the artificial variable has left the basis, it can never reenter the basis. Therefore, there is no need to carry throughout the calculation any columns of a simplex tableau that represent a nonbasic artificial variable. The value of the optimization function at this solution is $-1 \times -25 = 25$.

THE DUAL SIMPLEX METHOD

In the previous examples of the ordinary simplex method, the basis is kept feasible and iterated in order to achieve optimality. A method that does not follow this route is the *dual simplex method*, which differs from the ordinary simplex method in the way the variables leaving and entering the basis are chosen. The steps of the dual simplex method are summarized as follows:

1. Remove from the basis the variable with the most negative current basis value Bi.

2. Introduce into the basis the variable with the least Cj/Arj for Arj < 0 to keep the solution optimal and ensure that the incoming variable will have a nonnegative current value. (Note that r refers to the row and A signifies the specific element within the A matrix.)

3. Compute the tableaux as in the ordinary simplex method.

For example

Equation Set 2.9   Minimize $6\,Y1 + 7\,Y2 + 12\,Y3$

subject to $Y1 + Y2 + Y3 \geqslant 4$

$Y1 + 2\,Y2 + 4\,Y3 \geqslant 5$

or

Equation Set 2.10   Minimize $6\,Y1 + 7\,Y2 + 12\,Y3$

subject to $Y1 + Y2 + Y3 - Y4 \quad\quad = 4$

$Y1 + 2\,Y2 + 4\,Y3 \quad\quad - Y5 = 5$

Basis	B	Y1	Y2	Y3	Y4	Y5		Tableau 2.2
Y4	−4	−1	−1	−1	1	0		
Y5	−5	−1	−2	−4	0	1	**	
Obj	0	6	7	12	0	0		
				*				
Y4	−2.75	−0.75	−0.5	0	1	−0.25**		
Y3	1.25	0.25	0.5	1	0	−0.25		
Obj	−15	3	1	0	0	3		
			*					
Y2	5.5	1.5	1	0	−2	0.5		
Y3	−1.5	−0.5	0	1	1	−0.5	**	
Obj	−20.5	1.5	0	0	2	2.5		
		*						
Y2	1	0	1	3	1	−1		
Y1	3	1	0	−2	−2	1		
Obj	−25	0	0	3	5	1		

This is the first feasible solution, which is also the optimal solution. Notice the similarity of the problem formulation just solved,

Minimize $6 Y1 + 7 Y2 + 12 Y3$ **Equation Set 2.11**
subject to $Y1 + Y2 + Y3 \geq 4$
                     $Y1 + 2 Y2 + 4 Y3 \geq 5$

with the problem solved in Chapter 1:

Maximize $4 X1 + 5 X2$ **Equation Set 2.12**
subject to $X1 + X2 \leq 6$
                     $X1 + 2 X2 \leq 7$
                     $X1 + 4 X2 \leq 12$

Compare the solutions

Basis	B	Y1	Y2	Y3	Y4	Y5	Y6	Y7
Y1	3							
Y2	1							
Profit	25							
$Z_j - C_j$		0	0	3	5	1	—	—
		Structural			Slack		Artificial	

Basis	B	X1	X2	X3	X4	X5
X5	3					
X1	5					
X2	1					
Profit	25					
$Z_j - C_j$		0	0	3	1	0
		Structural		Slack		

The first thing that is apparent is that one problem formulation is roughly equivalent to the other formulation turned on its side: Rows are related to columns, and the constraint boundary values of one problem are related to the optimization function coefficients of the other. Furthermore, the solutions contain the same values. Values in the basis column of one solution are represented in the shadow values of the other solution. This is known as a *dual relationship*. A linear programming problem always has a dual formulation, known as its *dual*. The original formulation is known as the primal formulation, hence the dual of a dual is the primal. A closer examination of these problems is fruitful.

Consider the original problem confronting the edible fats manufacturer, which we shall now refer to as the primal formulation. We approached the solution by deciding on the number of products to manufacture by allocating the products among the scarce resources. The basis of the primal solution therefore defines the optimum number of products to produce and the marginal values of extra units of resource at the optimum solution.

Now look at the problem from the point of view of the director of the parent corporation of which the manufacturer of edible fats is a subsidary. He (or she, naturally) examines the costs of sunflower oil, plant capacity, and manpower. He defines the unit costs of the sunflower oil, plant capacity, and manpower to be Y1, Y2, and Y3 respectively. His total cost is therefore 6 Y1 + 7 Y2 + 12 Y3, which he wishes to minimize. To do this he considers the net revenue he will achieve by subdividing his resources. For example, margarine produces $4 per ton and each ton requires one unit of sunflower oil, one of plant capacity, and one of manpower. Thus, taking into account unused capacities within the system, he knows that for margarine

$$Y1 + Y2 + Y3 \geqslant 4$$

Also, with unused capacities, the relationship for cooking oil yields

$$Y1 + 2 Y2 + 4 Y3 \geqslant 5$$

Hence, by allocating resources to products in order to minimize costs, the corporate director arrives at the dual of the manager's problem of allocating products to resources to maximize profit.

The solution to the dual problem is, therefore, the unit costs of the resources *at the optimal solution*. These are equivalent to the shadow prices of the primal solution.

It should be stressed that these prices hold only at, or near, the optimal solution. They may not be used as control prices for, say, budgetary control purposes, since their range of applicability is limited. Furthermore, many cost variables that must be controlled in practice do not have shadow prices; manpower is an example in this problem. This point is examined and developed further in Chapter 3, on parametric programming.

FINDING THE DUAL

It may be observed that every constrained profit maximization problem has its associated dual problem of cost minimization. It is irrelevant whether the functions describing these are linear or not. The handling of nonlinear duals can be complex, but in the linear case there exists a convenient relationship between the primal and its dual. This may be summarized below:

Primal	Dual
Maximizes	Minimizes
Objective function	Constraint boundaries
Constraint boundaries	Objective function
j^{th} column of coefficients	j^{th} row of coefficients
i^{th} row of coefficients	i^{th} column of coefficients
i^{th} relationship $\leqslant$	i^{th} variable nonnegative
j^{th} variable nonnegative	j^{th} relationship $\geqslant$
j^{th} variable unrestricted in sign	j^{th} relationship $=$
i^{th} relationship $=$	i^{th} variable unrestricted in sign
i^{th} relationship $\geqslant$	i^{th} variable nonpositive

The three relationships below the line perhaps require elaboration. The standard relationships are conveniently depicted in a Tucker diagram:

Figure 2.1

PRIMAL

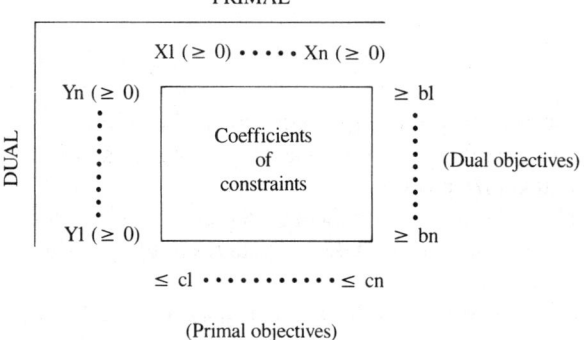

(Primal objectives)

Any linear programming problem can be expressed in this way; for example,

Equation Set 2.13

Maximize $4X_1 + 5X_2 + 3X_3$

subject to $X_1 + X_2 + 4X_3 \leq 6$

$X_1 + 2X_2 + X_3 = 7$

$X_1 + 4X_2 + 2X_3 \geq 12$

where X_1 is nonpositive

X_2 is nonnegative

X_3 is unrestricted in sign

Manipulating this in order to achieve a standard format, first by replacing the variable with only nonnegative variables, we get

Equation Set 2.14

Maximize $-4X_1' + 5X_2 + 3X_3' - 3X_3''$

subject to $-1X_1' + 1X_2 + 4X_3' - 4X_3'' \leq 6$

$-1X_1' + 2X_2 + 1X_3' - 1X_3'' = 7$

$-1X_1' + 4X_2 + 2X_3' - 2X_3'' \geq 12$

where X_1' is nonnegative

X_2 is nonnegative

X_3' is nonnegative

X_3'' is nonnegative

The next stage is to obtain a standard Tucker form:

Minimize $\quad\quad$ $4\,X1' - 5\,X2 - 3\,X3' + 3\,X3''$ $\hspace{3cm}$ **Equation Set 2.15**

subject to $\quad\quad$ $+ 1\,X1' - 1\,X2 - 4\,X3' + 4\,X3'' \geqslant -6$

$\quad\quad\quad\quad\quad - 1\,X1' + 2\,X2 + 1\,X3' - 1\,X3'' \geqslant \quad 7$

$\quad\quad\quad\quad\quad + 1\,X1' - 2\,X2 - 1\,X3' + 1\,X3'' \geqslant -7$

$\quad\quad\quad\quad\quad - 1\,X1' + 4\,X2 + 2\,X3' - 2\,X3'' \geqslant \quad 12$

Notice that

$$-1\,X1' + 2\,X2 + 1\,X3' - 1\,X3'' = 7$$

is equivalent to the pair

$$-1\,X1' + 2\,X2 + 1\,X3' - 1\,X3'' \leqslant 7$$
$$-1\,X1' + 2\,X2 + 1\,X3 \ - 1\,X3'' \geqslant 7$$

The standard form is then converted to its dual thus:

Maximize $\quad\quad$ $- 6\,Y1 + 7\,Y2 - 7\,Y3 + 12\,Y4$ $\hspace{2.5cm}$ **Equation Set 2.16**

subject to $\quad\quad$ $+ 1\,Y1 - 1\,Y2 + 1\,Y3 - \quad 1\,Y4 \leqslant \quad 4$

$\quad\quad\quad\quad\quad - 1\,Y1 + 2\,Y2 - 2\,Y3 + \quad 4\,Y4 \leqslant -5$

$\quad\quad\quad\quad\quad - 4\,Y1 + 1\,Y2 - 1\,Y3 + \quad 2\,Y4 \leqslant -3$

$\quad\quad\quad\quad\quad + 4\,Y1 - 1\,Y2 + 1\,Y3 - \quad 2\,Y4 \leqslant \quad 3$

$\quad\quad\quad\quad$ where Y1 is nonnegative

$\quad\quad\quad\quad\quad\quad\quad$ Y2 is nonnegative

$\quad\quad\quad\quad\quad\quad\quad$ Y3 is nonnegative

$\quad\quad\quad\quad\quad\quad\quad$ Y4 is nonnegative

This can be manipulated to its equivalent:

Maximize $\quad\quad$ $6\,Y1' + 7\,Y2' + 12\,Y4$ $\hspace{3cm}$ **Equation Set 2.17**

subject to $\quad\quad$ $- 1\,Y1' - 1\,Y2' - \quad 1\,Y4 \leqslant 4$

$\quad\quad\quad\quad\quad - 1\,Y1' - 2\,Y2' - \quad 4\,Y4 \geqslant 5$

$\quad\quad\quad\quad\quad + 4\,Y1' + 1\,Y2' + \quad 2\,Y4 = 3$

$\quad\quad\quad\quad$ where Y1' is nonpositive

$\quad\quad\quad\quad\quad\quad\quad$ Y2' is unrestricted in sign

$\quad\quad\quad\quad\quad\quad\quad$ Y4 $\,$ is nonnegative

A practical point: if the primal formulation is unbounded (the optimal solution may be improved to infinity), then the dual formulation will be inconsistent. The practical user should not waste time attempting to optimize such systems.

THE PRACTICAL USEFULNESS OF THE DUAL

As was observed earlier, the solution to a primal problem also generated the solution to the dual formulation. In this case, is there any use in considering the dual form? As we saw earlier, in Chapter 1, computer programs that are used in practice to solve these optimization problems use the product form of the revised simplex algorithms. Solution times for these algorithms depend on the number of rows; where the number of nonbasic variables is larger than the number of basic variables, the solution is more efficient. Thus savings of time may be significant in the format used to solve a particular problem. Alternatively, a formulation that has a large number of variables with a few constraints can be difficult to edit because of the limited screen width of the terminal. It is much more convenient to turn the formulation on its side, generating a long file of several equations that fit the screen nicely, even if it is computationally less efficient.

COMPUTER SOLUTIONS

In order to process a dual formulation or a formulation involving other than "less-than" constraints, it is only necessary to supply these instructions. For example, suppose that the dual of the three-variable problem solved in Chapter 1 has been prepared and stored in a permanent file labeled TAPE33. Log on the system, and once you have the operating system prompt, type

```
/ GET,TAPE33
```

Then, after you receive the "/" prompt again, in order to find out that it is indeed the required file, type

```
/ COPY,TAPE33
```

This will result in the output

Equation Set 2.18

```
MIN 6 Y1 + 7 Y2 + 12 Y3
ST
Y1 + Y2 + Y3 > 4
Y1 + 2 Y2 + 4 Y3 > 5
Y1 + Y2 + 2 Y3 > 3
END
GO
EOI ENCOUNTERED.
```

It is not necessary to put spaces in the expressions; in fact, it is quicker not to do so. It is a matter of preference (and storage).
        The file markers will now be at the end of this file, so in order to be able to use it again you must reset these using the command

```
/ REWIND,TAPE33
```

The system will respond with

```
REWIND,TAPE33
/
```

When this has been achieved; we are ready to process this file through LINDO. Type

```
/ FIND,LINDO
```

The response is

```
1 LINDO PROGRAM (CSU JAN-10-84)
:
```

Your reply,

```
:  TAKE
```

prompts the response

```
FILE NUMBER:
>
```

You enter "33." Processing then proceeds with this output:

```
LP OPTIMUM FOUND AT STEP          3

OBJECTIVE FUNCTION VALUE

1)              25.000000
VARIABLE          VALUE         REDUCED COST
Y1             3.000000              .000000
Y2             1.000000              .000000
Y3              .000000             3.000000


ROW         SLACK OR SURPLUS    DUAL PRICES
2)                  .000000        -5.000000
3)                  .000000        -1.000000
4)                10.000000          .000000
NO  OF  ITERATIONS=          3

DO  RANGE  (SENSITIVITY)  ANALYSIS ?  >
```

You reply

```
>  N
:  QUIT
/  BYE
```

Note that in this solution, the dual prices are negative. In the cases considered in Chapter 1, a negative dual price $(Z_j - C_j)$ would imply a nonoptimal solution. However, since the constraints on this solution are of the "greater than or equal to" form, this implies the use of a negative slack variable. The negative dual price reported in this solution represents this. It also implies that, as a minimization problem has been solved, the optimization function can be reduced by that dual price by the tightening of the particular constraint. It is, as discussed previously, the marginal cost of the constraint.

CHAPTER 2 PROBLEMS

The solutions to questions 1 through 10 will be found at the end of Chapter 2.

1. A manufacturer produces two types of plastic cladding. These have the trade names Ankalor and Beslite. One yard of Ankalor requires 8 lb of polyamine, 2.5 lb of diurethane and 2 lb of monomer. A yard of Beslite needs 10 lb of polyamine, 1 lb of diurethane, and 4 lb of monomer. The company has in stock 80,000 lb of polyamine, 20,000 lb of diurethane, and 30,000 lb of monomer. Both plastics can be produced by alternate parameter settings of the production plant, which is able to produce sheeting at the rate of 12 yards per hour. A total of 750 production plant hours are available for the next planning period. The contribution to profit on Ankalor is $10 per yard and $20 per yard on Beslite.

The company has a contract to deliver at least 3,000 yards of Ankalor. What production plan should be implemented in order to maximize the contribution to the firm's profit from this product division?

Should the company wish to avoid the contract on Ankalor, what will the solution indicate as a reasonable price ceiling for the company's negotiators?

2. The criterion for success in a particular organization is return on capital employed. The organization has two divisions employing different amounts of capital, each resulting in different returns. The activity levels of the divisions are described in transactions, which incidentally may span more than one accounting period per year. The amount of capital required by Division A is $2,500 per transaction, while that of Division B is $1,500 per transaction. There is also a headquarters overhead allocated at $2,500 per year. Profits for the divisions are $4,000 and $3,000 per transaction, respectively, but there is a franchise charge of $4,000 that must be met annually. Federal policy restricts the number of allowable transactions each year to five. Each transaction by Division A requires a headquarters support staff of 3, while transactions by B need a staff of 5 per transaction. There are 20 head-quarters support staff available to the divisions.

Calculate the optimal transaction policy for the divisions for next year and the return on capital employed assuming steady-state conditions.

3. R. Borist and Sons are formulating a new preparation to treat trees susceptible to Polish Palm-Tree disease. There are three active ingredients to the preparation. These are listed as Bi-cylo-p-dalamate, Benzoil, and T.P. The company has determined that there must be between 10 and 15 grams of the preparation in each gallon of solution used to spray the trees in order for the treatment to be effective. Due to interactions in the ingredients there must be more Bi-cylo-p-dalamate than Benzoil for effectiveness. This must not be increased too far, however, for it must not exceed twice the amount of T.P. in the preparation.

Now that R. Borist and Sons have these data, they can formulate the most cost-effective preparation for treatment of this troublesome disease. The costs of the base ingredients are $1.00 per gram of Bi-cyclo-p-dalamate, 30¢ per gram of Benzoil, and 90¢ per gram of T.P. What is the minimum cost preparation that will be effective?

4. I. Makecombe is a jobbing manufacturer who has con-tracted with the International Bottling Machine Company to supply IBM with machines for manufacturing Klein bottles.

The following delivery schedule has been agreed upon:

Month	Units to be Delivered
June	5
July	7
August	10
September	15

Makecombe believes that he might be better off making some of the units ahead of schedule, for these can be stored at a cost of $100 per month per unit. He is able to make ten machines per month in normal working hours.

Because of summer activities, many of his skilled bottle machinists decide to surf rather than work. He can, however, employ students in the summer. Thus his pay scale during the summer months varies with the surf conditions, but it has settled at the following table of rates:

Month	Cost to Produce a Bottle Machine at	
	Regular Pay	Overtime Pay
June	$100	$200
July	$200	$400
August	$300	$550
September	$150	$200

The maximum number of units that can be produced in overtime is four per month. What production schedule minimizes Makecombe's costs?

5. Blockwell Industries is a government aerospace contractor and is currently considering the production of some nonsensitive machined parts. There are four parts in question, each of which can be made in-house. The time needed to process each part using the company machines is detailed below:

Part No.	Machine (minutes)				
	Cutting	Milling	Grinding	Polishing	Finishing
DRNBG	18	—	12	25	20
RDL	—	40	32	75	34
SDR2	20	80	53	54	—
OICDR	24	—	—	30	10

Each machine operates on a two-shift production run of 80 hours per week. Demand for each piece is 2,000 units per week. Blockwell can subcontract out its work at the following rates:

Part No.	In-house Cost ($)	Subcontracted Cost ($)
DRNBG	40.00	63.50
RDL	65.00	99.00
SDR2	37.50	55.20
OICDR	32.00	45.00

How should Blockwell plan its operations?

6. A-Front, manufacturer of gentlemen's underwear, has asked its packaging and distribution manager, Ophelia Close, to determine the maximum tonnage of goods that she can move from the East Coast factories to the West Coast distributors to plan for the launch of their new product, Legal Briefs. A-Front has a network of distribution centers across the nation. These are shown on the following schematic map, along with the tonnage that can be moved between each pair of points:

Figure 2.2

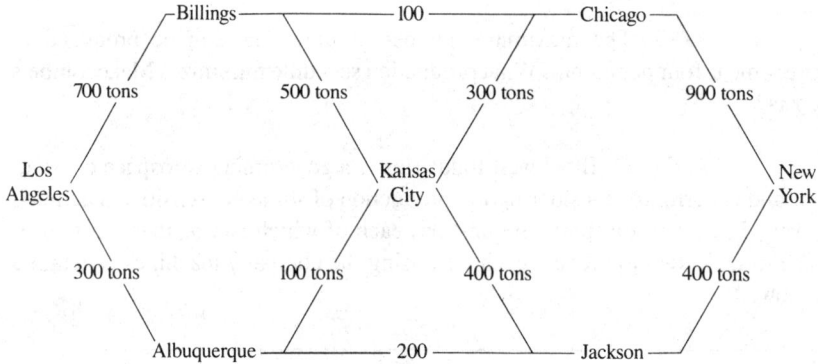

There is time for only one shipment.

What is the maximum tonnage that Close can ship between New York and Los Angeles for the product launch?

7. New Sprint has just taken control of Pay-per-Pros and has decided to use one of their paper-making machines for the manufacture of fan-fold computer paper. The machine was formerly used for the production of newsprint and manufactures paper in rolls of 1,000 yards long, 30 inches wide.

Computer paper is required to be supplied in widths of 9.5, 12, and 15 inches. A standard carton of fan-fold computer paper contains 3,600 sheets of 10-inch-long sheets. Orders have been received for the following:

Width	Cartons
9.5	130
12.0	120
15.0	85

What cutting policy minimizes the total production of paper and what is the waste produced? What policy minimizes the waste, and what quantity of paper is produced?

8. Tim Burr manufactures unfinished furniture. He wishes to expand his operation over the next four years by putting $500,000 into the business. Management consultants, Pye & Cohen, have identified three areas of investment:

Area	Maximum Effective Investment	Yield
Sales program	$25,000	25% after 1 yr.
2-year advertising program	$50,000	35% after 2 yrs.
3-year R&D program	$100,000	50% after 3 yrs.

How should Tim invest his $500,000?

9. Jack Uzi makes redwood spas and has contracted the following sales for the next six months:

Month	Contracted Sales ($)
April	25,000
May	30,000
June	50,000
July	55,000
August	60,000
Sept	40,000

Jack knows that his production at normal workload is 40 per month. He can produce an extra 10 per month on an overtime basis.

Unit cost per spa is $500 when manufactured on a normal basis and $800 when manufactured on an overtime basis. Inventory costs are $100 per month per unit. Cost to the customer, including shipping, is invoiced at $1,000 each. What is the optimal production schedule for Jack to minimize his cost of manufacture?

10. Find the dual of

Maximize $\qquad 4\,X1 + 3\,X2 - 6\,X3$

subject to $\qquad 2\,X1 - \quad X2 + 3\,X3 \leqslant 6$

$\qquad\qquad X1 + 7\,X2 - \quad X3 \geqslant 8$

$\qquad -3\,X1 \qquad\quad + 4\,X3 = 10$

where $X1 \geqslant 0$

$\qquad X2 \leqslant 0$

$\qquad X3$ is unrestricted

HOMEWORK PROBLEMS

Solutions to these problems do not appear in this text. Good luck!

11. The accounting firm of Adam Upp and B. Rich have determined their personnel requirements for the next three months to be 110, 125, and 140 accounting assistants, respectively. They anticipate hiring 95, 90, and 80 new employees during those three months, respectively. Of each group of new employees, one-tenth, one-fourth, and one-third have quit by the first, second, and third month of employment. During this time, productivity is 0.25, 0.5, and 0.75. It costs $500 to hire a new person and overtime premium is $1,500 per month. The limit on hires is 20 per month and overtime must not exceed 60 man-months. What policy should Adam Upp and B. Rich follow?

12. The portfolio management firm of I. D. Fawlt carries an investment portfolio on a set of stocks and bonds. They are currently considering new investment opportunities for $300,000. There are four stock options:

	A	B	C	D
Price per share	90	60	90	30
Annual rate of return	0.12	0.10	0.08	0.11
Risk factor	5	3	4	4

I. D. Fawlt insists on at least a 10 percent return on investments, and no one stock can account for more than 40 percent of the portfolio. What strategy maximizes income? What strategy minimizes risk?

13. Juan and A. Dew are owners of Going West Electronics, which manufactures microelectronic equipment and has assembly plants in the U.S.A. and in Taiwan. Their I.B.N. computer and A.P.T. phone systems have assembly costs as follows:

	Production Cost/Unit		Production Requirements
	Taiwan	U.S.A.	
I.B.N. computer	$600	$1,000	5,000
A.P.T. phones	$300	$ 870	2,500
Capacities	10,000	5,000	

What production schedules should the firm adopt?

14. The trash removal specialist firm M. T. Kahns has received a contract to transport 500,000 gallons of toxic waste to the Newport Back Bay landfill. It has apportioned $500,000 to the purchase of special vehicles for the contract. The data for these vehicles are as follows:

Vehicle	Capital Cost	Monthly Operating Cost	Payload Capacity
Rawls	$70,000	$750	8,000 gal
Murk	$50,000	$520	4,000 gal
Shevi	$30,000	$400	1,500 gal

Trip capacities of the vehicles are estimated at 20 per month for the Rawls, 25 per month for the Murk, and 30 per month for the Shevi. M. T. Kahns cannot add more than 14 vehicles to its fleet and has decided to buy at least 3 Shevi trucks for resale to the Bajha Cheese Co. at the completion of the contract. Also, the company is very patriotic, and wants at least 50 percent of its money to be spent in the U.S.A. (The Rawls is built in the U.K.; the others are assembled in the U.S.A.) What purchasing policy should M. T. Kahns adopt?

15. Electronic Village is a firm utilizing off-site workers who program computers at home, and submit their work to headquarters via telephone lines by downloading their production into the company computer.

Will B. Dunn is considering using two programming supervisors on two contracts. Because the supervisors must check each other's work, Dunn's policy is to balance their tasks. The two supervisors each have different skill levels in these areas, as follows:

	Supervisor 1	Supervisor 2
Standard APL segment	12 min	4 min
Standard ADA segment	6 min	8 min

It is Dunn's policy to schedule at least 10-hour segments of work to each operative. It is also his policy to ensure that the difference between any two operatives in amount of time scheduled cannot exceed one hour. The current contracts negotiated by Electronic Village return 50 percent more for ADA programs than APL. What is Dunn's best policy?

16. The P. Troll Oil Company is considering its blending strategies. It has crude oil available in its bunkers from three different sources of differing composition:

Type of Crude	Cost	Composition		
		A	B	C
1	$0.10	0.2	0.5	0.05
2	$0.15	0.4	0.3	0.1
3	$0.25	0.5	0.2	0

Type X gasoline must contain at least 40 percent of A and at most 0.5 percent of C. Type Y gasoline must contain at most 50 percent of B. Type Z gasoline must contain at least 20 percent of A and at most 40 percent of B. Demand for type X gasoline is 900,000 gal, for type Y is 400,000 gal, and for type Z is 10,000 gal. What blend minimizes the P. Troll's costs?

17. A manufacturer has forecast a demand of 5,000 units of its product next year. The product is assembled from three subassemblies, which previously have been manufactured in-house. The operation works an 8-hour day, 5-day week, 50 weeks per year. Each subassembly is processed by three sections in the factory. The following data apply:

Factory	Subassemblies (hours)		
Section	1	2	3
A	0.15	0.10	0.25
B	0.20	0.10	0.20
C	0.10	0.15	0.05

The following quotes have been obtained from outside manufacturers to subcontract the subassemblies:

Subassembly	In-house Cost	Subcontracted Cost
A	$3.00	$3.80
B	$1.60	$2.20
C	$4.40	$5.60

How should the manufacturer meet the projected demand at minimum cost?

18. Ms. Fun Kee operates the Chip Chop Shop where old microprocessors are recycled. She owns three chip dismounting machines, which differ in their speed of operation and success rate at salvaging microprocessors.

Machine	Success Rate (%)	Speed (units/hour)	Operating Cost (per hour)
A	86	300	$11.80
B	92	250	$10.40
C	78	400	$11.00

Kee has 5,000 microprocessors that she needs to salvage. She can only run the salvage machines for a single shift of 8 hours before maintenance. She also needs at least an 88 percent recovery rate. In addition, her employee is paid for 8 hours work, so Kee considers that there must be at least 8 hours salvage work done.

How should Ms. Kee organize her salvage operation?

19. Find the dual of the following:

Maximize $3 X1 + 6 X2 + 2 X3$
subject to $X1 + X2 - X3 \leqslant 10$
                  $- X1 + 2 X2 + 3 X3 \geqslant 4$
                  $4 X1 - 5 X2 - 7 X3 = 12$

where X1 is nonnegative
      X2 is nonpositive
      X3 is unrestricted

20. The Icon Sea, a property development company, has $1 million currently available for investment. It is considering four developments. These are summarized below, with negative entries signifying investment, positive entries being returns.

Year	Project ($K)			
	A	B	C	D
1	-50	-100	-30	-75
2	-150	-200	-40	-100
3	-300	-100	-90	-150
4	-40	300	10	-100
5	500	500	50	250
6	1,000	800	50	400
7	3,000	500	50	800
8	5,000	100	50	800
9	5,000	0	50	600
10	2,000	-5	50	100

In addition, Carl Icon has predicted his income stream over the next five years to be

Year	Income ($K)
1	250
2	200
3	350
4	300
5	150

Against this income, a tax rate of 30 percent will be levied.

The Icon Sea can borrow up to $1 million at 10 percent per annum, that is, the outstanding principal can never exceed $1 million. Surplus funds can be invested at 8 percent per annum.

Assuming that Icon Sea will be liable for a 50 percent tax payment, and that 75 percent of losses can be carried forward to the next tax period, what proportion of each project should Icon invest in? Solve it first ignoring tax, then taking into account tax.

CHAPTER 2 SOLUTIONS

1. Let

A = Number of yards produced of Ankalor
B = Number of yards produced of Beslite

Maximize	$10\ A + 20\ B$	
subject to	$8\ A + 10\ B \leq 80{,}000$	Polyamine
	$2.5\ A + 1\ B \leq 20{,}000$	Diurethane
	$2\ A + 4\ B \leq 30{,}000$	Monomer
	$A + B \leq 9{,}000$	Plant capacity
	$A \geq 3{,}000$	Contract

Maximize $10\ A + 20\ B + 0\ X1 + 0\ X2 + 0\ X3 + 0\ X4 + 0\ X5 - M\ X6$

subject to

$$8\ A + 10\ B + X1 = 80{,}000$$
$$2.5\ A + B + X2 = 20{,}000$$
$$2\ A + 4\ B + X3 = 30{,}000$$
$$A + B + X4 = 9{,}000$$
$$A - X5 + X6 = 3{,}000$$

Simplex tableaux:

C_i	Basis	B	C_j 10 A	20 B	0 X1	0 X2	0 X3	0 X4	0 X5	$-M$ X6	Q
0	X1	80,000	8	10	1	0	0	0	0	0	10,000
0	X2	20,000	2.5	1	0	1	0	0	0	0	8,000
0	X3	30,000	2	4	0	0	1	0	0	0	15,000
0	X4	9,000	1	1	0	0	0	1	0	0	9,000
$-M$	X6	3,000	1	0	0	0	0	0	-1	1	3,000*
	Z_j	$(-3{,}000M)$	$-M$	0	0	0	0	0	M	$-M$	
	$Z_j - C_j$		$(-M-10)$ **	-20	0	0	0	0	M	0	

(continued)

Ci	Cj Basis	B	10 A	20 B	0 X1	0 X2	0 X3	0 X4	0 X5	$-M$ X6	Q
0	X1	56,000	0	10	1	0	0	0	8	-8	5,600*
0	X2	12,500	0	1	0	1	0	0	2.5	-2.5	12,500
0	X3	24,000	0	4	0	0	1	0	2	-2	6,000
0	X4	6,000	0	1	0	0	0	1	1	-1	6,000
10	A	3,000	1	0	0	0	0	0	-1	1	—
	Zj	30,000	10	0	0	0	0	0	-10	10	
	Zj$-$Cj		0	-20 **	0	0	0	0	-10	$(10+M)$	
20	B	5,600	0	1	0.1	0	0	0	0.8	-0.8	
0	X2	6,900	0	0	-0.1	1	0	0	1.7	-1.7	
0	X3	1,600	0	0	-0.4	0	1	0	-1.2	1.2	
0	X4	400	0	0	-0.1	0	0	1	0.2	-0.2	
10	A	3,000	1	0	0	0	0	0	-1	1	
	Zj	142,000	10	20	2	0	0	0	6	-6	
	Zj$-$Cj		0	0	2	0	0	0	6	$-6+M$	

Solution: Make 3,000 yards of Alkalon and 5,600 yards of Beslite.

Contribution to profit from this operation is $142,000
There are 69,000 lb of diurethane unused
16,000 lb of spare monomer
and   400 yd of plant capacity unused.

The shadow price of polyamine is $2 per lb. The marginal cost of the Alkalon contract is $6 per yard. Solving without the contract constraint:

Ci	Cj Basis	B	10 A	20 B	0 X1	0 X2	0 X3	0 X4	Q
0	X1	80,000	8	10	1	0	0	0	8,000
0	X2	20,000	2.5	1	0	1	0	0	20,000
0	X3	30,000	2	4	0	0	1	0	7,500*
0	X4	9,000	1	1	0	0	0	1	9,000
	Zj	0	0	0	0	0	0	0	
	Zj$-$Cj		-10	-20	0	0	0	0	

Ci	Cj Basis	B	10 A	20 B	0 X1	0 X2	0 X3	0 X4	Q
0	X1	5,000	3	0	1	0	−2.5	0	
0	X2	12,500	2	0	0	1	−0.25	0	
20	B	7,500	0.5	1	0	0	0.25	0	
0	X4	1,500	0.5	0	0	0	−0.25	1	
	Zj	150,000	10	20	0	0	5	0	
	Zj − Cj		0	0	0	0	5	0	

The best policy is to manufacture 7,500 yards of Beslite, the total contribution to profit being $150,000. It will therefore be in the company's interest to tell its negotiators to have a ceiling of $150,000 − $142,000 = . $8,000 to negotiate out of the contract to supply the Ankalor cladding.

Computer solution:

```
/FIND,LINDO
:MAX 10A+20B
>ST
>8A+10B<80000
>2.5A+B<20000
>2A+4B<30000
>A+B<9000
>A>3000
>END
:GO
          LP OPTIMUM FOUND AT STEP        2
               OBJECTIVE FUNCTION VALUE
  1)              142000.000
VARIABLE              VALUE          REDUCED COST
  A              3000.000000             .000000
  B              5600.000000             .000000
ROW
               SLACK OR SURPLUS    DUAL PRICES
  2)                    .000000       2.000000
  3)               6900.000000        .000000
  4)               1600.000000        .000000
  5)                400.000000        .000000
  6)                    .000000      -6.000000
NO. OF ITERATIONS=        2
```

2. Let

X = Number of transactions performed by Division A
Y = Number of transactions performed by Division B

Maximize
subject to

$$\frac{4\ X + 3\ Y -\ 4}{2.5\,X + 1.5\,Y +\ 2.5}$$

$$X +\quad Y \leqslant\ 5$$
$$3\ X + 5\quad Y \leqslant 20$$
$$X, Y \geqslant 0$$

Let $\dfrac{X1}{Z} = X \qquad \dfrac{Y1}{Z} = Y$

then

Maximize
subject to

$$4\,X1 + 3\,Y1 -\ 4\,Z$$
$$5\,X1 + 3\,Y1 +\ 5\,Z = 2$$
$$X1 +\quad Y1 -\ 5\,Z \leqslant 0$$
$$3\,X1 + 5\,Y1 - 20\,Z \leqslant 0$$

The solution is

$$X1 = 0.2$$
$$Y1 = 0.2$$

Objective function = 1.08

3. For R. Borist's Polish Palm-Tree spray, let

X1 = Grams of Bi-cyclo-p-dalamate used in the preparation
X2 = Grams of Benzoil used
X3 = Grams of T.P. incorporated

Then

Minimize
subject to

$$1.00\,X1 + 0.30\,X2 + 0.90\,X3$$
$$1\quad X1 + 1\quad X2 + 1\quad X3 \geqslant 10$$
$$1\quad X1 + 1\quad X2 + 1\quad X3 \leqslant 15$$
$$1\quad X1 - 1\quad X2 \qquad\qquad \geqslant\ 0$$
$$-1\quad X1 \qquad\qquad 2\quad X3 \geqslant\ 0$$
$$X1, X2, X3, \geqslant 0$$

The solution is

$$X1 = 4 \text{ grams}$$
$$X2 = 4 \text{ grams}$$
$$X3 = 2 \text{ grams}$$
$$\text{Minimum cost of preparation} = \$7 \text{ per gallon}$$

4. For Mr. Makecombe's bottle machines:

$X1$ = Number of bottle machines produced in normal time in June

$X2$ = Number of bottle machines produced in normal time in July

$X3$ = Number of bottle machines produced in normal time in August

$X4$ = Number of bottle machines produced in normal time in September

$X5$ = Number of bottle machines produced in overtime in June

$X6$ = Number of bottle machines produced in overtime in July

$X7$ = Number of bottle machines produced in overtime in August

$X8$ = Number of bottle machines produced in overtime in September

$Y1$ = Number of machines stored from June to July

$Y2$ = Number of machines stored from July to August

$Y3$ = Number of machines stored from August to September

Minimize
$$100\,X1 + 200\,X2 + 300\,X3 + 150\,X4 + 200\,X5 + 400\,X6 + 550\,X7 + 200\,X8$$
$$+ 100\,Y1 + 100\,Y2 + 100\,Y3$$
subject to
$$
\begin{aligned}
X1 &&&&& \leq 10 \\
X2 &&&&& \leq 10 \\
X3 &&&&& \leq 10 \\
X4 &&&&& \leq 10 \\
X5 &&&&& \leq 4 \\
X6 &&&&& \leq 4 \\
X7 &&&&& \leq 4 \\
X8 &&&&& \leq 4 \\
X1 &+ X5 &- Y1 && &= 5 \\
X1 + X2 &+ X5 + X6 &- Y2 && &= 12 \\
X1 + X2 + X3 &+ X5 + X6 + X7 &- Y3 &&&= 22 \\
X1 + X2 + X3 + X4 &+ X5 + X6 + X7 + X8 &&&&= 37 \\
\end{aligned}
$$
$$X1, X2, X3, X4, X5, X6, X7, X8, Y1, Y2, Y3 \geq 0$$

Solution:

Minimum cost of operation = $ 7,600
Produce in

June,	5 machines in normal working hours
July,	10 machines in normal working hours
August,	8 machines in normal working hours
Sept,	10 machines in normal working hours
	4 machines in overtime

Store

July – August	3 machines
August – September	1 machine

5. For Blockwell let

X_i = Number of unit i produced in-house
Y_i = Number of unit i subcontracted

Then

Minimize
$$40\,X_1 + 65\,X_2 + 37.5\,X_3 + 32\,X_4 + 63.5\,Y_1 + 99\,Y_2 + 55.2\,Y_3 + 45\,Y_4$$
subject to
$$18\,X_1 \qquad\qquad + 20\ X_3 + 24\,X_4 \qquad\qquad\qquad\qquad\quad \le 4{,}800$$
$$\qquad\qquad 40\,X_2 + 80\ X_3 \qquad\qquad\qquad\qquad\qquad\qquad = 4{,}800$$
$$12\,X_1 + 32\,X_2 + 53\ X_3 \qquad\qquad\qquad\qquad\qquad\qquad = 4{,}800$$
$$25\,X_1 + 75\,X_2 + 54\ X_3 + 30\,X_4 \qquad\qquad\qquad\qquad = 4{,}800$$
$$20\,X_1 + 34\,X_2 + \qquad\quad 10\,X_4 \qquad\qquad\qquad\qquad\qquad = 4{,}800$$
$$X_1 + \qquad\qquad\qquad\qquad\qquad Y_1 \qquad\qquad\qquad\qquad = 2{,}000$$
$$\qquad X_2 \qquad\qquad\qquad\qquad\qquad + \ Y_2 \qquad\qquad\qquad = 2{,}000$$
$$\qquad\qquad\qquad X_3 \qquad\qquad\qquad\qquad + \ Y_3 \qquad\qquad = 2{,}000$$
$$\qquad\qquad\qquad\qquad X_4 \qquad\qquad\qquad\qquad\qquad + \ Y_4 = 2{,}000$$
$X_i, Y_i \ge 0$

Solution:

$X_1 =$	192	DRNBG	units produced in-house
$X_2 =$	0	RDL	units produced in-house
$X_3 =$	0	SDR2	units produced in-house
$X_4 =$	0	OICDR	units produced in-house

X5 = 1,808   DRNBG units subcontracted
X6 = 2,000   RDL   units subcontracted
X7 = 2,000   SDR2   units subcontracted
X8 = 2,000   OICDR units subcontracted
Objective function value = $520,888 minimum production cost.

6. For A-Front, label sites as follows:

Los Angeles   1
Billings   2
Albuquerque   3
Kansas   4
Chicago   5
Jackson   6
New York   7

Then the amount of goods shipped between two centers is X12 (amount shipped to Los Angeles from Billings), X13 (to Los Angeles from Albuquerque), etc.
        Then

Maximize   X12 + X13
subject to   X12 $\leq$ 700
                X13 $\leq$ 300
                X25 $\leq$ 100
                X24 $\leq$ 500
                X34 $\leq$ 100
                X36 $\leq$ 200
                X45 $\leq$ 300
                X46 $\leq$ 400
                X57 $\leq$ 900
                X67 $\leq$ 400
                X12 − X25 − X24   =   0
                X13 − X34 − X36   =   0
                X24 + X34 − X45 − X46 =   0
                X25 + X45 − X57   =   0
                X46 + X36 − X67   =   0
                Xij $\geq$ 0

Solution:

To	From	Amount Shipped
LA	Billings	500
LA	Albuquerque	300
Billings	Chicago	100
Billings	Kansas	400
Albuquerque	Kansas	100
Albuquerque	Jackson	200
Kansas	Chicago	300
Kansas	Jackson	200
Chicago	New York	400
Jackson	New York	400

Total amount shipped: 800 tons

7. Since one roll of paper is 1,000 yds long, it will produce 3,600 10-inch long sheets for one carton.

Let X_i = Number of rolls processed by cutting alternatives 1 through 5:

Alternative	Number of Rolls			Waste
	9.5 inch	12 inch	15 inch	
1	3	0	0	1.5
2	0	2	0	6
3	0	1	1	3
4	0	0	2	0
5	1	0	1	5.5

To minimize total paper production:

$$
\begin{aligned}
\text{Minimize} \quad & X_1 + X_2 + X_3 + X_4 + X_5 \\
\text{subject to} \quad & 3\,X_1 \hspace{4.2cm} + X_5 \geq 130 \\
& \hspace{1.3cm} 2\,X_2 + X_3 \hspace{2.3cm} \geq 120 \\
& \hspace{2.8cm} X_3 + 2\,X_4 + X_5 \geq 85 \\
& X_1, X_2, X_3, X_4, X_5 \geq 0
\end{aligned}
$$

Solution:

Total number of rolls used $= 145.833$

43.3 cut on alternative 1

60   cut on alternative 2

42.5 cut on alternative 4

$$
\begin{aligned}
\text{Waste} = \quad &43.33 \times 1000 \times 3 \times 12 \times 1.5 \\
+\ &60 \quad\ \times 1000 \times 3 \times 12 \times 6 \\
+\ &42.5 \ \ \times 1000 \times 3 \times 12 \times 0 \\
= \ &(2{,}339{,}982) + (12{,}960{,}000) \\
= \ &15{,}299{,}982 \text{ square inches}
\end{aligned}
$$

To minimize waste:

Minimize

$1.5\, X1 + 6\, X2 + 3\, X3 \qquad + 5.5\, X5$

subject to

$$
\begin{aligned}
3\, X1 \qquad\qquad\quad &+ \quad X5 \geqslant 130 \\
2\, X2 + \ X3 \qquad\qquad &\geqslant 120 \\
X3 + 2\, X4 + \quad X5 &\geqslant \ 85
\end{aligned}
$$

$X1, X2, X3, X4, X5 \geqslant 0$

The solution is the same as the first solution.

8. For Tim Burr, let

$Xij =$ Amount invested in alternative i in year j

$Yk =$ Amount held over

Then

Maximize $Y4$

subject to

$$
\begin{aligned}
X11 &\qquad\qquad\qquad\qquad\qquad\qquad\qquad\qquad \leqslant \ 25{,}000 \\
X21 &\qquad\qquad\qquad\qquad\qquad\qquad\qquad\qquad \leqslant \ 50{,}000 \\
X31 &\qquad\qquad\qquad\qquad\qquad\qquad\qquad\qquad \leqslant 100{,}000 \\
X11 + \quad X21 \qquad\qquad &+ \quad X31 + Y1 \qquad = 500{,}000 \\
X12 + \quad X22 - 1.25\, X11 + \quad &X32 + Y2 - Y1 = \qquad 0 \\
X13 + \quad X23 - 1.25\, X12 - 1.35\, X21 & \\
&+ \quad X33 + Y3 - Y2 = \qquad 0 \\
X14 - 1.35\, X22 - 1.25\, X13 - 1.5\ &X31 + Y4 - Y3 = \qquad 0
\end{aligned}
$$

Solution: At the end of the planning period, Burr's $500,000 will be worth $603,750. He will invest

$25,000 each year into sales programs,
$50,000 in year 1 and 2 into advertising,
and
$100,000 in year 1 into R&D.

9. For Jack Uzi, let

W_i = Regular production in month i
X_i = Overtime production in month i
Y_i = Inventory units in month i

Then

Minimize
$$500 (W_1 + W_2 + W_3 + W_4 + W_5 + W_6)$$
$$+ 800 (X_1 + X_2 + X_3 + X_4 + X_5 + X_6)$$
$$+ 100 (Y_1 + Y_2 + Y_3 + Y_4 + Y_5 + Y_6)$$

subject to
$$W_i \leq 40$$
$$X_i \leq 10$$
$$W_1 + X_1 - Y_1 \qquad = 25$$
$$W_2 + X_2 + Y_1 - Y_2 = 30$$
$$W_3 + X_3 + Y_2 - Y_3 = 50$$
$$W_4 + X_4 + Y_3 - Y_4 = 55$$
$$W_5 + X_5 + Y_4 - Y_5 = 60$$
$$W_6 + X_6 + Y_5 \qquad = 40$$
$$W_i, X_i, Y_i \geq 0$$

Solution: Jack should produce

40 units in standard time April through September
10 units in overtime in July
10 units in overtime in August

He should store

15 units in April
25 units in May
15 units in June
10 units in July

10. The dual is

Minimize
subject to

$$6\,Y1 + 8\,Y2 + 10\,Y3$$
$$2\,Y1 + 1\,Y2 -\ \ 3\,Y3 \geqslant 4$$
$$-\,1\,Y1 + 7\,Y2 \qquad\quad \leqslant 3$$
$$-\,3\,Y1 + 1\,Y2 -\ \ 4\,Y3 = 6$$
$$Y1 \geqslant 0$$
$$Y2 \leqslant 0$$

Y3 unrestricted

3

Parametric Programming

In Chapter 2, the duality of linear programming solutions and the significance of shadow values, or shadow prices, were examined. These shadow values represent the difference between the slope of the objective function and the slope of the constraint that is a constituent of the solution. It was pointed out that these shadow values are marginal values, existing only in the region of the optimum solution. Although the marginal price of a resource constraint might be indicated as, say, $3 per ton, there was no clue as to the number of tons one should seek to obtain. The shadow price might hold for several tons, or it might not hold over the range of even one full ton. Parametric programming enables us to perform sensitivity analyses on the solutions we obtain, and to investigate the ranges over which these hold. There are essentially two areas of particular interest: the sensitivity of our solutions when we vary the coefficients of the objective function and the sensitivity of our solutions when we vary the constraint capacities, or right-hand sides of the constraint expressions. The two forms of parametric programming are therefore known as parametric programming on the objective function and the other as parametric programming on the right-hand side.

PARAMETRIC PROGRAMMING ON THE OBJECTIVE FUNCTION

In the example depicted in Chapter 1, the profits on margarine and cooking oil were given as \$4 and \$5 per ton, respectively. Now we want to examine how the solution varies as the two profits rise simultaneously, with the profit on margarine rising twice as fast as that on cooking oil. The original profit function $f(x)$ was given by

$$f(x) = 4\,X1 + 5\,X2$$

This original function will now be modified by another function of the form $Q \cdot f'(x)$. The new composite function will therefore be of the form

$$F(x) = f(x) + Q \cdot f'(x)$$

The modifying function describes how the objective function is changing; in this case it is $(2\,X1 + X2)$. The scalar multiplier Q will then signify the rise in cooking-oil profit. This is often expressed as the change in the parameters being so much per year, so that Q will often signify the years of change in many applications. The new composite optimization function will therefore be

$$F(x) = 4\,X1 + 5\,X2 + Q(2\,X1 + X2)$$

When $Q = 0$, the original solution is obtained at $X1 = 5$, $X2 = 1$, with a profit of \$25. This is represented by the original simplex solution in equation format as

			X4	X3
X0	=	− 25	1	3
X5	=	3	3	− 2
X1	=	5	1	− 2
X2	=	1	− 1	1

Equation Set 3.1

This arose from the initial simplex form of

			X1	X2
X0	=	0	− 4	− 5
X3	=	6	− 1	− 1
X4	=	7	− 1	− 2
X5	=	12	− 1	− 4

Equation Set 3.2

Since the new parametric optimization function is a version of the original function modified by having an additional part, the initial set of equations can be conveniently represented as

Equation Set 3.3

			X1	X2
X0	=	0	-4	-5
X0'	=	0	$-2Q$	$-Q$
X3	=	6	-1	-1
X4	=	7	-1	-2
X5	=	12	-1	-4

In this case, to update the composite optimization function to point C in Figure 1.1, it is necessary merely to rewrite X0' in terms of X4 and X3:

Equation Set 3.4

		X4	X3	X3
X0	=	-25	1	3
X0'	=	$-11Q$	$-Q$	3Q
X5	=	3	3	-2
X1	=	5	1	-2
X2	=	1	-1	1

Note: X0' $= 0 - 2Q(5 + X4 - 2 X3) - Q(1 - X4 + X3)$
$\qquad\quad = -11Q - Q X4 + 3Q X3$

The values of the coefficients of the profit function and the actual values of profit for any value of Q are found by adding the two elements together. Thus, while Q lies between 0 and 1, the solution will be as depicted above, and will remain optimal. The profit of the solution will be \$(25 + 11Q). When Q exceeds 1, however, the combined coefficient of X4 becomes negative, which indicates that the solution may be improved by exchanging X4 in the same manner as the optimization of Chapter 1. This results in point B in Figure 1.1.

Equation Set 3.5

		X2	X3	X3
X0	=	-24	-1	4
X0'	=	$-12Q$	Q	2Q
X5	=	6	-3	1
X1	=	6	-1	-1
X4	=	1	-1	1

This solution remains optimal for all values of $Q > 1$, with a profit given by the expression $(24 + 12Q)$.

The above analysis is presented in the form of a functional modification of the objective function. Comparing the equation format with the graphical representaion of Figure 1.1 shows that as the objective function is modified, its slope with reference to the feasible region is altered. As one or more coefficients have an increasing or decreasing weight, the slope of the objective function changes. This causes the line, plane, or hyperplane representing the optimization function to roll around the surface enclosing the feasible region. Here, with the change represented as a function, an increase of the scalar multiplier Q will cause the slope of the objective function to tend more and more toward 2 as the modifier function becomes more dominant.

When using a computer package, it is likely that the package will generate the range over which the coefficient of each variable in the objective function may be varied, *with the other coefficients remaining at the values used to obtain the optimum.* If the package does not allow functions to be substituted directly, a simple device may be used as follows. Suppose that a manager is interested in discovering the solution to a linear problem that optimizes profit, and furthermore is also interested in the solution that optimizes market share. The two functions might be quite dissimilar; for example, they might contain variables from different groupings:

$$\text{Profit function} = X1 + 5\,X4 + 7\,Y2 - 3\,A1$$
$$\text{Market share} = X4 - 3\,V5 + 4\,X3 + 2\,A1$$

The formulation constraints are set up as usual, and then two extra constraints are added:

$$R - X1 - 5\,X4 - 7\,Y2 + 3\,A1 = 0$$
$$S - X4 + 3\,V5 - 4\,X3 - 2\,A1 = 0$$

Optimize profit by providing the total constraint set with the optimization function

Maximize R

To change to the market share criterion, simply change the optimization function to

Maximize S

PARAMETRIC PROGRAMMING ON THE RIGHT-HAND SIDE

Continuing with the basic problem faced by the edible fats manufacturer, let us examine how the problem varies as the constraints are altered. In particular, let us examine what happens when the plant capacity is increased. It was shown earlier that, under the conditions of solution of our basic problem, one extra unit of plant capacity would be worth \$1. Therefore, the purchase of extra capacity at less than \$1 would result in increased profit. Remember, however, that these shadow prices were marginal values existing only in the region of the optimum solution and could not be used for decision making over an extended range of values. This section examines how much capacity may be profitably purchased.

Consider the equation for plant capacity:

$$X1 + 2\,X2 + X4 = 7$$

Let the plant capacity be increased by Q units from 7 to $7 + Q$. This changes the equation representing the plant capacity to

$$X1 + 2\,X2 + X4 = 7 + Q$$

or

$$X1 + 2\,X2 + (X4 - Q) = 7$$

Replacing X4 in the original solution by $(X4 - Q)$, we obtain

Equation Set 3.6

$$X0 = -25 + (X4 - Q) + 3\,X3$$

$$
\begin{aligned}
X5 &= 3 + 3(X4 - Q) - 2\,X3 \\
X1 &= 5 + (X4 - Q) - 2\,X3 \\
X2 &= 1 - (X4 - Q) + X3
\end{aligned}
$$

Gathering terms together, this is

$$X0 = (-25 - Q) + X4 + 3\,X3$$

$$
\begin{aligned}
X5 &= (\ 3 - 3Q) + 3\,X4 - 2\,X3 \\
X1 &= (\ 5 - Q) + X4 - 2\,X3 \\
X2 &= (\ 1 + Q) - X4 + X3
\end{aligned}
$$

or

			X4	X3	
					Equation Set 3.7
X0	=	$(-25 - Q)$	1	3	
X5	=	$(\ \ 3 - 3Q)$	3	-2	
X1	=	$(\ \ 5 - Q)$	1	-2	
X2	=	$(\ \ 1 + Q)$	-1	1	

Examination of these equations shows that the profit increases from $25 as Q is increased. However, the value taken by the variable X5, which is currently in the basis of the solution, is steadily decreasing. This variable will decrease to zero when $Q = 1$ and will become negative for $Q > 1$. This represents an infeasible solution and is not allowed. Feasibility must be restored to the solution by pivoting between X5 and some other nonbasic variable. Looking at the equation for X5, we see that when $Q > 1$, the only other variable that can restore feasibility is X4. Pivoting then between X5 and X4 produces the result

			X5	X3	
					Equation Set 3.8
X0	=	-26	0.33	3.67	
X4	=	$(Q - 1)$	0.33	0.67	
X1	=	4	0.33	-1.33	
X2	=	2	-0.33	0.33	

This is represented by point K in Figure 1.1. The interpretation is that only one extra unit of plant capacity may be purchased, at a price of up to $1 per unit, when the extra profit made will be $26. The solution is that 4 units of X1 (4 tons of cooking oil) should be made, and 2 units of X2 (2 tons of margarine).

When plant capacity exceeds 8 units, the solution remains stable, the excess plant capacity over 8 units being shown as unused plant capacity in the basis. For increases in plant capacity between 0 and 1 unit, the solution is depicted by Equation Set 3.7. For example, with half a unit of extra plant capacity, $Q = 0.5$, then the solution becomes

$$X1 = 4.5, X2 = 1.5, X5 = 1.5$$

The profit is $25.50.

Consideration of Equation Set 3.7, alongside Figure 1.1, shows that as the resources for a constraint are increased, and the right-hand side of the equation is increased, the line or plane representing that constraint is replaced by a parallel line or plane further from the origin. This allows the intersection of the constraints forming the basis to travel along the other, unmodified constraint until it eventually reaches another constraint boundary, here the labor constraint. Further addition of resource to the right-hand side merely causes the constraint boundary to leave the feasible region.

In parametric programming on the right-hand side, the shadow prices remain constant over the range of applicability of the perturbed value, but the values of the variables in the basis change. In parametric programming on the objective function, the values of the variables in the basis remain constant over the range of value perturbation, while the shadow prices change.

Note here that the shadow price for sunflower oil has increased to $3.67 from its previous level, which held over the range of Equation Set 3.7. It may be that expansion in a way that was previously uneconomic may become economic following expansion in an alternative direction.

COMPUTER PACKAGE IMPLEMENTATION

In Chapters 1 and 2, after we submitted a linear programming problem to LINDO, we left the program without assenting to the question

```
DO RANGE (SENSITIVITY) ANALYSIS ? >
```

Let us rerun the linear program for the edible fats manufacturer's problem, which we now should have on file.

```
MAX 4X1+5X2
ST
X1+X2<6
X1+2X2<7
X1+4X2<12
END
GO
```

Now in response to the prompt above we answer "Y". The computer will respond with

```
RANGES IN WHICH THE BASIS IS UNCHANGED

                        OBJ COEFFICIENT RANGES
VARIABLE          CURRENT        ALLOWABLE        ALLOWABLE
                   COEF          INCREASE         DECREASE
   X1             4.000000       1.000000         1.500000
   X2             5.000000       3.000000         1.000000
                     RIGHTHAND SIDE RANGES
   ROW            CURRENT        ALLOWABLE        ALLOWABLE
                   RHS           INCREASE         DECREASE
    2             6.000000       1.000000         1.500000
    3             7.000000       1.000000         1.000000
    4            12.000000       INFINITY         3.000000
EOI ENCOUNTERED.
```

The computer reports that, concerning the objective function, the profit figure of $4 per ton for X1 may increase by $1, or may decrease by $1.50, all other variables remaining constant, with the solution found for this problem remaining optimal. Furthermore, labor, for which we have an oversupply, may be increased indefinitely, since it will always be in oversupply. The increase from 7 to 8 units of plant capacity is accurately reported, together with bounds for its reduction. The package has now indicated the ranges of applicability of the solution, but has not indicated the changes that will occur if any of these bounds are exceeded.

      LINDO allows the user, after considering the output from a run, to make alterations to the model. Some additional commands can be useful here. Suppose that we have just solved the three-variable problem in Chapter 1. In order to refresh our memory, we use the command LOOK ALL to see the current formulation held in the workspace. The response is

```
MAX 4 X1 + 5 X2 + 3 X3
SUBJECT TO
2)      X1 +    X2 + 4 X3 <=  6
3)      X1 + 2 X2 +    X3 <=  7
4)      X1 + 4 X2 + 2 X3 <= 12
END
```

Now let us investigate the change produced by a different element in the body of the matrix formulation, by using the command ALTER. In response, the computer asks for the location of the element and how to change it, and the user inserts this information after the ">" prompt:

```
ROW ?
> 3
VAR ?
> X1
NEW COEFFICIENT ?
> 2
```

In order to check your new formulation, use the LOOK ALL command again to see the current formulation held in the workspace:

```
: LOOK ALL
MAX 4 X1 + 5 X2 + 3 X3
SUBJECT TO
2)     X1 +    X2 + 4 X3 <=   6
3)   2 X1 + 2 X2 +    X3 <=   7
4)     X1 + 4 X2 + 2 X3 <= 12
END

: GO
             LP OPTIMUM FOUND AT STEP 2
               OBJECTIVE FUNCTION VALUE
 1)  17.1904762
        VARIABLE            VALUE         REDUCED COST
           X1              .666667          .000000
           X2             2.476190          .000000
           X3              .714286          .000000
        ROW        SLACK OR SURPLUS     DUAL PRICES
        2)              .000000            .142857
        3)              .000000           1.761905
        4)              .000000            .333333
NO. ITERATIONS= 2
DO RANGE (SENSITIVITY) ANALYSIS? > Y
```

```
RANGES IN WHICH THE BASIS IS UNCHANGED
                OBJ COEFFICIENT RANGES
VARIABLE        CURRENT        ALLOWABLE       ALLOWABLE
                COEFF          INCREASE        DECREASE
   X1           4.000000       1.000000        2.642857
   X2           5.000000       1.000000        1.000000
   X3           3.000000      12.333333         .500000

                RIGHTHAND SIDE RANGES
   ROW          CURRENT        ALLOWABLE       ALLOWABLE
                RHS            INCREASE        DECREASE
   2            6.000000      17.333333        2.500000
   3            7.000000       5.000000        1.000000
   4           12.000000       2.000000        7.428571
```

We see that in this new formulation, the right-hand side of the second constraint (row 3) is stable through a 5-unit increase from its current value of 7, up to a new value of 12 units. We can see what happens to the solution when the right-hand side value exceeds this figure by use of the PARARHS command

```
: PARARHS
 ROW  ?
> 3
 NEW RHS VAL=
> 13
```

This sequence has now set the range of the right-hand side value and the computer's response is

VAR OUT	VAR IN	PIVOT ROW	RHS VAL	DUAL VARIABLE	OBJ VAL
X3	SLK 3	2	12.0000	1.76190	26.0000
ART	SLK 3	0	12.0000	1.76190	26.0000

Note: ART designates artificial.

This indicates that the X3 variable leaves the basis when the right-hand-side value reaches 12. Further increases in supply of the resource represented by constraint 2 (row 3) merely increase the value of the slack variable associated with that resource, that is, the amount of unused resource increases directly.

We can also investigate the objective function coefficient changes using the ALTER command

```
: ALTER
 ROW ?
 > 1
 VAR ?
 > X1
 NEW COEFFICIENT ?
 > 6
```

Using LOOK ALL to see the current formulation held in the workspace, the response is

```
MAX 6 X1 + 5 X2 + 3 X3
SUBJECT TO
    2)     X1 +   X2 + 4 X3 <=   6
    3)   2 X1 + 2 X2 +   X3 <=   7
    4)     X1 + 4 X2 + 2 X3 <=  12
END
: GO
```

The computer solves this new formulation:

```
             LP OPTIMUM FOUND AT STEP 2
               OBJECTIVE FUNCTION VALUE
  1) 21.0000000
       VARIABLE           VALUE          REDUCED COST
          X1            3.142857             .000000
          X2             .000000            1.000000
          X3             .714286             .000000
       ROW        SLACK OR SURPLUS      DUAL PRICES
          2)            .000000             .000000
          3)            .000000            3.000000
          4)           7.428571             .000000
  NO. ITERATIONS= 2
  DO RANGE (SENSITIVITY) ANALYSIS? > Y
```

```
RANGES IN WHICH THE BASIS IS UNCHANGED
                OBJ COEFFICIENT RANGES
VARIABLE        CURRENT       ALLOWABLE      ALLOWABLE
                COEFF         INCREASE       DECREASE
   X1           6.000000       .000000       1.000000
   X2           5.000000      1.000000       INFINITY
   X3           3.000000     21.000000        .000000

                   RIGHTHAND SIDE RANGES
   ROW          CURRENT       ALLOWABLE      ALLOWABLE
                RHS           INCREASE       DECREASE
    2           6.000000     17.333333       2.500000
    3           7.000000      5.000000       1.000000
    4          12.000000      INFINITY       7.428571
```

Further modification may be made as desired.

As the formulation currently held in the LINDO workspace has been considerably modified from the file with which we started the session, it might be a good idea to retain a copy, using the command SAVE and assigning a new tape number to this file. The file will be accessible from the LINDO package by use of the command RETRIEVE.

Some additional commands help in the sensitivity analysis of linear programming formulations. If large problems are being investigated, it is sometimes confusing to identify just where a coefficient lies within the matrix, especially if the line wraps around the screen. Condensed formats are available to help you visualize element patterns. Let us return to the three-variable problem:

```
:TAKE
 FILE NUMBER?
> 33
: LOOK
 ROW:
> ALL
MAX 4 X1 + 5 X2 + 3 X3
SUBJECT TO
2)     X1 +   X2 + 4 X3 <=  6
3)     X1 + 2 X2 +   X3 <=  7
4)     X1 + 4 X2 + 2 X3 <= 12
END
```

This LOOK command has given us the full picture of the matrix currently in the formulation workspace. Because the problem is small, it is conveniently presented, but this would be much more confusing if the variables were not approximately aligned as above. A more condensed picture is obtained by the command PIC:

```
: PIC
      X X X
      1 2 3
 1:   4 5 3 MAX
 2:   1 1 4 < 6
 3:   1 2 1 < 7
 4:   1 4 2 < B
```

Notice that each element is assigned only a single character. The element that takes up two characters, here the 12 of the right-hand-side value of the third constraint, is assigned a value of B in order to maintain the pattern. This indicates that it is a number between 10 and 100. The full code will be listed by the command HELP PIC.

Step-by-step investigation is possible within LINDO by using the commands TABL and PIVOT. Returning to the three-variable problem, TABL produces the response

```
THE  TABLEAU
ROW (BASIS)      X1      X2      X3  SLK  2  SLK  3  SLK  4
1 ART          -4.0    -5.0    -3.0      .0      .0      .0      .0
2 SLK  2        1.0     1.0     4.0     1.0      .0      .0     6.0
3 SLK  3        1.0     2.0     1.0      .0     1.0      .0     7.0
4 SLK  4        1.0     4.0     2.0      .0      .0     1.0    12.0
```

The tableau presented here has been compressed in order to conserve space. With an 80-column display, the last two columns of the tableau will be truncated and displayed below the others.

LINDO uses the product form of the inverse method to perform simplex iterations, and thus does not normally generate the tableaux. They are generated as required.

You can investigate the development of a tableau with the command PIVOT. This command may be used on its own, when the next best nonbasic variable is chosen to enter the basis, or with a variable (PIVOT X3)

to investigate a variable entering the basis that is not the most desirable by the objective function criterion. In our example, PIVOT X2 produces the response

```
X2 ENTERS AT VALUE 3.0000 IN ROW 4 OBJ. VALUE= 15.000
```

In order to check on the new tableau, use TABL, which produces the response

```
THE TABLEAU
ROW (BASIS)      X1    X2   X3  SLK 2  SLK 3  SLK 4
1 ART         -2.75   .0  -.5    .0     .0   1.25  15.0
2 SLK 2         .75   .0  3.5   1.0     .0   -.25   3.0
3 SLK 3         .5    .0   .0    .0    1.0   -.5    1.0
4 X2            .25  1.0   .5    .0     .0    .25   3.0
```

There are two points of interest in using this approach:

   1. The PIVOT command may be used on its own or with a variable name, but *not* with a slack variable.

   2. Students who use the PIVOT command to generate simplex tableaux for checking longhand solutions should use it with care. Unexpected iterations might result, because the computer package uses a matrix operation set rather than the longhand simplex tableau iterations. For example, take the tableau

```
THE TABLEAU
ROW (BASIS)    X1    X2    X3  SLK 2  SLK 3  SLK 4
1 ART         .0    .0   -.5    .0     5.5   -1.5  20.5
2 SLK 2       .0    .0   3.5   1.0    -1.5    .5    1.5
3 X1         1.0    .0    .0    .0     2.0   -1.0   2.0
4 X2          .0   1.0    .5    .0     -.5    .5    2.5
```

One would expect that PIVOT would cause the entry with the largest $Zj - Cj$ equivalent entry (SLK 4) to enter the basis. However, PIVOT results in

```
SLK 4 ENTERS AT VALUE .42857 IN ROW 2 OBJ. VALUE= 20.714
```

TABL produces the response

```
THE TABLEAU
ROW (BASIS)  X1    X2    X3   SLK 2   SLK 3    SLK 4
1 ART        ,0    ,0    ,0   ,143   5,286   -1,429  20,0
2 X3         ,0    ,0   1,0   ,286   -,429     ,143    ,429
3 X1        1,0    ,0    ,0    ,0     2,0     -1,0     2,0
4 X2         ,0   1,0    ,0  -,143   -,286     ,429   2,286
```

It takes a further PIVOT command to obtain the expected final tableau.

In cases where the tableau is too large to be displayed conveniently on either a terminal or microcomputer screen, the command SHOCOLUMN will display a specified column of the problem.

Other commands are of use in investigating the solution of a problem formulation. EXT extends the problem by adding constraints. DEL deletes a specified constraint. APPC appends a new column to the problem. Upon receiving these commands, the computer will prompt you for appropriate specific information.

CHAPTER 3 PROBLEMS

The solutions to questions 1 through 21 will be found at the end of the chapter.

1. Gnomore Inc. produces three products for the new high-tech industries. Their data are listed below:

Production Data

	Product			Available
	A	B	C	
Raw material	3	1	4	30 tons
Manufacturing time	3	1	2	25 hours
Finishing time	4	3	3	17 hours

Financial Data	
Raw material cost per ton	$2
Direct labor expenses per ton	
A	$6
B	$4
C	$5
Selling price per ton	
A	$20
B	$12
C	$22
Fixed cash expenses per period	$5
Depreciation per period	$3

Balance Sheet			
	$		$
Cash assets	60	Bank loan	80
Accounts receivable	100	Long-term loans	20
Inventory	0	Plant & equipment	40
Equity	100		200
	200		

Further Data	
Dividends paid per period	$2
New plant purchases per period	$3

Taxes are paid from profits, at 50 percent of profit at the end of the period. All accounts receivable incurred in one period are received by the start of the next period. All payments are made as they accrue.

Analyze the situation facing Gnomore Inc. with regard to its production of the three high-tech items.

2. The managers of Gnomore Inc. have decided that, rather than facing each financial period as it occurs, they will plan for a 3-year horizon. It is anticipated that production and raw material costs will remain constant over the first 2 years and then rise. In the third year, raw material costs will rise by 10 percent and labor cost will increase by 8 percent. It is further anticipated that Gnomore will be able to increase the selling price of product B to $15 in year 2 and increase the selling prices of products A and C to $25 and $30 respectively in year 3. The firm has contracted to store any inventory at $1 per ton per period, payable at the end of each period.

As consultant to Gnomore Inc., you should advise the board on its available options.

3. Advise the Ding-he Boat Company (Problem 1, Chapter 1) on these two questions:

      a. How much can the price of sailboats rise before a new production schedule becomes necessary?

      b. How much may the price of rowboats fall before a new production schedule becomes necessary?

4. Advise Clone Computers (Problem 2, Chapter 1) how much they should pay for an increased supply of commercial ICs, and how many of these devices they should purchase.

5. A sorority (see Problem 3, Chapter 1) has asked the fraternity if it will sell 20 percent of its passion-fruit supply. Would you sell it, and what would be the price?

6. Mill Key Industries (Problem 4, Chapter 1) has been advised that it will be able to sell yogurt and cheese at the same contribution to profit. What effect does this have on the solution?

7. New C Foods (Problem 5, Chapter 1) has learned that, due to cancellation of a contract, four more new vessels will be available. Should New C Foods be interested in these vessels, and if so, what price should they pay?

8. What is the most economical storage cost per cubic foot of Fred's Superstores operation (Problem 6, Chapter 1)? Should Fred accept an offer of an additional 500 cubic feet of storage at 50 per cubic foot?

9. The bank has offered Eileen Dover (Problem 7, Chapter 1) an extra $30,000 at the same terms as her previous $25,000. Should she accept?

10. If home-loan interest rates drop by 2 percent, will this alter Robin Banks' (Problem 8, Chapter 1) investment strategy?

11. With falling interest rates, by how much will each investment interest rate need to fall before Rich Hustle (Problem 9, Chapter 1) needs to re-evaluate his strategy?

12. How much increase in profit margin on the floppy disc drive card will B. Quick be able to achieve (Problem 10, Chapter 1) before the production schedule is changed?

13. What should the plastics manufacturer in Problem 1, Chapter 2, be prepared to pay for extra polyamine, and how much will it wish to purchase at that price?

14. The organization described in Problem 2, Chapter 2, is considering increasing its franchise charge by $1,000. How does this affect the solution?

15. R. Borist has been offered $1.25 for 5 grams of Benziol (Problem 3, Chapter 2). Should he accept?

16. The operatives working for Mr. Makecombe (Problem 4, Chapter 2) have offered to work overtime in August at the July overtime rates. Should Makecombe take advantage of their offer, and how much overtime should he allocate?

17. By how much will the cost of in-house fabrication of DRNBG units have to increase (Problem 5, Chapter 2) before Blockwell Industries should change its production plans?

18. Ophelia Close (Problem 6, Chapter 2) has just been advised that, due to an accident on the way to Kansas City, the routes from Jackson to Kansas and from Kansas to Billings will now be reduced in capacity by 100 tons each. How does this affect Close's plans?

19. New Sprint (Problem 7, Chapter 2) has just found out that the order for 130 cartons of 9.5-inch computer paper should have been 120 cartons. How does this error affect the two solutions to minimize paper used or minimize waste?

20. If Tim Burr (Problem 8, Chapter 2) decides that he wishes to use part of his $500,000 to buy a yacht, how much of a yacht can he invest in without spoiling his business?

21. Jack Uzi (Problem 9, Chapter 2) has been advised that, in one of the next six months, one of the sales contract figures will be increased by 10 units. (The contractor does not, at this stage, know which month). What should Jack's response be?

HOMEWORK PROBLEMS

The answers to Problems 22 through 41 do not appear in this text.

22. Due to a breakdown in supply of three input parts, one of your machines will be idle next week. You may use the machine to fabricate the missing parts and in so doing you will save 10¢ on each input part A, 20¢ on each input part B, and 30¢ on each part C.

You do not normally fabricate these parts because it is more economical to use the machine time in the manufacture of your standard products X, Y, and Z, which return a contribution to profit of $5, $6, and $7 respectively.

Fabrication of X, Y, Z, A, B, and C takes 3, 4, 5, 1, 2, and 2 minutes, respectively, to process each unit. Your factory works a week of 40 hours, 50 minutes. Each X unit needs two A subassemblies and one B subassembly in its completion. Each Y unit needs three Bs and one C for completion, while each Z requires an A, a B, and one C.

You anticipate that supplies will be back to normal after this one week. What will be your optimum schedule of operations on the machine for next week?

What will be your economic subcontract rate for the subassemblies, and over what range will this hold?

23–41. Investigate the range of applicability of solutions to the homework problems in Chapters 1 and 2.

CHAPTER 3 SOLUTIONS

1. For Gnomore Inc.:

Raw material constraint	$3A + 1B + 4C \leq 30$
Manufacturing time constraint	$3A + 1B + 2C \leq 25$
Finishing time constraint	$4A + 3B + 3C \leq 17$

Cash constraint

$$5 + 3 + 2 + (6 + 6)A + (2 + 4)B + (8 + 5)C \leq 160$$

that is, $12A + 6B + 13C \leq 150$

Profit function

$$(20 - 12)A + (12 - 6)B + (22 - 13)C$$

that is, $8A + 6B + 9C$

Solution via LINDO:

```
look all
MAX 8 A + 6 B + 9 C
SUBJECT TO
2)  3 A + B + 4 C <= 30
3)  3 A + B + 2 C <= 25
4)  4 A + 3 B + 3 C <= 17
5)  12 A + 6 B + 13 C <= 150
END

: GO
LP OPTIMUM FOUND AT STEP 1
             OBJECTIVE FUNCTION VALUE
1)  51.0000000
     VARIABLE          VALUE        REDUCED COST
        A             .000000         4.000000
        B             .000000         3.000000
        C            5.666667          .000000

        ROW      SLACK OR SURPLUS    DUAL PRICES
        2)          7.333333           .000000
        3)         13.666667           .000000
        4)           .000000          3.000000
        5)         76.333333           .000000
NO. ITERATIONS= 1
DO RANGE (SENSITIVITY) ANALYSIS? > Y
```

(continued)

```
RANGES IN WHICH THE BASIS IS UNCHANGED
                   OBJ COEFFICIENT RANGES
VARIABLE          CURRENT        ALLOWABLE      ALLOWABLE
                   COEF          INCREASE       DECREASE
   A              8.000000       4.000000       INFINITY
   B              6.000000       3.000000       INFINITY
   C              9.000000       INFINITY       3.000000


                   RIGHTHAND SIDE RANGES
   ROW            CURRENT        ALLOWABLE      ALLOWABLE
                   RHS           INCREASE       DECREASE
   2             30.000000       INFINITY       7.333333
   3             25.000000       INFINITY       13.666667
   4             17.000000       5.500000       17.000000
   5            150.000000       INFINITY       76.333333
```

Interpreting this solution:

Produce 5.667 tons of C only, with a total value of $51, leaving 7.333 tons of spare raw material, 13.667 hours of spare manufacturing capacity, and $76.33 spare cash.

Should extra finishing time be available, it should be contracted out at a value of less than $3 per hour, until an extra 5.5 hours have been purchased. At this stage, the company will have exhausted its stocks of raw materials and will have produced 7.5 tons of C at a profit of $67.50 minus the price paid for the subcontracted finishing time.

Now the use of extra raw material is investigated. The value of this extra raw material to the company is linked to the price paid for the extra finishing time. This will be less than $3. Say that the price paid is $X.

The price paid for raw material will be less than $0.25(9 - X) = ($2.25 - 0.25X)$ in order for this to be profitable to the firm. This extra raw material price will hold until 16.15 extra tons have been purchased and a total of 29.15 hours finishing time has been subcontracted. At this point, the firm will have exhausted its cash limits and a new loan must be negotiated.

At the original solution where only 5.667 tons of product C is produced, should the company wish to produce some product A, it will reduce the company profit by $4 per ton of A produced. Similarly, each ton of B produced will reduce the company profit by $3.

Hence, Gnomore Inc. should investigate the possibility of increasing the prices of products A and B by at least $4 per ton and $3 per ton, respectively.

2. Let

Aij = Amount of product A made in period i and sold in period j

Bij = Amount of product B made in period i and sold in period j

Cij = Amount of product C made in period i and sold in period j

The constraints change as follows:

Raw material constraints

Period 1

$3\,A11 + 3\,A12 + 3\,A13 + B11 + B12 + B13 + 4\,C11 + 4\,C12 + 4\,C13 \leq 30$

Period 2

$3\,A22 + 3\,A23 \quad + B22 + B23 \quad + 4\,C22 + 4\,C23 \leq 30$

Period 3

$3\,A33 \quad + B33 \quad + 4\,C33 \leq 30$

Manufacturing time constraints

Period 1

$3\,A11 + 3\,A12 + 3\,A13 + B11 + B12 + B13 + 2\,C11 + 2\,C12 + 2\,C13 \leq 25$

Period 2

$3\,A22 + 3\,A23 \quad + B22 + B23 \quad + 2\,C22 + 2\,C23 \leq 25$

Period 3

$3\,A33 \quad + B33 \quad + 2\,C33 \leq 25$

Finishing time constraints

Period 1

$4\,A11 + 4\,A12 + 4\,A13 + 3\,B11 + 3\,B12 + 3\,B13 + 3\,C11 + 3\,C12 + 3\,C13 \leq 17$

Period 2

$4\,A22 + 4\,A23 \quad + 3\,B22 + 3\,B23 \quad + 3\,C22 + 3\,C23 \leq 17$

Period 3

$4\,A33 \quad + 3\,B33 \quad + 3\,C33 \leq 17$

Cash constraints

Period 1

$12\,A11 + \mathbf{13\,A12} + \mathbf{13\,A13}$

$+ \ 6\,B11 + \ \mathbf{7\,B12} + \ \ \mathbf{7\,B13}$

$+ \ 13\,C11 + \mathbf{14\,C12} + \mathbf{14\,C13} + X1 = 150$

The second and third columns (shown in boldface) are increased by \$1 to cover storage charge.

Period 2

$$\mathbf{-4\,A11 - 3\,B11 - 4.5\,C11 - X1} + 10 + 12\,A22$$
$$+ 13\,A23 + 6\,B22 + 7\,B23 + 13\,C22 + 14\,C23 + X2 = 0$$

The first four figures (shown in boldface) are profit for the first year's trading and subsequent years less tax.

Period 3

$$-4\,A22 - 4\,A12 - \mathbf{4.5\,B22 - 4.5\,B12} - 4.5\,C22$$
$$-4.5\,C12 - X2 + 10 + 13.08\,A33 + 6.52\,B33$$
$$+ 14.2\,C33 + X3 = 0$$

There is a price rise in year 2 on product B (shown in boldface).

Objective function:

Maximize $13\,A13 + 13\,A23 + 11.92\,A33 + 9\,B13 + 9\,B23$
$$+ 8.48\,B33 + 17\,C13 + 17\,C23 + 15.8\,C33 + X3$$

Solution by LINDO:

```
look all
 MIN - 13 A13 - 13 A23 - 11.92 A33 - 9 B13 - 9 B23
  - 8.48 B33 - 17 C13 - 17 C23 - 15.8 C33 - X3
SUBJECT TO
2)  3 A13 + B13 + 4 C13 + 3 A11 + 3 A12 + B11 + B12
  + 4 C11 + 4 C12 <= 30
3)  3 A23 + B23 + 4 C23 + 3 A22 + B22 + 4 C22 <= 30
4)  3 A33 + B33 + 4 C33 <= 30
5)  3 A13 + B13 + 2 C13 + 3 A11 + 3 A12 + B11 + B12
  + 2 C11 + 2 C12 <= 25
6)  3 A23 + B23 + 4 C23 + 3 A22 + B22 + 4 C22 <= 25
7)  3 A33 + B33 + 4 C33 <= 25
8)  4 A13 + 3 B13 + 3 C13 + 4 A11 + 4 A12 + 3 B11 +
  3 B12 + 3 C11 + 3 C12 <= 17
```

```
9) 4 A23 + 3 B23 + 3 C23 + 4 A22 + 3 B22 + 3 C22
 <= 17
10) 4 A33 + 4 B33 + 4 C33 <= 17
11) 13 A13 + 7 B13 + 14 C13 + 12 A11 + 13 A12 + 6
 B11 + 7 B12 + 13 C11 + 14 C12 + X1 = 150
12) - 13 A23 - 7 B23 - 14 C23 + 4 A11 + 3 B11 +
 4.5 C11 - 12 A22 - 6 B22 - 13 C22 + X1 - X2 = 10
13) - 13.08 A33 - 6.52 B33 - 14.2 C33 - X3 + 4 A12
 + 4.2 B12 + 4.2 C12 + 4 A22 + 4.2 B22 + 4.2 C22 +
 X2 = 10
END


: GO
LP OPTIMUM FOUND AT STEP 7
              OBJECTIVE FUNCTION VALUE
1) -164.276190
      VARIABLE           VALUE          REDUCED COST
         A13            .000000           3.190476
         A23            .000000           3.190476
         A33            .000000           4.057143
         B13           2.387143            .000000
         B23           5.666667            .000000
         B33           4.250000            .000000
         C13           3.279524            .000000
         C23            .000000            .000000
         C33            .000000           1.457143
          X3            .000000            .142857
         A11            .000000          10.476190
         A12            .000000          11.619048
         B11            .000000           4.428571
         B12            .000000           4.200000
         C11            .000000          10.714286
         C12            .000000          12.200000
         A22            .000000          10.476190
         B22            .000000           3.057143
         C22            .000000          11.057143
          X1          87.376667            .000000
          X2          37.710000            .000000


         ROW       SLACK OR SURPLUS    DUAL PRICES
          2)          14.494762            .000000
          3)          24.333333            .000000
          4)          25.750000            .000000
          5)          16.053810            .000000
          6)          19.333333            .000000
          7)          20.750000            .000000
          8)            .000000            .333333
```

(continued)

```
        9)              .000000            .333333
       10)              .000000            .257143
       11)              .000000           1.142857
       12)              .000000          -1.142857
       13)              .000000          -1.142857
NO. ITERATIONS= 7
DO RANGE (SENSITIVITY) ANALYSIS? > Y

RANGES IN WHICH THE BASIS IS UNCHANGED
                   OBJ COEFFICIENT RANGES
VARIABLE        CURRENT          ALLOWABLE        ALLOWABLE
                 COEF            INCREASE         DECREASE
   A13        -13.000000         INFINITY         3.190476
   A23        -13.000000         INFINITY         3.190476
   A33        -11.920000         INFINITY         4.057143
   B13         -9.000000          .500000          .000000
   B23         -9.000000          .000000         INFINITY
   B33         -8.480000         1.028571         INFINITY
   C13        -17.000000          .000000         1.000000
   C23        -17.000000         INFINITY          .000000
   C33        -15.800000         INFINITY         1.457143
   X3          -1.000000         INFINITY          .142857
   A11          .000000          INFINITY        10.476190
   A12          .000000          INFINITY        11.619048
   B11          .000000          INFINITY         4.428571
   B12          .000000          INFINITY         4.200000
   C11          .000000          INFINITY        10.714286
   C12          .000000          INFINITY        12.200000
   A22          .000000          INFINITY        10.476190
   B22          .000000          INFINITY         3.057143
   C22          .000000          INFINITY        11.057143
   X1           .000000           .142857          .000000
   X2           .000000           .157756          .142857

                  RIGHTHAND SIDE RANGES
ROW             CURRENT          ALLOWABLE        ALLOWABLE
                 RHS             INCREASE         DECREASE
    2         30.000000         INFINITY        14.494762
    3         30.000000         INFINITY        24.333333
    4         30.000000         INFINITY        25.750000
    5         25.000000         INFINITY        16.053810
    6         25.000000         INFINITY        19.333333
    7         25.000000         INFINITY        20.750000
    8         17.000000         9.838571         3.580714
    9         17.000000         9.838571         7.161429
   10         17.000000        14.083845        10.251534
   11        150.000000        16.710000        22.956667
   12         10.000000        22.956667        16.710000
   13         10.000000        22.956667        16.710000
```

The interpretation of this solution is that Gnomore Inc. should produce

      2.387 tons of B in year 1 and sell in year 3
      5.667 tons of B in year 2 and sell in year 3
      4.250 tons of B in year 3 and sell in year 3
      3.279 tons of C in year 1 and sell in year 3

      $87.37 should be carried over from year 1 to year 2
      $37.71 should be carried over from year 2 to year 3

      Total profit produced = $ 164.27

      Gnomore Inc. will have

      14.49 tons of spare raw material in year 1
      24.33 tons of spare raw material in year 2
      25.75 tons of spare raw material in year 3

      16.05 hours of spare manufacturing capacity in year 1
      19.33 hours of spare manufacturing capacity in year 2
      20.75 hours of spare manufacturing capacity in year 3

Gnomore Inc. will use up all its finishing capacity in years 1 and 2. Finishing capacity might be subcontracted at up to 33¢ per hour; 9.83 extra units of finishing time would be of use to the firm during these two periods. In year 3 the profitable subcontract rate drops to 25.5¢ per hour, when they can use an extra 14.08 hours.

      Again, Gnomore could profitably employ more cash. Here, each $1 extra invested in any one of the three years will increase the total profit produced at the horizon time by $1.14. This is the economic value of cash to the firm. Note the correspondance between X3 and the dual prices of rows 11, 12, and 13.

The management of Gnomore Inc. might also find the following information useful:

Producing One Unit of	In Year	Selling in Year	Reduces the Horizon-Time Profit by
A	1	1	$10.47
A	1	2	$11.61
A	1	3	$ 3.19
B	1	1	$ 4.42
B	1	2	$ 4.20
C	1	1	$10.71
C	1	2	$12.20
A	2	2	$10.47
A	2	3	$ 3.19
B	2	2	$ 3.06
C	2	2	$11.06
C	2	3	$ 0
A	3	3	$ 4.06
C	3	3	$ 1.46

The cost of capital to the enterprise, previously stated at 14 percent, is shown in that each $1 retained in the business, to increase liquidity for example, reduces the horizon-time profit by 14¢.

3. For the Ding-he Boat Company, solution via LINDO is

```
look all
MAX 10 A + 7.5 B
SUBJECT TO
2) 2 A + B <= 1000
3) A + B <= 800
4) A <= 400
5) B <= 700
END
```

```
? GO
LP OPTIMUM FOUND AT STEP 3
               OBJECTIVE FUNCTION VALUE
1)  6500.00000
       VARIABLE              VALUE          REDUCED COST
          A               200.000000           .000000
          B               600.000000           .000000


          ROW         SLACK OR SURPLUS     DUAL PRICES
           2)               .000000          2.500000
           3)               .000000          5.000000
           4)             200.000000           .000000
           5)             100.000000           .000000
NO. ITERATIONS= 3
DO RANGE (SENSITIVITY) ANALYSIS? ? Y

RANGES IN WHICH THE BASIS IS UNCHANGED
                 OBJ COEFFICIENT RANGES
VARIABLE        CURRENT       ALLOWABLE       ALLOWABLE
                 COEF         INCREASE        DECREASE
   A          10.000000      5.000000        2.500000
   B           7.500000      2.500000        2.500000

                   RIGHTHAND SIDE RANGES
   ROW          CURRENT       ALLOWABLE       ALLOWABLE
                  RHS         INCREASE        DECREASE
    2        1000.000000    200.000000      100.000000
    3         800.000000     50.000000      200.000000
    4         400.000000     INFINITY       200.000000
    5         700.000000     INFINITY       100.000000
```

The solution indicates that the price of sailboats may be increased by $500 before production schedules should be altered, and the price of rowboats may fall by up to $250 before rescheduling production.

4. The solution for Clone Computers is:

```
look all
MAX 1.9 G + C
SUBJECT TO
2)  450 G + 250 C <= 2500000
3)  100 G <= 100000
4)  175 G + 350 C <= 1050000
5)  3 G + 10 C <= 21000
END
```

(continued)

```
: GO
LP OPTIMUM FOUND AT STEP  2
OBJECTIVE FUNCTION VALUE
1)  3700.00000
        VARIABLE                VALUE          REDUCED COST
            G            1000.000000             .000000
            C            1800.000000             .000000

        ROW          SLACK OR SURPLUS       DUAL PRICES
        2)            1600000.000000           .000000
        3)                  .000000           .016000
        4)             245000.000000           .000000
        5)                  .000000           .100000
NO. ITERATIONS=  2
DO RANGE (SENSITIVITY) ANALYSIS? > Y


RANGES IN WHICH THE BASIS IS UNCHANGED
                   OBJ COEFFICIENT RANGES
VARIABLE        CURRENT        ALLOWABLE        ALLOWABLE
                COEF           INCREASE         DECREASE
    G          1.900000        INFINITY         1.600000
    C          1.000000        5.333333         1.000000

                   RIGHTHAND SIDE RANGES
    ROW         CURRENT        ALLOWABLE        ALLOWABLE
                RHS            INCREASE         DECREASE
    2      2500000.000000      INFINITY    1600000.000000
    3       100000.000000 350000.000000  100000.000000
    4      1050000.000000      INFINITY   245000.000000
    5        21000.000000   7000.000000    18000.000000
```

The indicated shadow price of secret ICs is 0.016 units, that is, 1.6 percent of the price of a civilian computer. Clone should purchase up to 350,000 more secret ICs if it can get them at a price below 1.6 percent of the sales price of a civilian computer.

5. The LINDO solution for the fraternity is:

```
look all
MAX 7.6 F + 9.6 L
SUBJECT TO
2)  34 F + 19 L <= 646
3)  13 F + 33 L <= 429
4)  L <= 10
END
```

```
: GO
LP OPTIMUM FOUND AT STEP 3
               OBJECTIVE FUNCTION VALUE
1)  182.256000
        VARIABLE              VALUE          REDUCED COST
           F              15.048000            .000000
           L               7.072000            .000000


          ROW         SLACK OR SURPLUS       DUAL PRICES
          2)             .000000              .144000
          3)             .000000              .208000
          4)            2.928000              .000000
NO. ITERATIONS= 3
DO RANGE (SENSITIVITY) ANALYSIS? > Y

RANGES IN WHICH THE BASIS IS UNCHANGED
               OBJ COEFFICIENT RANGES
VARIABLE        CURRENT       ALLOWABLE       ALLOWABLE
                 COEF         INCREASE        DECREASE
   F           7.600000       9.578947        3.818182
   L           9.600000       9.692308        5.352941

                  RIGHTHAND SIDE RANGES
   ROW          CURRENT       ALLOWABLE       ALLOWABLE
                  RHS         INCREASE        DECREASE
    2         646.000000     476.000000     197.076923
    3         429.000000      75.352941     182.000000
    4          10.000000      INFINITY        2.928000
```

The solution indicates that 29.2 percent of the passion-fruit supply is unused. They should therefore sell the 20 percent for whatever price they can get.

6. For Mill Key Industries, the LINDO solution is:

```
look all
MAX 190 X1 + 250 X2
SUBJECT TO
2)  X1 + X2 <= 2000
3)  2 X1 + 3 X2 <= 5400
4)  2 X1 + X2 <= 3000
END
```

(continued)

```
: GO
LP OPTIMUM FOUND AT STEP 2
                OBJECTIVE FUNCTION VALUE
1)  464000.000

        VARIABLE            VALUE          REDUCED COST
           X1            600.000000           .000000
           X2           1400.000000           .000000
          ROW        SLACK OR SURPLUS      DUAL PRICES
           2)             .000000          70.000000
           3)             .000000          60.000000
           4)          400.000000            .000000
NO. ITERATIONS= 2
DO RANGE (SENSITIVITY) ANALYSIS? > Y

RANGES IN WHICH THE BASIS IS UNCHANGED
                OBJ COEFFICIENT RANGES
  VARIABLE        CURRENT        ALLOWABLE       ALLOWABLE
                   COEF          INCREASE        DECREASE
     X1          190.000000      60.000000      23.333333
     X2          250.000000      35.000000      60.000000

                RIGHTHAND SIDE RANGES
     ROW          CURRENT        ALLOWABLE       ALLOWABLE
                   RHS           INCREASE        DECREASE
      2          2000.000000    100.000000     200.000000
      3          5400.000000    600.000000     400.000000
      4          3000.000000     INFINITY      400.000000
```

The solution indicates that the contribution to profit on yogurt can rise by $60 to the same level as the profit on cheese before the solution will be affected. Also, the contribution to profit on cheese may drop by $60 to the level of yogurt without affecting the solution parameters.

7. The LINDO solution for the New C Foods situation is:

```
look all
MAX 110 X1 + 55 X2
SUBJECT TO
2)  4 X1 + X2 <= 20
3)  4 X1 + 3 X2 <= 24
4)  6 X1 + 13 X2 <= 78
END
```

```
: GO
LP OPTIMUM FOUND AT STEP 2
OBJECTIVE FUNCTION VALUE
1) 605.000000
        VARIABLE            VALUE           REDUCED COST
           X1            4.500000               .000000
           X2            2.000000               .000000

        ROW          SLACK OR SURPLUS     DUAL PRICES
         2)               .000000          13.750000
         3)               .000000          13.750000
         4)             25.000000            .000000
NO. ITERATIONS= 2
DO RANGE (SENSITIVITY) ANALYSIS? > Y

RANGES IN WHICH THE BASIS IS UNCHANGED
              OBJ COEFFICIENT RANGES
VARIABLE         CURRENT       ALLOWABLE      ALLOWABLE
                 COEF          INCREASE       DECREASE
   X1         110.000000     110.000000      36.666667
   X2          55.000000      27.500000      27.500000

              RIGHTHAND SIDE RANGES
   ROW           CURRENT       ALLOWABLE      ALLOWABLE
                 RHS           INCREASE       DECREASE
    2          20.000000       4.000000       5.882353
    3          24.000000       4.347826       4.000000
    4          78.000000       INFINITY      25.000000
```

The solution indicates that 4.347 vessels can be used profitably. (Clearly 0.347 of a boat would be unseaworthy.) New C should take the four vessels if they can be obtained at a price below the indicated shadow price of 13.75 units of production. This will be 13.75 divided by 605, and then multiplied by the dollar return of the crop.

8. Fred's Superstores solution:

```
look all
MAX 2 X1 + 3 X2 + 4 X3
SUBJECT TO
2) 100 X1 + 120 X2 + 180 X3 <= 122000
3) X1 + 2.5 X2 + 5 X3 <= 2000
END
```

```
? GO
LP OPTIMUM FOUND AT STEP 3
OBJECTIVE FUNCTION VALUE
1)  2800.00000
        VARIABLE              VALUE          REDUCED COST
            X1           500.000000              .000000
            X2           600.000000              .000000
            X3              .000000             1.076923

        ROW         SLACK OR SURPLUS      DUAL PRICES
         2)              .000000             .015385
         3)              .000000             .461538
NO. ITERATIONS= 3
DO RANGE (SENSITIVITY) ANALYSIS? ? Y

RANGES IN WHICH THE BASIS IS UNCHANGED
                OBJ COEFFICIENT RANGES
VARIABLE        CURRENT        ALLOWABLE       ALLOWABLE
                COEF           INCREASE        DECREASE
    X1          2.000000        .500000         .800000
    X2          3.000000       2.000000         .437500
    X3          4.000000       1.076923         INFINITY

                RIGHTHAND SIDE RANGES
    ROW         CURRENT        ALLOWABLE       ALLOWABLE
                RHS            INCREASE        DECREASE
     2      122000.000000   78000.000000   26000.000000
     3        2000.000000     541.666667     780.000000
```

This indicates that 50¢ per cubic foot is too high a price to pay. Fred could, however, use an extra 541 cubic feet at a price lower than 46¢.

9. The solution for Eileen Dover is:

```
look all
MAX 250000 X1 + 25000 X2 + 100000 X3
SUBJECT TO
2)  2000 X1 + 250 X2 + 500 X3 <= 25000
3)  X1 <= 25
4)  X2 <= 25
5)  X3 <= 25
END
```

```
? GO
LP OPTIMUM FOUND AT STEP 2
                OBJECTIVE FUNCTION VALUE
1) 4062500.00
        VARIABLE           VALUE          REDUCED COST
           X1            6.250000            .000000
           X2             .000000         6250.000000
           X3           25.000000            .000000

        ROW         SLACK OR SURPLUS    DUAL PRICES
          2)             .000000         125.000000
          3)           18.750000            .000000
          4)           25.000000            .000000
          5)             .000000       37500.000000
NO. ITERATIONS= 2
DO RANGE (SENSITIVITY) ANALYSIS? ? Y

RANGES IN WHICH THE BASIS IS UNCHANGED
                OBJ COEFFICIENT RANGES
VARIABLE      CURRENT        ALLOWABLE       ALLOWABLE
               COEF          INCREASE        DECREASE
    X1    250000.000000  150000.000000   50000.000000
    X2     25000.000000    6250.000000      INFINITY
    X3    100000.000000      INFINITY     37500.000000

                RIGHTHAND SIDE RANGES
  ROW         CURRENT        ALLOWABLE       ALLOWABLE
               RHS           INCREASE        DECREASE
   2      25000.000000   37500.000000   12500.000000
   3         25.000000      INFINITY      18.750000
   4         25.000000      INFINITY      25.000000
   5         25.000000      25.000000      25.000000
```

Yes, Eileen should accept the bank's offer, because she can profitably use an extra $37,500. In fact, she might be well advised to take the above output to her enlightened bank manager to argue for more funds.

10. Robin Banks' solution by LINDO:

```
look all
MAX 0.12 X1 + 0.14 X2 + 0.165 X3 + 0.1525 X4
+ 0.09 X5
SUBJECT TO
2)  X5 <= 3000000
3)  X1 + X2 - X5 <= 0
4)  - 0.49 X1 - 0.49 X2 - 0.49 X3 + 0.51 X4
    - 0.49 X5 <= 0
5)  - 0.5 X1 - 0.5 X2 + X3 <= 0
6)  X1 + X2 + X3 + X4 + X5 <= 9000000
END


? GO
LP OPTIMUM FOUND AT STEP 5
                OBJECTIVE FUNCTION VALUE
1)  1246275.00
        VARIABLE           VALUE         REDUCED COST
           X1            .000000          .020000
           X2        1836000.000000       .000000
           X3         918000.000000       .000000
           X4        4410000.000000       .000000
           X5        1836000.000000       .000000


         ROW      SLACK OR SURPLUS    DUAL PRICES
          2)      1164000.000000        .000000
          3)            .000000         .035000
          4)            .000000         .027500
          5)            .000000         .040000
          6)            .000000         .138475
NO. ITERATIONS= 5
DO RANGE (SENSITIVITY) ANALYSIS? ? Y


RANGES IN WHICH THE BASIS IS UNCHANGED
                OBJ COEFFICIENT RANGES
VARIABLE CURRENT        ALLOWABLE        ALLOWABLE
         COEF           INCREASE         DECREASE
   X1      .120000       .020000        INFINITY
   X2      .140000       .068750         .020000
   X3      .165000       .137500         .050000
   X4      .152500      INFINITY         .027500
   X5      .090000       .058333         .678799
```

```
                RIGHTHAND SIDE RANGES
  ROW     CURRENT         ALLOWABLE       ALLOWABLE
            RHS           INCREASE        DECREASE
   2  3000000.000000      INFINITY       1164000.000000
   3        .000000   3060000.000000     1940000.000000
   4        .000000   4590000.000000     2910000.000000
   5        .000000   4590000.000000     1147500.000000
   6  9000000.000000   5705882.352941    9000000.000000
```

All of the entries in the objective function are stable over a decrease of 2 percentage points of the interest rate. Thus Robin does not need to reevaluate his strategy.

11. In light of the falling interest rates, Rich Hustle's solution is as follows:

```
look all
MAX 0.1 X11 + 0.1 X12 + 0.13 X21 + 0.18 X32
SUBJECT TO
2) X11 + X21 <= 150000
3) - 0.1 X11 + X12 - 0.13 X21 + X32 <= 150000
4) X21 <= 100000
5) X32 <= 50000
END

? GO
LP OPTIMUM FOUND AT STEP 4
            OBJECTIVE FUNCTION VALUE
1)  38800.0000
      VARIABLE          VALUE          REDUCED COST
        X11          50000.000000        .000000
        X12         118000.000000        .000000
        X21         100000.000000        .000000
        X32          50000.000000        .000000

        ROW       SLACK OR SURPLUS     DUAL PRICES
        2)              .000000          .110000
        3)              .000000          .100000
        4)              .000000          .033000
        5)              .000000          .080000
NO. ITERATIONS= 4
DO RANGE (SENSITIVITY) ANALYSIS? ? Y
```

(continued)

```
RANGES IN WHICH THE BASIS IS UNCHANGED
                    OBJ COEFFICIENT RANGES
VARIABLE        CURRENT         ALLOWABLE       ALLOWABLE
                COEF            INCREASE        DECREASE
   X11             .100000         .033000         .110000
   X12             .100000         .080000         .100000
   X21             .130000        INFINITY         .033000
   X32             .180000        INFINITY         .080000


                    RIGHTHAND SIDE RANGES
  ROW           CURRENT         ALLOWABLE       ALLOWABLE
                RHS             INCREASE        DECREASE
    2       150000.000000      INFINITY      50000.000000
    3       150000.000000      INFINITY     118000.000000
    4       100000.000000   50000.000000    100000.000000
    5        50000.000000  118000.000000     50000.000000
```

Rich is in good shape. The 10 percent rate needs to drop by 11 percentage points and 10 percentage points, respectively, over the two time periods. The 13 percent rate can decrease by 3 percentage points and the 18 percent rate by 8 percentage points before his strategy is affected.

12. B. Quick's data are:

```
look all
MAX 25 X1 + 12 X2 + 19.375 X3 + 9.615 X4
SUBJECT TO
2) 125 X1 + 53 X2 <= 5000
3) 125 X3 + 53 X4 <= 15000
4) - 25 X1 - 12 X2 + 45.9375 X3 + 18.0775 X4
   <= 5000
END

? GO
LP OPTIMUM FOUND AT STEP 4
          OBJECTIVE FUNCTION VALUE
1) 3853.30189
      VARIABLE           VALUE        REDUCED COST
         X1               .000000        3.301887
         X2             94.339623         .000000
         X3               .000000        3.301887
         X4            283.018868         .000000
```

```
     ROW          SLACK OR SURPLUS    DUAL PRICES
      2)                .000000        .226415
      3)                .000000        .181415
      4)            1015.801887        .000000
NO. ITERATIONS= 4
DO RANGE (SENSITIVITY) ANALYSIS? ? Y

RANGES IN WHICH THE BASIS IS UNCHANGED
               OBJ COEFFICIENT RANGES
VARIABLE      CURRENT        ALLOWABLE      ALLOWABLE
               COEF          INCREASE       DECREASE
   X1        25.000000       3.301887       INFINITY
   X2        12.000000       INFINITY       1.400000
   X3        19.375000       3.301887       INFINITY
   X4         9.615000       INFINITY       1.400000

               RIGHTHAND SIDE RANGES
  ROW         CURRENT        ALLOWABLE      ALLOWABLE
               RHS           INCREASE       DECREASE
   2        5000.000000      INFINITY       4486.458333
   3       15000.000000      2978.149634    15000.000000
   4        5000.000000      INFINITY       1015.801887
```

The solution indicates that B. Quick is able to push the price of his floppy disc drive cards as high as he wishes, because there are no equations limiting demand in the solution.

13. For the plastics manufacturer, the LINDO solution is:

```
look all
MAX 10 A + 20 B
SUBJECT TO
2) 8 A + 10 B <= 80000
3) 2.5 A + B <= 20000
4) 2 A + 4 B <= 30000
5) A + B <= 9000
6) A >= 3000
END
```

(continued)

```
? GO
LP OPTIMUM FOUND AT STEP 2
               OBJECTIVE FUNCTION VALUE
1)  142000.000
         VARIABLE              VALUE           REDUCED COST
            A              3000.000000             .000000
            B              5600.000000             .000000


            ROW         SLACK OR SURPLUS      DUAL PRICES
            2)                 .000000          2.000000
            3)             6900.000000           .000000
            4)             1600.000000           .000000
            5)              400.000000           .000000
            6)                 .000000         -6.000000
NO. ITERATIONS=  2
DO RANGE (SENSITIVITY) ANALYSIS?  ?  Y

RANGES IN WHICH THE BASIS IS UNCHANGED
                    OBJ COEFFICIENT RANGES
VARIABLE        CURRENT        ALLOWABLE        ALLOWABLE
                 COEF          INCREASE         DECREASE
   A           10.000000       6.000000         INFINITY
   B           20.000000       INFINITY         7.500000


                 RIGHTHAND SIDE RANGES
   ROW          CURRENT        ALLOWABLE        ALLOWABLE
                 RHS           INCREASE         DECREASE
   2         80000.000000    4000.000000     56000.000000
   3         20000.000000      INFINITY       6900.000000
   4         30000.000000      INFINITY       1600.000000
   5          9000.000000      INFINITY        400.000000
   6          3000.000000    2000.000000      1333.333333
```

The company can therefore use an extra 4,000 lb of polyamine, and they would be prepared to pay up to $2 per pound for it.

14. For the company, the LINDO solution is:

```
look all
MAX 4 X1 + 3 Y1 - 4 Z
SUBJECT TO
2)  5 X1 + 3 Y1 + 5 Z = 2
3)  X1 + Y1 - 5 Z <= 0
4)  3 X1 + 5 Y1 - 20 Z <= 0
END
```

```
? GO
LP OPTIMUM FOUND AT STEP 3
              OBJECTIVE FUNCTION VALUE
1)  1.08000000
       VARIABLE            VALUE          REDUCED COST
          X1              .200000           .000000
          Y1              .200000           .000000
           Z              .080000           .000000


          ROW         SLACK OR SURPLUS    DUAL PRICES
          2)              .000000           .540000
          3)              .000000          1.180000
          4)              .000000           .040000
NO. ITERATIONS= 3
DO RANGE (SENSITIVITY) ANALYSIS? ? Y

RANGES IN WHICH THE BASIS IS UNCHANGED
              OBJ COEFFICIENT RANGES
VARIABLE        CURRENT       ALLOWABLE        ALLOWABLE
                 COEF         INCREASE         DECREASE
     X1        4.000000        .100000          .694118
     Y1        3.000000        .513043          .066667
      Z       -4.000000       3.687500         1.000000

                 RIGHTHAND SIDE RANGES
   ROW          CURRENT       ALLOWABLE        ALLOWABLE
                 RHS          INCREASE         DECREASE
    2          2.000000       INFINITY         2.000000
    3           .000000        .086957          .117647
    4           .000000        .500000          .333333
```

The current franchise charge of $4,000 can be increased to a limit of $5,000 without disturbing the solution strategy.

15. R. Borist's complete LINDO output is:

```
look all
MIN X1 + 0.3 X2 + 0.9 X3
SUBJECT TO
2)  X1 + X2 + X3 >= 10
3)  X1 + X2 + X3 <= 15
4)  X1 - X2 >= 0
5)  - X1 + 2 X3 >= 0
END
```

(continued)

```
? GO
LP OPTIMUM FOUND AT STEP 3
                OBJECTIVE FUNCTION VALUE
1)  7.00000000
        VARIABLE           VALUE          REDUCED COST
          X1             4.000000           .000000
          X2             4.000000           .000000
          X3             2.000000           .000000

        ROW        SLACK OR SURPLUS     DUAL PRICES
         2)              .000000          -.700000
         3)             5.000000           .000000
         4)              .000000          -.400000
         5)              .000000          -.100000
NO. ITERATIONS= 3
DO RANGE (SENSITIVITY) ANALYSIS? ? Y

RANGES IN WHICH THE BASIS IS UNCHANGED
                    OBJ COEFFICIENT RANGES
VARIABLE        CURRENT        ALLOWABLE        ALLOWABLE
                 COEF          INCREASE         DECREASE
    X1         1.000000        .500000         1.000000
    X2          .300000        .500000         1.750000
    X3          .900000        INFINITY         .250000

                    RIGHTHAND SIDE RANGES
    ROW         CURRENT        ALLOWABLE        ALLOWABLE
                 RHS           INCREASE         DECREASE
     2         10.000000       5.000000        10.000000
     3         15.000000       INFINITY         5.000000
     4          .000000        6.666667        10.000000
     5          .000000       20.000000         5.000000
```

R. Borist has 5 grams of Benzoil unused at his optimal strategy. He should therefore sell it for whatever he can get (even though it cost him more than the offer price).

16. Mr. Makecombe's solution is:

```
look all
MIN 100 X1 + 200 X2 + 300 X3 + 150 X4 + 200 X5 +
400 X6 + 550 X7 + 200 X8 + 100 Y1 + 100 Y2 + 100
Y3
SUBJECT TO
2)  X1 <= 10
3)  X2 <= 10
4)  X3 <= 10
5)  X4 <= 10
6)  X5 <= 4
7)  X6 <= 4
8)  X7 <= 4
9)  X8 <= 4
10) X1 + X5 - Y1 = 5
11) X1 + X2 + X5 + X6 - Y2 = 12
12) X1 + X2 + X3 + X5 + X6 + X7 - Y3 = 22
13) X1 + X2 + X3 + X4 + X5 + X6 + X7 + X8 = 37
END


? GO
LP OPTIMUM FOUND AT STEP 7
              OBJECTIVE FUNCTION VALUE
1)  7600.00000
       VARIABLE           VALUE         REDUCED COST
          X1            5.000000            .000000
          X2           10.000000            .000000
          X3            8.000000            .000000
          X4           10.000000            .000000
          X5             .000000         100.000000
          X6             .000000         200.000000
          X7             .000000         250.000000
          X8            4.000000            .000000
          Y1             .000000            .000000
          Y2            3.000000            .000000
          Y3            1.000000            .000000
```

(continued)

```
        ROW        SLACK OR SURPLUS    DUAL PRICES
          2)          5.000000           .000000
          3)           .000000           .000000
          4)          2.000000           .000000
          5)           .000000        250.000000
          6)          4.000000           .000000
          7)          4.000000           .000000
          8)          4.000000           .000000
          9)           .000000        200.000000
         10)           .000000        100.000000
         11)           .000000        100.000000
         12)           .000000        100.000000
         13)           .000000       -400.000000
NO. ITERATIONS= 7
DO RANGE (SENSITIVITY) ANALYSIS? ? Y


RANGES IN WHICH THE BASIS IS UNCHANGED
                  OBJ COEFFICIENT RANGES
VARIABLE         CURRENT       ALLOWABLE       ALLOWABLE
                  COEF         INCREASE        DECREASE
    X1         100.000000     100.000000        .000000
    X2         200.000000       .000000        INFINITY
    X3         300.000000       .000000         .000000
    X4         150.000000     250.000000       INFINITY
    X5         200.000000      INFINITY       100.000000
    X6         400.000000      INFINITY       200.000000
    X7         550.000000      INFINITY       250.000000
    X8         200.000000     200.000000       INFINITY
    Y1         100.000000      INFINITY         .000000
    Y2         100.000000       .000000         .000000
    Y3         100.000000      INFINITY       200.000000


                  RIGHTHAND SIDE RANGES
    ROW          CURRENT       ALLOWABLE       ALLOWABLE
                  RHS          INCREASE        DECREASE
     2          10.000000      INFINITY        5.000000
     3          10.000000      8.000000        2.000000
     4          10.000000      INFINITY        2.000000
     5          10.000000      1.000000        2.000000
     6           4.000000      INFINITY        4.000000
     7           4.000000      INFINITY        4.000000
     8           4.000000      INFINITY        4.000000
     9           4.000000      1.000000        2.000000
    10           5.000000      5.000000        2.000000
    11          12.000000      3.000000        INFINITY
    12          22.000000      1.000000        INFINITY
    13          37.000000      2.000000        1.000000
```

Makecombe should decline the offer, but indicate interest in overtime paid at less than $300.

17. Blockwell's solution via LINDO:

```
look all
MIN 40 X1 + 65 X2 + 37.5 X3 + 32 X4 + 63.5 Y1 + 99
Y2 + 55.2 Y3 + 45 Y4
SUBJECT TO
2)  18 X1 + 20 X3 + 24 X4 <= 4800
3)  40 X2 + 80 X3 <= 4800
4)  12 X1 + 32 X2 + 53 X3 <= 4800
5)  25 X1 + 75 X2 + 54 X3 + 30 X4 <= 4800
6)  20 X1 + 34 X2 + 10 X4 <= 4800
7)  X1 + Y1 = 2000
8)  X2 + Y2 = 2000
9)  X3 + Y3 = 2000
10) X4 + Y4 = 2000
END

? GO
LP OPTIMUM FOUND AT STEP 6
OBJECTIVE FUNCTION VALUE
1)  520888.000
        VARIABLE           VALUE           REDUCED COST
           X1          192.000000             .000000
           X2            .000000           36.500000
           X3            .000000           33.060000
           X4            .000000           15.200000
           Y1         1808.000000             .000000
           Y2         2000.000000             .000000
           Y3         2000.000000             .000000
           Y4         2000.000000             .000000

        ROW        SLACK OR SURPLUS    DUAL PRICES
          2)          1344.000000          .000000
          3)          4800.000000          .000000
          4)          2496.000000          .000000
          5)            .000000           .940000
          6)           960.000000          .000000
          7)            .000000         -63.500000
          8)            .000000         -99.000000
          9)            .000000         -55.200000
         10)            .000000         -45.000000
NO. ITERATIONS= 6
DO RANGE (SENSITIVITY) ANALYSIS? ? Y
```

(continued)

```
RANGES IN WHICH THE BASIS IS UNCHANGED
                    OBJ COEFFICIENT RANGES
VARIABLE        CURRENT         ALLOWABLE       ALLOWABLE
                COEF            INCREASE        DECREASE
    X1          40.000000       12.166667       INFINITY
    X2          65.000000       INFINITY        36.500000
    X3          37.500000       INFINITY        33.060000
    X4          32.000000       INFINITY        15.200000
    Y1          63.500000       INFINITY        12.166667
    Y2          99.000000       36.500000       INFINITY
    Y3          55.200000       33.060000       INFINITY
    Y4          45.000000       15.200000       INFINITY


                    RIGHTHAND SIDE RANGES
    ROW         CURRENT         ALLOWABLE       ALLOWABLE
                RHS             INCREASE        DECREASE
     2          4800.000000     INFINITY        1344.000000
     3          4800.000000     INFINITY        4800.000000
     4          4800.000000     INFINITY        2496.000000
     5          4800.000000     1200.000000     4800.000000
     6          4800.000000     INFINITY        960.000000
     7          2000.000000     INFINITY        1808.000000
     8          2000.000000     INFINITY        2000.000000
     9          2000.000000     INFINITY        2000.000000
    10          2000.000000     INFINITY        2000.000000
```

The cost of in-house fabrication of DRNBG units must rise by $12.17 before changes are needed.

18. The LINDO solution to Close's problem is:

```
look all
MAX  X12 + X13
SUBJECT TO
2)   X12 <= 700
3)   X13 <= 300
4)   X25 <= 100
5)   X24 <= 500
6)   X34 <= 100
7)   X36 <= 200
8)   X45 <= 300
9)   X46 <= 400
10)  X57 <= 900
11)  X67 <= 400
12)  X12 - X25 - X24 = 0
13)  X13 - X34 - X36 = 0
14)  X24 + X34 - X45 - X46 = 0
15)  X25 + X45 - X57 = 0
16)  X36 + X46 - X67 = 0
END


? GO
LP OPTIMUM FOUND AT STEP 10
OBJECTIVE FUNCTION VALUE
1)  800.000000
        VARIABLE              VALUE            REDUCED COST
            X12           500.000000               .000000
            X13           300.000000               .000000
            X25           100.000000               .000000
            X24           400.000000               .000000
            X34           100.000000               .000000
            X36           200.000000               .000000
            X45           300.000000               .000000
            X46           200.000000               .000000
            X57           400.000000               .000000
            X67           400.000000               .000000
```

(continued)

```
        ROW         SLACK OR SURPLUS      DUAL PRICES
         2)            200.000000            .000000
         3)                 .000000          .000000
         4)                 .000000         1.000000
         5)            100.000000            .000000
         6)                 .000000          .000000
         7)                 .000000          .000000
         8)                 .000000         1.000000
         9)            200.000000            .000000
        10)            500.000000            .000000
        11)                 .000000         1.000000
        12)                 .000000         1.000000
        13)                 .000000         1.000000
        14)                 .000000         1.000000
        15)                 .000000          .000000
        16)                 .000000         1.000000
```

NO. ITERATIONS= 10
DO RANGE (SENSITIVITY) ANALYSIS? ? Y

RANGES IN WHICH THE BASIS IS UNCHANGED

	OBJ COEFFICIENT RANGES		
VARIABLE	CURRENT COEF	ALLOWABLE INCREASE	ALLOWABLE DECREASE
X12	1.000000	.000000	1.000000
X13	1.000000	INFINITY	.000000
X25	.000000	INFINITY	1.000000
X24	.000000	.000000	1.000000
X34	.000000	.000000	.000000
X36	.000000	INFINITY	.000000
X45	.000000	INFINITY	1.000000
X46	.000000	.000000	1.000000
X57	.000000	INFINITY	1.000000
X67	.000000	INFINITY	1.000000

```
                     RIGHTHAND SIDE RANGES
    ROW        CURRENT         ALLOWABLE       ALLOWABLE
                 RHS           INCREASE        DECREASE
     2        700.000000       INFINITY       200.000000
     3        300.000000        .000000       100.000000
     4        100.000000      200.000000      100.000000
     5        500.000000       INFINITY       100.000000
     6        100.000000       INFINITY         .000000
     7        200.000000      100.000000        .000000
     8        300.000000      100.000000      300.000000
     9        400.000000       INFINITY       200.000000
    10        900.000000       INFINITY       500.000000
    11        400.000000      100.000000      200.000000
    12          .000000       200.000000      500.000000
    13          .000000       100.000000        .000000
    14          .000000       100.000000      400.000000
    15          .000000       400.000000      500.000000
    16          .000000       100.000000      200.000000
```

Fortunately, Billings-Kansas has 100 tons of spare capacity, and Jackson-Kansas has 200 tons of spare capacity, so Close just makes it.

19. With regard to New Sprint's first problem, the minimizing paper used is:

```
look all
MIN X1 + X2 + X3 + X4 + X5
SUBJECT TO
2)  3 X1 + X5 >= 130
3)  2 X2 + X3 >= 120
4)  X3 + 2 X4 + X5 >= 85
END

? GO
LP OPTIMUM FOUND AT STEP 3
OBJECTIVE FUNCTION VALUE
1)  145.833333
      VARIABLE           VALUE           REDUCED COST
         X1            43.333333           .000000
         X2            60.000000           .000000
         X3              .000000           .000000
         X4            42.500000           .000000
         X5              .000000           .166667
```

(continued)

```
       ROW          SLACK OR SURPLUS    DUAL PRICES
        2)               .000000          -.333333
        3)               .000000          -.500000
        4)               .000000          -.500000
NO. ITERATIONS= 3
DO RANGE (SENSITIVITY) ANALYSIS? ? Y


RANGES IN WHICH THE BASIS IS UNCHANGED
                   OBJ COEFFICIENT RANGES
VARIABLE       CURRENT        ALLOWABLE       ALLOWABLE
                COEF          INCREASE        DECREASE
   X1         1.000000        .500000        1.000000
   X2         1.000000        .000000        1.000000
   X3         1.000000        INFINITY        .000000
   X4         1.000000        .000000        1.000000
   X5         1.000000        INFINITY        .166667


                   RIGHTHAND SIDE RANGES
   ROW         CURRENT        ALLOWABLE       ALLOWABLE
                RHS           INCREASE        DECREASE
    2        130.000000       INFINITY      130.000000
    3        120.000000       INFINITY      120.000000
    4         85.000000       INFINITY       85.000000
```

With regard to the problem of minimizing waste:

```
look all
MIN 1.5 X1 + 6 X2 + 3 X3 + 5.5 X5
SUBJECT TO
2)  3 X1 + X5 >= 130
3)  2 X2 + X3 >= 120
4)  X3 + X5 + 2 X4 >= 85
END

: GO
LP OPTIMUM FOUND AT STEP 3
           OBJECTIVE FUNCTION VALUE
1)  425.000000
     VARIABLE           VALUE          REDUCED COST
        X1            43.333333           .000000
        X2            60.000000           .000000
        X3             .000000            .000000
        X4             .000000           5.000000
        X4            42.500000           .000000
```

```
        ROW           SLACK OR SURPLUS    DUAL PRICES
         2)                .000000         -.500000
         3)                .000000         -3.000000
         4)                .000000          .000000
NO. ITERATIONS= 3
DO RANGE (SENSITIVITY) ANALYSIS? > Y

RANGES IN WHICH THE BASIS IS UNCHANGED
                  OBJ COEFFICIENT RANGES
VARIABLE       CURRENT        ALLOWABLE       ALLOWABLE
                COEF          INCREASE        DECREASE
   X1         1.500000       15.000000        1.500000
   X2         6.000000         .000000        6.000000
   X3         3.000000       INFINITY          .000000
   X5         5.500000       INFINITY         5.000000
   X4          .000000         .000000         .000000


                  RIGHTHAND SIDE RANGES
   ROW         CURRENT        ALLOWABLE       ALLOWABLE
                RHS           INCREASE        DECREASE
    2        130.000000      INFINITY       130.000000
    3        120.000000      INFINITY       120.000000
    4         85.000000      INFINITY        85.000000
```

The decreased order does not affect the solution in either case.

20. For Tim Burr:

```
look all
MAX Y4
SUBJECT TO
2)  X11 <= 25000
3)  X12 <= 25000
4)  X13 <= 25000
5)  X21 <= 50000
6)  X22 <= 50000
7)  X23 <= 50000
8)  X31 <= 100000
9)  X32 <= 100000
10)  X33 <= 100000
11)  X11 + X21 + X31 + Y1 = 500000
12)  - 1.25 X11 + X12 + X22 + X32 - Y1 + Y2 = 0
13)  - 1.25 X12 + X13 - 1.35 X21 + X23 + X33 - Y2 +
Y3 = 0
14)  Y4 - 1.25 X13 - 1.35 X22 - 1.5 X31 - Y3 + X14
 = 0
END

: GO
LP OPTIMUM FOUND AT STEP 14
            OBJECTIVE FUNCTION VALUE
1)  603750.000
      VARIABLE            VALUE          REDUCED COST
         Y4         603750.000000         .000000
        X11          25000.000000         .000000
        X12          25000.000000         .000000
        X13          25000.000000         .000000
        X21          50000.000000         .000000
        X22          50000.000000         .000000
        X23               .000000        1.000000
        X31         100000.000000         .000000
        X32               .000000        1.000000
        X33               .000000        1.000000
         Y1         325000.000000         .000000
         Y2         281250.000000         .000000
         Y3         355000.000000         .000000
        X14               .000000        1.000000
```

ROW	SLACK OR SURPLUS	DUAL PRICES
2)	.000000	.250000
3)	.000000	.250000
4)	.000000	.250000
5)	.000000	.350000
6)	.000000	.350000
7)	50000.000000	.000000
8)	.000000	.500000
9)	100000.000000	.000000
10)	100000.000000	.000000
11)	.000000	1.000000
12)	.000000	1.000000
13)	.000000	1.000000
14)	.000000	1.000000

NO. ITERATIONS= 14
DO RANGE (SENSITIVITY) ANALYSIS? > Y

RANGES IN WHICH THE BASIS IS UNCHANGED
OBJ COEFFICIENT RANGES

VARIABLE	CURRENT COEF	ALLOWABLE INCREASE	ALLOWABLE DECREASE
Y4	1.000000	INFINITY	1.000000
X11	.000000	INFINITY	.250000
X12	.000000	INFINITY	.250000
X13	.000000	INFINITY	.250000
X21	.000000	INFINITY	.350000
X22	.000000	INFINITY	.350000
X23	.000000	1.000000	INFINITY
X31	.000000	INFINITY	.500000
X32	.000000	1.000000	INFINITY
X33	.000000	1.000000	INFINITY
Y1	.000000	.250000	INFINITY
Y2	.000000	.250000	1.000000
Y3	.000000	.250000	1.000000
X14	.000000	1.000000	INFINITY

(continued)

	RIGHTHAND SIDE RANGES		
ROW	CURRENT RHS	ALLOWABLE INCREASE	ALLOWABLE DECREASE
2	25000.000000	325000.000000	25000.000000
3	25000.000000	281250.000000	25000.000000
4	25000.000000	355000.000000	25000.000000
5	50000.000000	281250.000000	50000.000000
6	50000.000000	281250.000000	50000.000000
7	50000.000000	INFINITY	50000.000000
8	100000.000000	281250.000000	100000.000000
9	100000.000000	INFINITY	100000.000000
10	100000.000000	INFINITY	100000.000000
11	500000.000000	INFINITY	281250.000000
12	.000000	INFINITY	281250.000000
13	.000000	INFINITY	355000.000000
14	.000000	INFINITY	603750.000000

This indicates that Burr will not be using all of his cash and can spend $281,250 on his yacht.

21. Jack Uzi's solution is:

```
look all
MIN 500 X1 + 500 X2 + 500 X3 + 500 X4 + 500 X5 +
 500 X6 + 800 Y1 + 800 Y2 + 800 Y3 + 800 Y4 + 800
 Y5 + 800 Y6 + 100 Z1 + 100 Z2 + 100 Z3 + 100 Z4 +
 100 Z5 + 100 Z6
SUBJECT TO
2) X1 <= 40
3) X2 <= 40
4) X3 <= 40
5) X4 <= 40
6) X5 <= 40
7) X6 <= 40
8) Y1 <= 10
9) Y2 <= 10
10) Y3 <= 10
11) Y4 <= 10
12) Y5 <= 10
13) Y6 <= 10
14) X1 + Y1 - Z1 = 25
15) X2 + Y2 + Z1 - Z2 = 30
16) X3 + Y3 + Z2 - Z3 = 50
17) X4 + Y4 + Z3 - Z4 = 55
18) X5 + Y5 + Z4 - Z5 = 60
19) X6 + Y6 + Z5 = 40
END
```

(continued)

```
: GO
LP OPTIMUM FOUND AT STEP 14
OBJECTIVE FUNCTION VALUE
1) 142500,000
        VARIABLE           VALUE          REDUCED COST
            X1          40,000000              ,000000
            X2          40,000000              ,000000
            X3          40,000000              ,000000
            X4          40,000000              ,000000
            X5          40,000000              ,000000
            X6          40,000000              ,000000
            Y1            ,000000           200,000000
            Y2            ,000000           100,000000
            Y3            ,000000              ,000000
            Y4          10,000000              ,000000
            Y5          10,000000              ,000000
            Y6            ,000000           300,000000
            Z1          15,000000              ,000000
            Z2          25,000000              ,000000
            Z3          15,000000              ,000000
            Z4          10,000000              ,000000
            Z5            ,000000           600,000000
            Z6            ,000000           100,000000

        ROW       SLACK OR SURPLUS       DUAL PRICES
          2)            ,000000           100,000000
          3)            ,000000           200,000000
          4)            ,000000           300,000000
          5)            ,000000           400,000000
          6)            ,000000           500,000000
          7)            ,000000              ,000000
          8)          10,000000              ,000000
          9)          10,000000              ,000000
         10)          10,000000              ,000000
         11)            ,000000           100,000000
         12)            ,000000           200,000000
         13)          10,000000              ,000000
         14)            ,000000          -600,000000
         15)            ,000000          -700,000000
         16)            ,000000          -800,000000
         17)            ,000000          -900,000000
         18)            ,000000         -1000,000000
         19)            ,000000          -500,000000
NO, ITERATIONS= 14
DO RANGE (SENSITIVITY) ANALYSIS? > Y
```

```
RANGES IN WHICH THE BASIS IS UNCHANGED
                OBJ COEFFICIENT RANGES
VARIABLE       CURRENT       ALLOWABLE      ALLOWABLE
               COEF          INCREASE       DECREASE
   X1        500.000000     100.000000      INFINITY
   X2        500.000000     200.000000      INFINITY
   X3        500.000000     300.000000      INFINITY
   X4        500.000000     400.000000      INFINITY
   X5        500.000000     500.000000      INFINITY
   X6        500.000000     300.000000      INFINITY
   Y1        800.000000      INFINITY      200.000000
   Y2        800.000000      INFINITY      100.000000
   Y3        800.000000     100.000000     100.000000
   Y4        800.000000     100.000000      INFINITY
   Y5        800.000000     200.000000      INFINITY
   Y6        800.000000      INFINITY      300.000000
   Z1        100.000000     100.000000     200.000000
   Z2        100.000000     100.000000     100.000000
   Z3        100.000000      INFINITY      100.000000
   Z4        100.000000      INFINITY      200.000000
   Z5        100.000000      INFINITY      600.000000
   Z6        100.000000      INFINITY      100.000000


                RIGHTHAND SIDE RANGES
  ROW          CURRENT       ALLOWABLE      ALLOWABLE
               RHS           INCREASE       DECREASE
   2          40.000000       .000000      10.000000
   3          40.000000       .000000      10.000000
   4          40.000000       .000000      10.000000
   5          40.000000       .000000      10.000000
   6          40.000000       .000000      10.000000
   7          40.000000      INFINITY       .000000
   8          10.000000      INFINITY      10.000000
   9          10.000000      INFINITY      10.000000
  10          10.000000      INFINITY      10.000000
  11          10.000000       .000000      10.000000
  12          10.000000       .000000      10.000000
  13          10.000000      INFINITY      10.000000
  14          25.000000      10.000000      .000000
  15          30.000000      10.000000      .000000
  16          50.000000      10.000000      .000000
  17          55.000000      10.000000      .000000
  18          60.000000      10.000000      .000000
  19          40.000000       .000000      40.000000
```

The solution indicates (rows 14 through 18) that demand can be increased in any of these periods without affecting the plan.

4

The Transportation Problem

The transportation problem is a special case in linear programming. Here the constraints are all equality constraints; furthermore, only one basic commodity is involved. This basic commodity is available from several supply depots and must be redistributed to several consumer depots. The difficulty of the problem is that the redistribution costs vary between the various routes from supply depots to consumer depots.

Let us return briefly to the previous linear programming problem, which could be written mathematically as

Optimize
$$\sum_{i=1}^{n} A_i X_i$$

over
$$\sum_{j=1-m}^{n} C_{ij} X_i = B_j \qquad \text{usually } Xi \geq 0$$

In other words, the problem of the previous chapter was

Maximize $4 X1 + 5 X2 + 3 X3 + 0 X4 + 0 X5 + 0 X6$

subject to $1 X1 + 1 X2 + 4 X3 + 0 X4 + 0 X5 + 0 X6 = 6$

$1 X1 + 2 X2 + 1 X3 + 0 X4 + 1 X5 + 0 X6 = 7$

$1 X1 + 4 X2 + 2 X3 + 0 X4 + 0 X5 + 1 X6 = 12$

Thus

$$n = 6, \quad m = 3.$$
$$A_1 = 4, \quad A_2 = 5, \text{ etc.}$$
$$C_{11} = 1, \quad C_{12} = 1, \quad C_{13} = 4, \text{ etc.}$$
$$B_1 = 6, \quad B_2 = 7, \text{ etc.}$$

The transportation problem has the form

Optimize $$\sum_i \sum_j C_{ij} X_{ij}$$

subject to $$\sum_j X_{ij} = A_i$$

and $$\sum_i X_{ij} = B_j \qquad X_{ij} \geq 0$$

While the mathematical representation looks vaguely similar, it may not be immediately apparent how the two forms are linked. Actually, a few general linear programming problems may be rewritten as transportation problems with some ease in computation. However, *all* transportation problems can be solved via the simplex method, although not very efficiently. When considering computer solutions, one must decide whether to use a transportation algorithm implemented on a microcomputer (which has more steps but requires less data input), or to use the same microcomputer to access a simplex package on a mainframe computer (which requires more data input but allows for more post-optimal analysis). My own preference is to do the latter.

The best way to discuss the transportation problem is with an example. Consider the situation of a company with three warehouses and four retail shops. The cost of distributing a unit of merchandise from any particular warehouse to any particular shop is given on the next page:

Shop	Warehouse		
	a	b	c
A	8	7	9
B	4	6	2
C	2	1	1
D	5	7	7

(C_{ij})

The shops have demands of 5, 10, 12, and 3 units, respectively, and the warehouses have 10, 8, and 12 units in stock. Note that we are dealing here with a single commodity, unlike the previous problem with constraints arising from several commodities. Now, if the amounts transferred between warehouses and shops are depicted as X_{ij}, the problem may be written as

Minimize
$$8\,X11 + 7\,X12 + 9\,X13$$
$$+\ 4\,X21 + 6\,X22 + 2\,X23$$
$$+\ 2\,X31 + 1\,X32 + 1\,X33$$
$$+\ 5\,X41 + 7\,X42 + 7\,X43$$

subject to

$$X11 + X12 + X13 = 5$$
$$X21 + X22 + X23 = 10 \quad \text{(Shop constraints)}$$
$$X31 + X32 + X33 = 12$$
$$X41 + X42 + X43 = 3$$
$$X11 + X21 + X31 + X41 = 10 \quad \text{(Warehouse}$$
$$X12 + X22 + X32 + X42 = 8 \quad \text{constraints)}$$
$$X13 + X23 + X33 + X43 = 12$$
$$X_{ij} \geqslant 0$$

Some consideration of these equations will show that at least one X_{ij} in each equation must be positive. Furthermore, no more than two X_{ij}'s in each equation need be positive in order to have a feasible solution. In fact, if there are more than two positive X_{ij}'s in any equation, a better solution will be found by rearranging the solution to give, at most, only two positive X_{ij}'s. An examination of the transportation tableau may help to clarify this:

Warehouse

Shop	Split Costs	a	b	c	Total Requirements
A		8	7	9	5
B		4	6	2	10
C		2	1	1	12
D		5	7	7	3
Total Available		10	8	12	30

Tableau 4.1

       Tableau 4.1 conveniently depicts the problem. If we say that the problem has N requirement rows and M availability columns, we may choose a feasible solution that has $N + M - 1$ nonzero X_{ij} elements, known as used routes. (This is not always the case; the alternatives will be discussed later.)

       A solution, not necessarily the best solution, will be found by assigning the maximum possible value to X_{11} and then following out the rest of the X_{ij}'s from this point. The solution is known as the "northwest corner solution" (referring to the upper left-hand corner of the tableau) (see Tableau 4.2).

       This solution to the problem is not necessarily a good one. We can check on the unused routes in any row or column by comparing the costs implied by the proposed solution with the actual costs via a device known as *split costs*. The costs for the proposed routes used in the solution are split into a cost of dispatch and a cost of receipt, such that

Cost of Dispatch + Cost of Receipt = Cost of Transportation

		Warehouse			
Shop	Split Costs	a	b	c	Total Require-ments
A		5 8	7	9	5
B		5 4	5 6	2	10
C		2	3 1	9 1	12
D		5	7	3 7	3
	Total Available	10	8	12	30

Tableau 4.2

In Tableau 4.2, the solution calls for X11 = 5; in other words, the route aA is utilized. Thus the transportation cost of 8 must be split into two parts. There are many ways of adding two numbers to achieve a sum of 8; but the ambiguity is removed if we arbitrarily decide to set the cost of receipt by A at zero. (The method will still work with any other number, including a negative one.)

Having set the cost of receipt by A at 0, then the split cost for point a must be 8 for this used route aA. Since the proposed solution calls for the route aB to be used, the split cost of B must therefore be −4. As route bB is used in this solution, b's split cost must be 10 (because 10 + (−4) = 6), etc. *But only the used routes are utilized in calculating the split costs.* Tableau 4.3 shows these split costs.

The next stage in testing the solution is to check the unused routes, such as bA, cA, Bc, and so forth. The cost of transporting from b to A implied by the proposed solution is given by adding the split costs of b and A, in this case 10 cost units. The actual cost of this route is only 7 units, so an improvement of 3 units could be achieved by moving goods out of the proposed route into route bA. The difference between the actual and implied values (-3 for bA) is then written for convenience into the top left-hand corner of each unused route cell. These values are also depicted in Tableau 4.3, where examination of these top left-hand corner values indicates that maximum benefit is achieved by improving the solution to take into account route cB (for this has the largest negative coefficient). In fact, these top left-hand corner values are the $Z_j - C_j$ coefficients for the solution, which calls for the X_{ij} elements that are positive being in the basis of the linear programming solution.

Warehouse

Shop	Split Costs	a	b	c	Total Requirements
		8	10	10	
A	0	5 8	-3 7	-1 9	5
B	-4	5 4	5 6	-4 2	10
C	-9	3 2	3 1	9 1	12
D	-3	0 5	0 7	3 7	3
	Total Available	10	8	12	30

Tableau 4.3

How can route cB be incorporated? Tableau 4.4 shows the method. Let X be the amount of goods consigned via cB. These must be withdrawn from other routes in such a way that the constraints of the problem remain satisfied. The only way this can be achieved is to withdraw X from cC, add X to bC, and withdraw X from Bb. The best improvement is achieved with the maximum value of X transferred. The maximum value is 5 units, and Tableau 4.5 results from its transfer.

		Warehouse			
		a	b	c	
Shop	Split Costs	8	10	10	Total Requirements
A	0	5 8	−3 7	−1 9	5
B	−4	5 4	5−X... 6	−4 ...X 2	10
C	−9	3 2	3+X... 1	...9−X 1	12
D	−3	0 5	0 7	3 7	3
	Total Available	10	8	12	30

Tableau 4.4

To improve the result, we can analyze Tableau 4.5 again via split costs and the route aD is found to offer an improvement. The maximum amount of the commodity that may be transferred from the proposed solution without making any of the entries negative (thus proposing an infeasible solution) is 3 units (Tableau 4.6). The process of successive improvements is followed through to Tableau 4.7, where none of the unused routes offers an improvement over the proposed solution. This solution is therefore optimal.

Tableau 4.5

Warehouse

Shop	Split Costs	a (8)	b (6)	c (6)	Total Requirements
A	0	**5** (8)	^{1} (7)	^{3} (9)	**5**
B	−4	**5−X** (4)	^{4} (6)	**5+X** (2)	**10**
C	−5	$^{-1}$ (2)	**8** (1)	**4** (1)	**12**
D	1	$^{-4}$ **X** (5)	^{0} (7)	**3−X** (7)	**3**
Total Available		10	8	12	30

Tableau 4.6

Warehouse

Shop	Split Costs	a (8)	b (6)	c (6)	Total Requirements
A	0	**5** (8)	^{1} (7)	^{3} (9)	**5**
B	−4	**2−X** (4)	^{4} (6)	**8+X**	10
C	−5	$^{-1}$ **X** (2)	**8** (1)	**4−X** (1)	**12**
D	−3	**3** (5)	^{4} (7)	^{4} (7)	**3**
Total Available		10	8	12	30

Tableau 4.7

		Warehouse			
Shop	Split Costs	a	b	c	Total Requirements
		8	7	7	
A	0	0 **5** 8	0 7	2 9	5
B	−5	1 4	4 6	**10** 2	10
C	−6	**2** 2	**8** 1	**2** 1	12
D	−3	**3** 5	3 7	3 7	3
Total Available		10	8	12	30

       One point of interest is the zero entry in the top left-hand corner of the element bA of Tableau 4.7. This indicates that there is an alternative solution to the one depicted, with exactly the same optimization function value. The alternative is discovered by inserting X into the element bA and updating the tableau via the same procedure. Note that the process is self-correcting: A mistake made in generating Tableau 4.5 (for example aD) will result in Tableau 4.8. This error will be automatically corrected between Tableaux 4.8 and 4.11. Tableau 4.11 is identical to Tableau 4.7, and the process requires one extra iteration because of the error.

       Note the indication of equivalent optimal solutions by the zero element in the top left-hand corner of Ab. Up to 5 units of the commodity may be rearranged from the Tableau 4.7 solution into this route. This is shown in Tableaux 4.7A and 4.7B.

Warehouse   **Tableau 4.7A**

Shop	Split Costs	a	b	c	Total Requirements
		8	7	7	
A	0	**5 − X** 8	0 **X** 7	2 9	**5**
B	−5	1 4	4 6	**10** 2	**10**
C	−6	**2 + X** 2	**8 − X** 1	**2** 1	**12**
D	−3	**3** 5	3 7	3 7	**3**
Total Available		10	8	12	30

Warehouse   **Tableau 4.7B**

Shop	Split Costs	a	b	c	Total Requirements
		8	7	7	
A	0	0 8	**5** 7	2 9	**5**
B	−5	1 4	4 6	**10** 2	**10**
C	−6	**7** 2	**3** 1	**2** 1	**12**
D	−3	**3** 5	3 7	3 7	**3**
Total Available		10	8	12	30

Tableau 4.8

Shop	Split Costs	Warehouse a	b	c	Total Requirements
		8	6	6	
A	0	5 8	1 7	3 9	5
B	−4	5−X 4	4 6	5+X 2	10
C	−5	−1 X 2	8 1	4−X 1	12
D	1	4 5	0 7	3 7	3
Total Available		10	8	12	30

Note: This tableau has an error in cell aD.

Tableau 4.9

Shop	Split Costs	Warehouse a	b	c	Total Requirements
		8	7	6	
A	0	5 8	0 7	3 9	5
B	−4	1−X 4	3 6	9+X 2	10
C	−6	4 2	8 1	1 1	12
D	1	−4 X 5	−1 7	3−X 7	3
Total Available		10	8	12	30

Warehouse

Tableau 4.10

Shop	Split Costs	a — 8	b — 7	c — 10	Total Requirements
A	0	**5** (8)	0 — (7)	−1 — (9)	**5**
B	−8	4 — (4)	7 — (6)	**10** (2)	**10**
C	−6	**4−X** (2)	**8** (1)	−3 — **X** (1)	**12**
D	−3	**1+X** (5)	3 — (7)	**2−X** (7)	**3**
Total Available		10	8	12	30

Warehouse

Tableau 4.11

Shop	Split Costs	a — 8	b — 7	c — 7	Total Requirements
A	0	**5** (8)	0 — (7)	2 — (9)	**5**
B	−5	1 — (4)	4 — (6)	**10** (2)	**10**
C	−6	**2** (2)	**8** (1)	**2** (1)	**12**
D	−3	**3** (5)	3 — (7)	3 — (7)	**3**
Total Available		10	8	12	30

UNBALANCED SUPPLIES AND DEMAND

In the example illustrating the transportation technique just shown, the total supply of the commodity was conveniently equal to the total demand for the product. Should this not be the case, then a dummy supplier or a dummy receiver may be inserted into the tableau in order to ensure that the demand and supply balance, thus restoring equality constraints. The costs of transporting the commodity from or to this dummy may be set to any figure as long as it is equal throughout the row (or column). The unit cost figure for the dummy route is usually set at zero. This is convenient because use of the dummy route means that goods either remain in the warehouse at a transportation cost of zero, or are supplied by a dummy warehouse, again at a cost of zero. For example, Tableau 4.12 depicts a dummy warehouse and Tableau 4.13 depicts a dummy customer.

Tableau 4.12

Dummy Warehouse

		Warehouse				
	Shop / Split Costs	a	b	c	d	Total Requirements
A		8	7	9	0	**5**
B		4	6	2	0	**10**
C		2	1	1	0	**12**
D		5	7	7	0	**5**
Total Available		10	8	12	2	32

Warehouse

Shop	Split Costs	a	b	c	Total Requirements
A		8	7	9	5
B		4	6	2	10
C		2	1	1	12
D		5	7	7	3
E		0	0	0	2
Total Available		10	10	12	32

Tableau 4.13
Dummy Customer

DEGENERACY

Not all transportation solutions utilize exactly $M + N - 1$ used routes, as previously implied. For example, in the previous problem, had shop B's requirement been 8 units rather than 10, and warehouse a had only 8 units in stock rather than 10, then the second iteration from the northwest corner initial solution would have been changed from that depicted in Tableau 4.5 to that shown in Tableau 4.14. In this case, X would be 3 and two routes would disappear from the solution, with only one replacing them. This leaves only $M + N - 2$ used routes. Now it is not possible to discover a complete set of split costs for only $M + N - 2$ used routes. One way around this difficulty is to assign a value of zero to one of the unused routes and determine the split costs, with the solution transporting 0 units via that route. The updated tableau is shown in Tableau 4.15, optimization proceeding accordingly.

			Warehouse			
	Shop	Split Costs	a	b	c	Total Requirements
			8	6	6	
Tableau 4.14	A	0	5 〔8〕	1 〔7〕	3 〔9〕	5
	B	−4	3−X 〔4〕	4 〔6〕	5+X 〔2〕	8
	C	−5	−1 〔2〕	8 〔1〕	4 〔1〕	12
	D	1	−4 X 〔5〕	0 〔7〕	3−X 〔7〕	3
		Total Available	8	8	12	28

Warehouse

Shop	Split Costs	a	b	c	Total Requirements
		8	6	6	
A	0	1 **5** 8	7	3 9	5
B	−4	**0** 4	4 6	**8** 2	8
C	−5	−1 2	**8** 1	**4** 1	12
D	−3	**3** 5	4 7	4 7	3
Total Available		8	8	12	28

Tableau 4.15

There is no reason that the number of used routes should not drop even lower in some cases. The device may be used again as many times as necessary, with a zero appearing each time degeneracy occurs. It is also common to find that a solution will be improved by reallocating zero commodities among the routes.

This degenerate solution is exactly analogous to the degeneracy that can occur in the regular simplex algorithm, where one or more basic variables take the value zero.

VOGEL'S METHOD

The northwest corner rule produced a particularly inefficient initial solution to the transportation problem. The example required four successive improvements before the optimum solution was reached. Vogel's method, in contrast, regularly produces an extremely good solution which, if not optimal, is regularly almost optimal. The method uses the commonsense assumption that one wishes, if possible, to utilize the minimum cost routes. This might not be possible directly, because using the minimum cost route for one constraint might force a high-cost route in another constraint. Therefore, opportunity penalties are evaluated, as shown in Tableau 4.16, by noting the differences between the smallest and the next smallest element in each row and column.

Tableau 4.16

	Shop / Split Costs	Warehouse a	b	c	Total Requirements	Difference between smallest and next smallest in row
A		8	7	9	**5**	1
B		4	6	2	**10**	2
C		2	1	1	**12**	0
D		5	7	7	**3**	2
Total Available		10	8	12	30	
Difference between smallest and next smallest cost element in column		2	**5**	1		

The 5 is bold because this is the largest opportunity cost of taking the smallest cost element.

The largest element, 5, occurs in column b. Thus, if column b were not to supply as much as possible via bC (the smallest unit transport cost element), there would be an extra cost of at least 5 units for any other such deployment. We must therefore deploy as much as possible via bC. The largest deployment possible is 8 units of commodity. Eight units therefore exhausts the total stock in the warehouse and thus removes warehouse b from any further consideration.

The differences are then recomputed for the now modified problem depicted in Tableau 4.17. The largest opportunity cost now arises in column a, and also in rows B and D. Make a choice, say B, which allocates 10 units to cB. Reiterating gives an opportunity cost of 6 in column c (Tableau 4.18), producing the final tableau after two more iterations (Tableaux 4.19 and 4.20). This is identical to Tableau 4.11, the optimal solution. However, since the method is not guaranteed to produce an optimal solution, it must be checked via split costs and if necessary iterated, as previously, to an optimal solution.

		Warehouse				
Shop	Split Costs	a	b	c	Total Requirements	Difference between smallest and next smallest in row
A		8	*	9	**5**	1
B		4	*	2	**10**	2*
C		2	**8** *	1	**4**	1
D		5	*	7	**3**	2
Total Available		10	*	12	22	
Difference between smallest and next smallest cost element in column		2	*	1		

Tableau 4.17

The b column is now no longer considered a candidate for the minimum cost element, because we have exhausted the total in warehouse b by allocating all of it to route bC.

Warehouse

Tableau 4.18

Shop	Split Costs	a	b	c	Total Require-ments	Difference between smallest and next smallest in row
A		8	*	9	**5**	1
B		*	*	**10** *	*	*
C		2	**8** *	1	**4**	1
D		5	*	7	**3**	2
Total Available		10	*	2	12	

Difference between smallest and next smallest cost element in column: a = 3   b = *   c = **6**

Warehouse

Shop	Split Costs	a	b	c	Total Requirements	Difference between smallest and next smallest in row
A		8	*	*	**5**	1
B		*	*	**10** *	*	*
C		2	**8** *	**2** *	**2**	1
D		5	*	*	**3**	2
Total Available		10	*	*	10	
Difference between smallest and next smallest cost element in column		3	*	*		

Tableau 4.19

Warehouse

		a	b	c	Total Requirements
Shop	Split Costs	8	7	7	
A	0	**5** / 8	0 / 7	2 / 9	**5**
B	−5	1 / 4	4 / 6	**10** / 2	**10**
C	−6	**2** / 2	**8** / 1	**2** / 1	**12**
D	−3	**3** / 5	3 / 7	3 / 7	**3**
Total Available		10	8	12	30

Tableau 4.20

COMPUTER SOLUTION OF TRANSPORTATION PROBLEMS

It has been pointed out that the transportation problem is merely a special case of a linear programming (LP) problem. The LP formulation was written out for the example problem. For a computer solution via a generalized LP package, such as LINDO, it is merely necessary to input that formulation:

```
MIN
8X11+7X12+9X13
+4X21+6X22+2X23
+2X31+1X32+1X33
+5X41+7X42+7X43
ST
X11+X12+X13=5
X21+X22+X23=10
X31+X32+X33=12
X41+X42+X43=3
X11+X21+X31+X41=10
X12+X22+X32+X42=8
X13+X23+X33+X43=12
END
```

The LINDO package produces the output:

```
LP OPTIMUM FOUND AT STEP 6
          OBJECTIVE FUNCTION VALUE
1)  89.000000

     VARIABLE          VALUE          REDUCED COST
        X11           .000000           .000000
        X12          5.000000           .000000
        X13           .000000          2.000000
        X21           .000000          1.000000
        X22           .000000          4.000000
        X23         10.000000           .000000
        X31          7.000000           .000000
        X32          3.000000           .000000
        X33          2.000000           .000000
        X41          3.000000           .000000
        X42           .000000          3.000000
        X43           .000000          3.000000
```

(continued)

```
          ROW         SLACK OR SURPLUS      DUAL PRICES
           2)              .000000         -8.000000
           3)              .000000         -3.000000
           4)              .000000         -2.000000
           5)              .000000         -5.000000
           6)              .000000           .000000
           7)              .000000          1.000000
           8)              .000000          1.000000
NO. OF ITERATIONS= 6

DO RANGE (SENSITIVITY) ANALYSIS? > Y

RANGES IN WHICH THE BASIS IS UNCHANGED
```

```
                        OBJ COEFFICIENT RANGES
VARIABLE         CURRENT         ALLOWABLE        ALLOWABLE
                  COEF           INCREASE         DECREASE
   X11          8.000000         INFINITY          .000000
   X12          7.000000          .000000         INFINITY
   X13          9.000000         INFINITY         2.000000
   X21          4.000000         INFINITY         1.000000
   X22          6.000000         INFINITY         4.000000
   X23          2.000000         1.000000         INFINITY
   X31          2.000000          .000000         3.000000
   X32          1.000000         3.000000          .000000
   X33          1.000000         2.000000         1.000000
   X41          5.000000         3.000000         INFINITY
   X42          7.000000         INFINITY         3.000000
   X43          7.000000         INFINITY         3.000000
```

```
                        RIGHTHAND SIDE RANGES
   ROW           CURRENT         ALLOWABLE        ALLOWABLE
                  RHS            INCREASE         DECREASE
    2           5.000000          .000000         5.000000
    3          10.000000          .000000         7.000000
    4          12.000000          .000000         7.000000
    5           3.000000          .000000         3.000000
    6          10.000000         INFINITY          .000000
    7           8.000000         7.000000          .000000
    8          12.000000         7.000000          .000000
```

There are some interesting points concerning this output. First, it has generated a solution different from that generated by the long-hand method. Notice that both solutions have indicated a further optimal solution, the long-hand method by the generation of a zero entry in the top left-hand corner route evaluation of cell Ab (equivalent to X12). The computer solution indicates the existence of further solutions by the reported zero in both the value and the reduced cost of X11, equivalent to route Aa.

The next point of interest is that the reduced cost on the routes designated as "unused" by the computer solution are the same values as those generated in the top left-hand corners of the long-hand method. The dual prices indicated on the rows show the extra cost savings that might be made by having an extra unit available in the indicated warehouse, or not required by a customer (hence the difference in sign between the row duals and the column duals).

When parametric analysis is performed on this solution, the unused routes have a cost increase allowable to infinity before an alternate solution is indicated. This is hardly surprising. The parametric ranging (discussed in Chapter 3) indicated on (1) the cost coefficients for the unit cost of transportation via the routes and (2) the perturbation ranges of the customer and supply constraints could well be useful in practice.

One further point: In the case of unbalanced supply and demand, a dummy can be provided and equality constraints submitted for computer solution, or the imbalance can be handled with inequality constraints thus:

```
MIN
8X11+7X12+9X13
+4X21+6X22+2X23
+2X31+1X32+1X33
+5X41+7X42+7X43
ST
X11+X12+X13<5
X21+X22+X23<10
X31+X32+X33<12
X41+X42+X43<5
X11+X21+X31+X41=10
X12+X22+X32+X42=8
X13+X23+X33+X43=12
END
```

(continued)

```
LP OPTIMUM FOUND AT STEP 7

                OBJECTIVE FUNCTION VALUE
1)  83.000000
        VARIABLE           VALUE        REDUCED COST
            X11          3.000000          .000000
            X12           .000000          .000000
            X13           .000000         2.000000
            X21           .000000         1.000000
            X22           .000000         4.000000
            X23         10.000000          .000000
            X31          2.000000          .000000
            X32          8.000000          .000000
            X33          2.000000          .000000
            X41          5.000000          .000000
            X42           .000000         3.000000
            X43           .000000         3.000000


         ROW      SLACK OR SURPLUS      DUAL PRICES
          2)          2.000000           .000000
          3)           .000000          5.000000
          4)           .000000          6.000000
          5)           .000000          3.000000
          6)           .000000         -8.000000
          7)           .000000         -7.000000
          8)           .000000         -7.000000
```

This solution indicates the position of the imbalance in row 1, equivalent to customer A, who only gets three of the requested five units. The surplus is the unused resource of the requested five. Here the signs of the dual process appear to have been reversed from the previous, balanced formulation. This is because we now have positive slack variables associated with each row and negative slacks associated with the columns, consistent with the simplex formulation approach.

BASIC PROGRAM TO GENERATE A MATRIX TEMPLATE

Preparation of larger files can be quite time consuming and prone to error. The following BASIC program will allow you to generate a file that can conveniently be used in conjunction with an editor or word processor to represent any problem of the form used in the past two chapters.

```
10 OPEN "O",#1,"MATRIX"
100 I1 = 65
110 I2 = 90
120 PRINT#1,"MAX       "
130 K = 1
140 FOR I = I1 TO I2
150 FOR J = I1 TO I2
160 PRINT#1,"+       ";CHR$(I);CHR$(J);
170 K = K+1
180 IF K< 10 THEN 200
190 PRINT#1," ":K=1
200 NEXT J
210 NEXT I
300 PRINT#1," ":PRINT#1,"ST":PRINT#1," "
310 K=1
330 FOR I=I1 TO I2
340 FOR J=I1 TO I2
350 PRINT#1,"+       ";CHR$(I);CHR$(J);
360 K=K+1
370 IF K<10 THEN 400
380 PRINT #1," "
390 K=1
400 NEXT J
410 PRINT#1,"=       "
420 K=1
430 NEXT I
510 K=1
530 FOR J=I1 TO I2
540 FOR I=I1 TO I2
550 PRINT#1,"+       ";CHR$(I);CHR$(J);
560 K=K+1
570 IF K<10 THEN 600
580 PRINT #1," "
590 K=1
600 NEXT I
610 PRINT#1,"=       "
620 K=1
630 NEXT J
640 CLOSE #1:END
```

CHAPTER 4 PROBLEMS

The solutions to Problems 1 through 10 appear at the end of Chapter 4.

1. San Casas is a building firm with 1,000 employees. It is engaged in construction of a large housing development. The employees work in four different areas: Annapurna, Bahia, Copacabanya, and Dolrayme. For security and administration purposes, the firm has arranged for the employees to go to three different pay stations at the construction site on each Friday. The capacities of the three stations are as follows: pay station 1, 200 accounts; pay station 2, 350 accounts; pay station 3, 450 accounts. On Fridays, the employees are' entitled to attend one of these stations during working hours in order to collect their pay. The times spent, in minutes, in walking to the pay stations from their particular work areas are as follows:

	Pay Stations		
	1	2	3
Annapurna	10	6	5
Bahia	4	8	2
Copacabanya	12	7	15
Dolrayme	2	13	6

The number of employees working at each region of the development are as follows:

Annapurna	200
Bahia	450
Copacabanya	300
Dolrayme	50

San Casas wishes to minimize the time lost in the payment of wages. The supervisor in charge of wages and salaries has been requested to make the appropriate arrangements.

What is the optimal schedule for wage payment? Also, ascertain the number of working hours lost each Friday when this optimal schedule is followed.

2. C. Mentor Inc. owns three cement manufacturing plants. Currently it has four major customers, Red Eye Mix, Con Crete, Quick Pour, and C. Walls. The capacities of C. Mentor's plants are:

Plant 1	100 tons per week
Plant 2	220 tons per week
Plant 3	150 tons per week

The requirements from each customer are:

Red Eye Mix	80 tons per week
Con Crete	160 tons per week
Quick Pour	80 tons per week
C. Walls	150 tons per week

The distribution costs (dollars per ton) from each plant to each customer are:

	Plant		
	1	2	3
Red Eye Mix	5	1	3
Con Crete	3	5	4
Quick Pour	7	9	9
C. Walls	2	7	6

What is the best distribution policy and the total cost of distribution?

As the result of a fire, plant 2 has its output reduced to 180 tons per week. During this time, a quota system is initiated to restrict shortages to 15 tons per week for any particular customer. How should the allocation be modified? What is the cost of operating the quota system?

3. Fruit grown in three major growing areas is gathered into three packing stations, and then shipped to four major distribution centers. The shipping costs per ton between these stations and distribution centers are as follows:

		Packing Station			Demand
		a	b	c	
	A	8	10	13	260
Distribution	B	15	9	7	340
Center	C	11	5	16	200
	D	17	11	6	300
Supply		400	400	400	

State regulations forbid the shipment of fruit from packing station b to distribution center C. What is the optimal schedule of shipments?

4. Three wineries operate in the same valley. The wineries can process, at full capacity, the following amounts of wine:

	Number of Vats
E. & J. Garlic	20
Derringer	40
Buckthorn	30

The valley contains four major growers, who can produce grapes next season in the following amounts:

	Number of Vats
Mr. Sturgess	10
Mr. Silver	50
Mrs. Chef	40
Ms. Isreal	50

Ms. Isreal refuses to supply grapes to the Buckthorn winery, and the Derringer group considers Mr. Silver an old pirate and will not do business with him.

The wineries' managements have decided to form a cooperative, and at their first meeting discover that the growers have negotiated different prices for their crop. These are (dollars per vat):

	Garlic	Derringer	Buckthorn
Mr. Sturgess	7,000	8,000	5,000
Mr. Silver	6,000		6,000
Mrs. Chef	10,000	4,000	5,000
Ms. Isreal	3,000	9,000	

How can the cooperative minimize its purchasing policy cost in the valley?

5. Solve the following transportation problem (minimize):

From	a	b	c	d	e	Demand
to						
A	$63	45	54		99	200
B	36	39	51	6	87	150
C	84	60	66	30	105	250
D	51		30	15	81	300
E	18	36	24	3	69	250
Supply	250	300	200	350	250	

6. Solve the following as a transportation problem:

Minimize

$$18\,X11 + 10\,X12 + 16\,X13 + 11\,X21 + 15\,X22 + 10\,X23$$
$$+ 9\,X31 + 14\,X32 + 12\,X33 + 20\,X41 + 8\,X42 + 12\,X43$$

subject to

$$X11 + X12 + X13 = 140$$
$$X21 + X22 + X23 = 70$$
$$X31 + X32 + X33 = 180$$
$$X41 + X42 + X43 = 30$$
$$X11 + X21 + X31 + X41 = 200$$
$$X12 + X22 + X32 + X42 = 140$$
$$X13 + X23 + X33 + X43 = 80$$

7. Solve problem 6 as a LINDO LP problem.

8. How is the solution affected as the cost of route X11 is reduced?

9. Solve the transportation problem with the following cost matrix:

	a	b	c	d	e	f	g	h	i	j	k	l	m	n	o	p	q	r	s	t
A	6	9	6	4	6	9	2	3	1	2	7	8	6	1	2	1	6	8	2	3
B	3	8	9	9	1	2	6	4	7	8	2	3	3	8	6	7	3	6	8	3
C	4	5	3	7	2	5	6	2	8	9	4	6	3	5	2	5	6	3	2	4
D	2	5	8	2	4	2	4	3	5	7	2	9	9	6	7	4	5	3	5	2
E	2	3	5	3	4	5	6	3	4	2	6	7	8	5	4	2	3	5	2	7
F	3	5	1	3	5	7	3	8	6	9	4	3	6	7	8	4	1	2	1	3
G	3	5	3	2	5	4	5	3	8	7	6	9	6	7	3	6	4	6	2	3
H	7	6	8	7	9	6	7	7	8	6	5	9	4	7	3	6	7	9	6	3
I	5	6	3	4	5	3	7	8	9	4	5	3	2	6	7	5	8	9	3	5
J	6	7	8	9	5	4	3	5	7	5	8	9	4	3	5	7	8	9	6	4
K	5	6	4	7	8	9	5	4	8	6	7	7	4	9	9	3	3	3	10	5
L	6	7	9	12	3	4	8	7	9	6	7	4	3	6	7	3	6	4	2	1
M	5	6	7	4	3	7	8	9	4	5	3	5	2	1	2	8	9	5	4	5
N	4	5	6	3	7	3	4	7	8	9	3	4	2	3	12	5	6	7	8	3
O	6	7	4	3	5	6	4	7	8	5	6	7	3	4	5	2	7	8	4	5
P	2	3	4	2	5	7	8	4	3	2	4	3	5	6	6	7	8	4	5	3
Q	5	6	7	4	5	3	4	7	2	8	9	4	2	3	1	4	6	7	4	7
R	5	6	8	9	4	3	2	5	6	7	8	5	6	3	2	3	2	1	4	5
S	4	5	6	7	4	5	6	3	4	8	6	5	4	3	1	3	4	2	5	6
T	5	6	3	7	8	9	5	7	4	8	9	3	2	1	5	4	5	2	4	6
U	5	3	6	7	8	9	4	3	2	5	6	3	5	2	7	8	9	3	2	1
V	5	6	7	4	3	2	5	6	7	5	4	3	2	6	7	8	9	4	5	3
W	4	5	7	8	5	4	6	3	5	4	2	1	4	3	2	6	7	8	5	4
X	4	3	6	7	3	5	7	8	9	4	3	2	1	5	4	7	8	6	5	3

Supply		Demand	
a	12	A	10
b	24	B	5
c	20	C	12
d	35	D	8
e	25	E	10
f	20	F	30
g	24	G	13
h	28	H	25
i	13	I	14
j	50	J	30
k	25	K	25
l	60	L	53
m	23	M	9
n	30	N	10
o	36	O	34
p	32	P	30
q	25	Q	10
r	15	R	6
s	13	S	12
t	28	T	11
		U	13
		V	20
		W	5
		X	10

10. Solve problem 9 with the matrix interpreted as a profit matrix, and determine the range of applicability of the data for the solution.

HOMEWORK PROBLEMS

11. Bauxite is mined at three sites, with production capacities as follows:

Mine	Production
A	250
B	310
C	195

Smelters are sited conveniently close to the sites, each having a processing capacity of:

Smelter	Capacity
a	175
b	95
c	170
d	280

The following shipping costs have been determined per ton:

To	a	b	c	d
From				
A	$20	$12	$18	$14
B	11	9	10	16
C	15	14	16	23

Due to local highway problems, route bB is currently unavailable.
What transportation solution minimizes the shipping costs of the operation? What is the cost of the local difficulty restricting route bB? How much would you be prepared to pay to resolve it?

12. The Round Tuit Corporation has many orders for its product, since numerous people have finally decided to get a round tuit. These tuits are sold in standard, deluxe, and executive models. The company can manufacture tuits on three different machines. The orders for tuits are currently:

Standard	4,000
Deluxe	1,500
Executive	1,850

The machine processing capacities are:

Machine	Capacity (units)
A	3,000
B	2,800
C	1,800

The product costs per machine are:

	Standard	Deluxe	Executive
A	$1.00	1.30	1.40
B	1.30	1.40	1.20
C	1.15	1.00	1.20

Generate the minimum production cost schedule for the Round Tuit Corporation. If the production manager would like the minimum number of product changeovers on the machines, are there any alternatives?

13. The Balsam Air Conditioner company has obtained a contract to supply units to major condominium developments in four major cities. The company has plants situated near these developments. The contract calls for:

Supply	To city
100	A
150	B
130	C
190	D

The plant capacities are:

Plant	Capacity
a	200
b	200
c	180

The estimated profit per unit shipped is estimated by the accounts department as follows:

Produced By	Shipped To			
	A	B	C	D
a	$12	$20	$15	$25
b	40	35	26	21
c	15	28	25	21

Determine the optimal shipment schedule for the company.

14. Solve the following transportation problem (minimize):

From	a	b	c	d	e	Demand
To						
A	$63	$45	$54	$28	$99	$250
B	36	39	51		87	150
C	84	60	66	30	105	250
D	51	43	30		81	300
E	18	36	24	3	69	200
Supply	250	300	200	350	250	

15. Solve the following transportation problem (maximize):

From	a	b	c	d	e	Demand
To						
A	$63	$45	$54	$0	$99	$200
B	36	39	51	6	87	150
C	84	60	66	30	105	250
D	51		30	15	81	300
E	18	36	24	3	69	250
Supply	250	300	200	350	250	

16. Solve the following as a transportation problem:

Minimize

$$20\,X11 + 15\,X12 + 12\,X13 + 10\,X21 + 12\,X22 + 20\,X23$$
$$+ 19\,X31 + 14\,X32 + 16\,X33 + 15\,X41 + 12\,X42 + 16\,X43$$

subject to

$$X11 + X12 + X13 = 140$$
$$X21 + X22 + X23 = 110$$
$$X31 + X32 + X33 = 120$$
$$X41 + X42 + X43 = 50$$
$$X11 + X21 + X31 + X41 = 200$$
$$X12 + X22 + X32 + X42 = 140$$
$$X13 + X23 + X33 + X43 = 80$$

17. Solve problem 16 as a LINDO LP problem.

18. How is the solution of problem 16 (17) affected as the cost of route X23 is reduced?

19. Solve the transportation problem with the following cost matrix:

	a	b	c	d	e	f	g	h	i	j	k	l	m	n	o	p	q	r	s	t
A	16	9	6	4	6	9	2	3	1	2	7	8	6	1	2	1	6	8	2	3
B	3	8	9	9	1	12	6	4	7	8	2	3	3	8	6	7	3	6	8	3
C	4	5	3	17	2	5	6	2	8	9	4	16	3	5	2	5	6	3	2	4
D	2	5	8	2	4	2	4	3	5	7	2	9	9	6	7	14	5	3	5	2
E	2	3	5	3	4	5	6	3	4	2	6	7	8	5	4	2	3	5	2	7
F	3	5	1	3	5	7	3	8	6	9	4	3	6	7	8	4	1	2	1	3
G	3	5	3	2	5	4	5	3	8	7	6	9	6	7	3	6	4	6	2	3
H	7	6	8	7	9	6	7	7	8	6	5	9	4	7	3	6	7	9	6	3
I	5	6	3	4	5	13	7	8	9	4	5	3	2	6	7	5	8	9	3	5
J	6	7	8	9	5	4	3	5	7	5	8	9	4	3	5	7	8	9	6	4
K	5	6	4	7	8	9	5	4	8	6	7	17	4	9	9	3	3	3	10	5
L	6	7	9	12	3	4	8	7	9	6	7	4	3	6	7	3	6	4	2	1
M	5	6	7	4	3	7	8	9	4	5	3	5	2	1	2	8	9	5	4	5
N	4	5	6	3	7	3	4	7	8	9	3	4	2	3	12	5	6	7	8	3
O	6	7	4	3	5	6	4	7	8	5	6	7	3	4	5	2	7	8	4	5
P	2	3	4	2	5	7	8	4	3	2	4	3	5	6	6	7	8	4	5	3
Q	5	6	7	4	5	3	4	7	2	8	9	4	2	3	1	4	6	7	4	7
R	5	6	8	9	4	3	2	5	6	7	8	5	6	3	2	3	2	1	4	5
S	4	5	6	7	4	5	6	3	4	8	6	5	4	3	1	3	4	2	5	6
T	5	6	3	7	8	9	5	7	4	8	9	3	2	1	5	4	5	2	4	6
U	5	3	6	7	8	9	4	3	2	5	6	3	5	2	7	8	9	3	2	1
V	5	6	7	4	3	2	5	6	7	5	4	3	2	6	7	8	9	4	5	3
W	4	5	7	8	5	4	6	3	5	4	2	1	4	3	2	6	7	8	5	4
X	4	3	6	7	3	5	7	8	9	4	3	2	1	5	4	7	8	6	5	3

Supply		Demand	
a	22	A	10
b	24	B	15
c	20	C	12
d	35	D	18
e	25	E	10
f	20	F	30
g	24	G	13
h	28	H	25
i	13	I	14
j	50	J	30
k	25	K	25
l	60	L	53
m	23	M	19
n	30	N	10
o	36	O	34
p	32	P	30
q	25	Q	10
r	15	R	16
s	13	S	12
t	28	T	11
		U	13
		V	20
		W	5
		X	10

20. Solve problem 19 with the matrix interpreted as a profit matrix. Determine the range of applicability of the data for the solution.

CHAPTER 4 SOLUTIONS

1. San Casas's wages payment schedule:

Pay Stations

Area	Split Costs	1 0	2 −1	3 −2	Total Requirements
A	7	3 10	**50** 6	**150** 5	200
B	4	**150** 4	5 8	**300** 2	450
C	8	4 12	**300** 7	9 15	300
D	2	**50** 2	12 13	6 6	50
Total Available		200	350	50	1000

$$\text{Time lost} = (150 \times 4) + (50 \times 2) + (50 \times 6) + (300 \times 7)$$
$$+ (150 \times 5) + (300 \times 2) \text{ min}$$
$$= 74 \text{ hr } 10 \text{ min}$$

2. C. Mentor's distribution policy and distribution cost:

C. Mentor Plant

Customer	Split Costs	1	2	3	Total Requirements
		−4	1	0	
A	0	9 5	3 **80** 1	3 3	80
B	4	3 3	**10** 5	**150** 4	160
C	8	3 7	**80** 9	1 9	80
D	6	**100** 2	**50** 7	0 6	150
Total Available		100	220	150	470

Cost = (80 x 1) + (10 x 5) + (150 x 4) + (80 x 9)

       + (100 x 2) + (50 x 7)

   = $2,000

As a result of the fire, without a quota:

C. Mentor Plant

Customer	Split Costs	1 -4		2 1		3 0		Dummy -8		Total Requirements
A	0	9	5	**80** 3	1	3	3	0	0	80
B	4	3	3	**10**	5	**150** 4	4	4	0	160
C	8	3	7	**80**	9	1	9	**40** 4	0	80
D	6	**100**	2	**50**	7	0	6	2	0	150
Total Available		100		180		150		40		470

Cost = $1,640

As a result of the fire, with the quota:

C. Mentor Plant

Customer	Split Costs	1 (−4)	2 (1)	3 (−3)	Total Requirements
A	0	7 · · 5	**80** · 1	6 · · 3	80
B	7	**0** · 3	3 · · 5	**150** · 4	160
C	8	3 · · 7	**65** · 9	4 · · 9	80
D	6	**100** · 2	**35** · 7	3 · · 6	150
Total Available		100	180	150	

Cost = $1,710

Therefore,

Cost of quota = $1,710 − $1,640
                = $70

3. Optimal schedule of fruit shipments:

Packing Station

Distrib. Center	Split Costs	a 8	b 10	c 8	Total Requirements
A	0	**200** ⟨8⟩	**60** ⟨10⟩	5 ⟨13⟩	260
B	−1	8 ⟨15⟩	**240** ⟨9⟩	**100** ⟨7⟩	340
C	3	**200** ⟨11⟩	M ⟨M⟩	5 ⟨16⟩	200
D	−2	12 ⟨18⟩	3 ⟨11⟩	**300** ⟨6⟩	300
Dummy	−10	2 ⟨0⟩	**100** ⟨0⟩	2 ⟨0⟩	100
Total Available		400	400	400	1200

4. Minimum cost of grapes:

Winery

Grower	Split Costs	1	2	3	Dummy	Total Growers
		0	−4	3	−3	
A	0	7 7	12 8	**10** 5	3 0	10
B	3	3 6	M M	**20** 6	**30** 0	50
C	8	2 10	**40** 4	0 5	5 0	40
D	3	**10** 3	10 9	M M	**30** 0	50
Total Available		20	40	30	60	150

Note that this solution is degenerate.

5.

a – E	250
b – A	200
b – C	100
c – D	200
d – B	150
d – C	150
d – D	50
e – D	50
e – Dummy	100

This solution is degenerate.

6.

"X"	Split Costs	1 11	2 10	3 10	Total Requirements
A	0	7 18	**140** 10	6 16	140
B	0	**20** 11	5 15	**50** 10	70
C	−2	**180** 9	6 14	2 12	180
D	−2	11 20	**0** 8	**30** 12	30
Total Available		200	140	80	

7. LINDO solution to Problem 6:

```
look all
  MIN 18 X11 + 10 X12 + 16 X13 + 11 X21 + 15 X22
  + 10 X23 + 9 X31 + 14 X32 + 12 X33 + 20 X41
  + 8 X42 + 12 X43
  SUBJECT TO
  2)  X11 + X12 + X13 = 140
  3)  X21 + X22 + X23 = 70
  4)  X31 + X32 + X33 = 180
  5)  X41 + X42 + X43 = 30
  6)  X11 + X21 + X31 + X41 = 200
  7)' X12 + X22 + X32 + X42 = 140
  8)  X13 + X23 + X33 + X43 = 80
  END

: GO
  LP OPTIMUM FOUND AT STEP 6


            OBJECTIVE FUNCTION VALUE
  1)  4100.00000
      VARIABLE          VALUE         REDUCED COST
        X11            .000000          3.000000
        X12         140.000000           .000000
        X13            .000000          2.000000
        X21          20.000000           .000000
        X22            .000000          9.000000
        X23          50.000000           .000000
        X31         180.000000           .000000
        X32            .000000         10.000000
        X33            .000000          4.000000
        X41            .000000          7.000000
        X42            .000000           .000000
        X43          30.000000           .000000


        ROW        SLACK OR SURPLUS    DUAL PRICES
         2)            .000000        -15.000000
         3)            .000000        -11.000000
         4)            .000000         -9.000000
         5)            .000000        -13.000000
         6)            .000000           .000000
         7)            .000000          5.000000
         8)            .000000          1.000000
  NO. ITERATIONS= 6
```

8. Continuation of Problem 7:

```
DO RANGE (SENSITIVITY) ANALYSIS? > Y

RANGES IN WHICH THE BASIS IS UNCHANGED

                      OBJ COEFFICIENT RANGES
VARIABLE        CURRENT        ALLOWABLE        ALLOWABLE
                COEF           INCREASE         DECREASE
  X11          18.000000       INFINITY          3.000000
  X12          10.000000       2.000000         INFINITY
  X13          16.000000       INFINITY          2.000000
  X21          11.000000       3.000000          4.000000
  X22          15.000000       INFINITY          9.000000
  X23          10.000000       4.000000          3.000000
  X31           9.000000       4.000000         INFINITY
  X32          14.000000       INFINITY         10.000000
  X33          12.000000       INFINITY          4.000000
  X41          20.000000       INFINITY          7.000000
  X42           8.000000       9.000000          2.000000
  X43          12.000000       2.000000          9.000000


                      RIGHTHAND SIDE RANGES
  ROW           CURRENT        ALLOWABLE        ALLOWABLE
                RHS            INCREASE         DECREASE
    2          140.000000       .000000         20.000000
    3           70.000000       .000000         20.000000
    4          180.000000       .000000        180.000000
    5           30.000000       .000000         20.000000
    6          200.000000      INFINITY          .000000
    7          140.000000      20.000000         .000000
    8           80.000000      20.000000         .000000
```

The cost coefficient of X11 (route from 1 to A) may be reduced by 3 units before the solution is affected.

9. This problem illustrates the relative merits of Vogel's method in discovering an initial solution to a reasonably sized problem. However, if parameter range information is required, computer solution is advisable. Even this relatively modest problem, when presented in standard LP format, consists of 480 variables and 44 constraints.

A microcomputer vastly assists in the preparation of the input file. The file illustrated here, while taking a fairly long time to prepare initially, can be modified easily with any screen editor or word processor to accommodate any problem encountered.

Notice that upper and lower case have been used in the preparation of the objective function; these will all be translated into upper case by the mainframe. The use of upper and lower case reminds the user that the first character is the uppercase character and the second character refers to the original lower-case input.

Note too that the X_{ij} format might have been used, but this increases the transmission time over the two-character format.

This file takes approximately 10 minutes to load into a mainframe using a 300-baud phone link. The output of this problem takes about 18 minutes to capture over the same 300-baud phone link. When parametric ranges are called for, the total output will take about 40 minutes at 300 baud. Using a 1200-baud modem, this will be reduced to 10. (A proportionate reduction will probably not be achieved with a 2400-baud installation due to phone line quality, necessitating retransmission of characters.)

Pacific Bell's proposal for providing 9600-baud lines for residential users, under test at the time of writing, should enable the output of the large file option to be received in about one minute.

```
MIN
6Aa + 9Ab + 6Ac + 4Ad + 6Ae + 9Af + 2Ag + 3Ah + 1Ai + 2Aj + 7AK +
8Al + 6Am + 1An + 2Ao + 1Ap + 6Aq + 8Ar + 2As + 3At + 3Ba + 8Bb +
9Bc + 9Bd + 1Be + 2Bf + 6Bg + 4Bh + 7Bi + 8Bj + 2BK + 3Bl + 3Bm +
8Bn + 6Bo + 7Bp + 3Bq + 6Br + 8Bs + 3Bt + 4Ca + 5Cb + 3Cc + 7Cd +
2Ce + 5Cf + 6Cg + 2Ch + 8Ci + 9Cj + 4CK + 6Cl + 3Cm + 5Cn + 2Co +
5Cp + 6Cq + 3Cr + 2Cs + 4Ct + 2Da + 5Db + 8Dc + 2Dd + 4De + 2Df +
4Dg + 3Dh + 5Di + 7Dj + 2DK + 9Dl + 9Dm + 6Dn + 7Do + 4Dp + 5Dq +
3Dr + 5Ds + 2Dt + 2Ea + 3Eb + 5Ec + 3Ed + 4Ee + 5Ef + 6Eg + 3Eh +
4Ei + 2Ej + 6EK + 7El + 8Em + 5En + 4Eo + 2Ep + 3Eq + 5Er + 2Es +
7Et + 3Fa + 5Fb + 1Fc + 3Fd + 5Fe + 7Ff + 3Fg + 8Fh + 6Fi + 9Fj +
4FK + 3Fl + 6Fm + 7Fn + 8Fo + 4Fp + 1Fq + 2Fr + 1Fs + 3Ft + 3Ga +
5Gb + 3Gc + 2Gd + 5Ge + 4Gf + 5Gg + 3Gh + 8Gi + 7Gj + 6GK + 9Gl +
6Gm + 7Gn + 3Go + 6Gp + 4Gq + 6Gr + 2Gs + 3Gt + 7Ha + 6Hb + 8Hc +
7Hd + 9He + 6Hf + 7Hg + 7Hh + 8Hi + 6Hj + 5HK + 9Hl + 4Hm + 7Hn +
3Ho + 6Hp + 7Hq + 9Hr + 6Hs + 3Ht + 5Ia + 6Ib + 3Ic + 4Id + 5Ie +
3If + 7Ig + 8Ih + 9Ii + 4Ij + 5IK + 3Il + 2Im + 6In + 7Io + 5Ip +
8Iq + 9Ir + 3Is + 5It + 6Ja + 7Jb + 8Jc + 9Jd + 5Je + 4Jf + 3Jg +
5Jh + 7Ji + 5Jj + 8JK + 9Jl + 4Jm + 3Jn + 5Jo + 7Jp + 8Jq + 9Jr +
6Js + 4Jt + 5Ka + 6Kb + 4Kc + 7Kd + 8Ke + 9Kf + 5Kg + 4Kh + 8Ki +
6Kj + 7KK + 7Kl + 4Km + 9Kn + 9Ko + 3Kp + 3Kq + 3Kr+ 10Ks + 5Kt +
6La + 7Lb + 9Lc+ 12Ld + 3Le + 4Lf + 8Lg + 7Lh + 9Li + 6Lj + 7LK +
4Ll + 3Lm + 6Ln + 7Lo + 3Lp + 6Lq + 4Lr + 2Ls + 1Lt + 5Ma + 6Mb +
                                                          (continued)
```

```
7Mc + 4Md + 3Me + 7Mf + 8Mg + 9Mh + 4Mi + 5Mj + 3MK + 5M1 + 2Mm +
1Mn + 2Mo + 8MP + 9Mq + 5Mr + 4Ms + 5Mt + 4Na + 5Nb + 6Nc + 3Nd +
7Ne + 3Nf + 4Ng + 7Nh + 8Ni + 9Nj + 3NK + 4N1 + 2Nm + 3Nn+ 12No +
5NP + 6Nq + 7Nr + 8Ns + 3Nt + 6Oa + 7Ob + 4Oc + 3Od + 5Oe + 6Of +
4Og + 7Oh + 8Oi + 5Oj + 6OK + 7O1 + 3Om + 4On + 5Oo + 2OP + 7Oq +
8Or + 4Os + 5Ot + 2Pa + 3Pb + 4Pc + 2Pd + 5Pe + 7Pf + 8Pg + 4Ph +
3Pi + 2Pj + 4PK + 3P1 + 5Pm + 6Pn + 6Po + 7PP + 8Pq + 4Pr + 5Ps +
3Pt + 5Qa + 6Qb + 7Qc + 4Qd + 5Qe + 3Qf + 4Qg + 7Qh + 2Qi + 8Qj +
9QK + 4Q1 + 2Qm + 3Qn + 1Qo + 4QP + 6Qq + 7Qr + 4Qs + 7Qt + 5Ra +
6Rb + 8Rc + 9Rd + 4Re + 3Rf + 2Rg + 5Rh + 6Ri + 7Rj + 8RK + 5R1 +
6Rm + 3Rn + 2Ro + 3RP + 2Rq + 1Rr + 4Rs + 5Rt + 4Sa + 5Sb + 6Sc +
7Sd + 4Se + 5Sf + 6Sg + 3Sh + 4Si + 8Sj + 6SK + 5S1 + 4Sm + 3Sn +
1So + 3SP + 4Sq + 2Sr + 5Ss + 6St + 5Ta + 6Tb + 3Tc + 7Td + 8Te +
9Tf + 5Tg + 7Th + 4Ti + 8Tj + 9TK + 3T1 + 2Tm + 1Tn + 5To + 4TP +
5Tq + 2Tr + 4Ts + 6Tt + 5Ua + 3Ub + 6Uc + 7Ud + 8Ue + 9Uf + 4Ug +
3Uh + 2Ui + 5Uj + 6UK + 3U1 + 5Um + 2Un + 7Uo + 8UP + 9Uq + 3Ur +
2Us + 1Ut + 5Va + 6Vb + 7Vc + 4Vd + 3Ve + 2Vf + 5Vg + 6Vh + 7Vi +
5Vj + 4VK + 3V1 + 2Vm + 6Vn + 7Vo + 8VP + 9Vq + 4Vr + 5Vs + 3Vt +
4Wa + 5Wb + 7Wc + 8Wd + 5We + 4Wf + 6Wg + 3Wh + 5Wi + 4Wj + 2WK +
1W1 + 4Wm + 3Wn + 2Wo + 6WP + 7Wq + 8Wr + 5Ws + 4Wt + 4Xa + 3Xb +
6Xc + 7Xd + 3Xe + 5Xf + 7Xg + 8Xh + 9Xi + 4Xj + 3XK + 2X1 + 1Xm +
5Xn + 4Xo + 7XP + 8Xq + 6Xr + 5Xs + 3Xt
ST
AA + AB + AC + AD + AE + AF + AG + AH + AI + AJ + AK + AL + AM +
AN + AO + AP + AQ + AR + AS + AT = 10
BA + BB + BC + BD + BE + BF + BG + BH + BI + BJ + BK + BL + BM +
BN + BO + BP + BQ + BR + BS + BT = 5
CA + CB + CC + CD + CE + CF + CG + CH + CI + CJ + CK + CL + CM +
CN + CO + CP + CQ + CR + CS + CT = 12
DA + DB + DC + DD + DE + DF + DG + DH + DI + DJ + DK + DL + DM +
DN + DO + DP + DQ + DR + DS + DT = 8
EA + EB + EC + ED + EE + EF + EG + EH + EI + EJ + EK + EL + EM +
EN + EO + EP + EQ + ER + ES + ET = 10
FA + FB + FC + FD + FE + FF + FG + FH + FI + FJ + FK + FL + FM +
FN + FO + FP + FQ + FR + FS + FT = 30
GA + GB + GC + GD + GE + GF + GG + GH + GI + GJ + GK + GL + GM +
GN + GO + GP + GQ + GR + GS + GT = 13
HA + HB + HC + HD + HE + HF + HG + HH + HI + HJ + HK + HL + HM +
HN + HO + HP + HQ + HR + HS + HT = 25
IA + IB + IC + ID + IE + IF + IG + IH + II + IJ + IK + IL + IM +
IN + IO + IP + IQ + IR + IS + IT = 14
JA + JB + JC + JD + JE + JF + JG + JH + JI + JJ + JK + JL + JM +
JN + JO + JP + JQ + JR + JS + JT = 30
KA + KB + KC + KD + KE + KF + KG + KH + KI + KJ + KK + KL + KM +
KN + KO + KP + KQ + KR + KS + KT = 25
```

```
LA + LB + LC + LD + LE + LF + LG + LH + LI + LJ + LK + LL + LM +
LN + LO + LP + LQ + LR + LS + LT = 53
MA + MB + MC + MD + ME + MF + MG + MH + MI + MJ + MK + ML + MM +
MN + MO + MP + MQ + MR + MS + MT = 9
NA + NB + NC + ND + NE + NF + NG + NH + NI + NJ + NK + NL + NM +
NN + NO + NP + NQ + NR + NS + NT = 10
OA + OB + OC + OD + OE + OF + OG + OH + OI + OJ + OK + OL + OM +
ON + OO + OP + OQ + OR + OS + OT = 34
PA + PB + PC + PD + PE + PF + PG + PH + PI + PJ + PK + PL + PM +
PN + PO + PP + PQ + PR + PS + PT = 30
QA + QB + QC + QD + QE + QF + QG + QH + QI + QJ + QK + QL + QM +
QN + QO + QP + QQ + QR + QS + QT = 10
RA + RB + RC + RD + RE + RF + RG + RH + RI + RJ + RK + RL + RM +
RN + RO + RP + RQ + RR + RS + RT = 6
SA + SB + SC + SD + SE + SF + SG + SH + SI + SJ + SK + SL + SM +
SN + SO + SP + SQ + SR + SS + ST = 12
TA + TB + TC + TD + TE + TF + TG + TH + TI + TJ + TK + TL + TM +
TN + TO + TP + TQ + TR + TS + TT = 11
UA + UB + UC + UD + UE + UF + UG + UH + UI + UJ + UK + UL + UM +
UN + UO + UP + UQ + UR + US + UT = 13
VA + VB + VC + VD + VE + VF + VG + VH + VI + VJ + VK + VL + VM +
VN + VO + VP + VQ + VR + VS + VT = 20
WA + WB + WC + WD + WE + WF + WG + WH + WI + WJ + WK + WL + WM +
WN + WO + WP + WQ + WR + WS + WT = 5
XA + XB + XC + XD + XE + XF + XG + XH + XI + XJ + XK + XL + XM +
XN + XO + XP + XQ + XR + XS + XT = 10
AA + BA + CA + DA + EA + FA + GA + HA + IA + JA + KA + LA + MA +
NA + OA + PA + QA + RA + SA + TA + UA + VA + WA + XA < 12
AB + BB + CB + DB + EB + FB + GB + HB + IB + JB + KB + LB + MB +
NB + OB + PB + QB + RB + SB + TB + UB + VB + WB + XB < 24
AC + BC + CC + DC + EC + FC + GC + HC + IC + JC + KC + LC + MC +
NC + OC + PC + QC + RC + SC + TC + UC + VC + WC + XC < 20
AD + BD + CD + DD + ED + FD + GD + HD + ID + JD + KD + LD + MD +
ND + OD + PD + QD + RD + SD + TD + UD + VD + WD + XD < 35
AE + BE + CE + DE + EE + FE + GE + HE + IE + JE + KE + LE + ME +
NE + OE + PE + QE + RE + SE + TE + UE + VE + WE + XE < 25
AF + BF + CF + DF + EF + FF + GF + HF + IF + JF + KF + LF + MF +
NF + OF + PF + QF + RF + SF + TF + UF + VF + WF + XF < 20
AG + BG + CG + DG + EG + FG + GG + HG + IG + JG + KG + LG + MG +
NG + OG + PG + QG + RG + SG + TG + UG + VG + WG + XG < 24
AH + BH + CH + DH + EH + FH + GH + HH + IH + JH + KH + LH + MH +
NH + OH + PH + QH + RH + SH + TH + UH + VH + WH + XH < 28
AI + BI + CI + DI + EI + FI + GI + HI + II + JI + KI + LI + MI +
NI + OI + PI + QI + RI + SI + TI + UI + VI + WI + XI < 13
AJ + BJ + CJ + DJ + EJ + FJ + GJ + HJ + IJ + JJ + KJ + LJ + MJ +
NJ + OJ + PJ + QJ + RJ + SJ + TJ + UJ + VJ + WJ + XJ < 50
```

(continued)

```
AK + BK + CK + DK + EK + FK + GK + HK + IK + JK + KK + LK + MK +
NK + OK + PK + QK + RK + SK + TK + UK + VK + WK + XK < 25
AL + BL + CL + DL + EL + FL + GL + HL + IL + JL + KL + LL + ML +
NL + OL + PL + QL + RL + SL + TL + UL + VL + WL + XL < 60
AM + BM + CM + DM + EM + FM + GM + HM + IM + JM + KM + LM + MM +
NM + OM + PM + QM + RM + SM + TM + UM + VM + WM + XM < 23
AN + BN + CN + DN + EN + FN + GN + HN + IN + JN + KN + LN + MN +
NN + ON + PN + QN + RN + SN + TN + UN + VN + WN + XN < 30
AO + BO + CO + DO + EO + FO + GO + HO + IO + JO + KO + LO + MO +
NO + OO + PO + QO + RO + SO + TO + UO + VO + WO + XO < 36
AP + BP + CP + DP + EP + FP + GP + HP + IP + JP + KP + LP + MP +
NP + OP + PP + QP + RP + SP + TP + UP + VP + WP + XP < 32
AQ + BQ + CQ + DQ + EQ + FQ + GQ + HQ + IQ + JQ + KQ + LQ + MQ +
NQ + OQ + PQ + QQ + RQ + SQ + TQ + UQ + VQ + WQ + XQ < 25
AR + BR + CR + DR + ER + FR + GR + HR + IR + JR + KR + LR + MR +
NR + OR + PR + QR + RR + SR + TR + UR + VR + WR + XR < 15
AS + BS + CS + DS + ES + FS + GS + HS + IS + JS + KS + LS + MS +
NS + OS + PS + QS + RS + SS + TS + US + VS + WS + XS < 13
AT + BT + CT + DT + ET + FT + GT + HT + IT + JT + KT + LT + MT +
NT + OT + PT + QT + RT + ST + TT + UT + VT + WT + XT < 28
END
```

The output from this input file is found in Appendix 3.

   10. The solution is achieved simply by changing MIN to MAX in the previous problem. The output received is found in the Appendix.

5

Assignment

Just as the transportation problem is a special case of linear programming, the assignment or allocation problem is a special case of the transportation problem. All assignment problems are capable of solution as transportation problems as well as via the general simplex algorithm. An assignment problem arises when we have N units of commodity, each stored in a separate distinct location, and one unit is required by a separate customer at N separate locations. We thus have a maximally degenerate case of a transportation problem. If we use the transportation method to find a solution, we would have to fill in $N - 1$ zeros to test and modify each solution. Fortunately, there is a more efficient procedure, known as Konig's Method, or the Hungarian Procedure.

The mathematical representation of the problem is

Minimize
$$\sum_{i=1}^{N}\sum_{j=1}^{N} C_{ij} X_{ij}$$

subject to
$$\sum_{i=1}^{N} X_{ij} = 1 \qquad j = 1 \ldots N$$

$$\sum_{j=1}^{N} X_{ij} = 1 \qquad i = 1 \ldots N$$

and $\qquad\qquad X_{ij} = 0 \text{ or } 1$

This last might also be written as

$$X_{ij} = X_{ij}^2$$

The method of solution relies on two fairly obvious observations:

1. Any solution is unchanged if we add (or subtract) a constant to any row or column of the Cij matrix.

2. If all Cij $\geqslant$ 0, then if a set Xij can be discovered such that

$$\sum_{i=1}^{N} \sum_{j=1}^{N} X_{ij} C_{ij} = 0$$

then it is optimal.

Let us again consider an example. In this case, the terms Cij are the lengths of time taken by four operatives to perform four tasks (see Tableau 5.1). The problem is to assign the operatives to the tasks so that the total time is minimized.

Tableau 5.1

Task	\multicolumn{4}{c}{Operator}
	a	b	c	d
A	8	7	9	8
B	4	6	2	8
C	2	1	1	8
D	7	7	5	8

The solution entails the following steps:

1. Subtract the smallest number in each row from the other elements in each row. Note that, in so doing, the relative speed of each operator for the task is maintained (Observation 1): The quickest remains quickest and slowest remains slowest (see Tableau 5.2).

	Operator			
Task	a	b	c	d
A	1	0	2	1
B	2	4	0	6
C	1	0	0	7
D	2	2	0	3

Tableau 5.2

The assignment that minimizes the cost of Tableau 5.2 also minimizes the cost of Tableau 5.1 (but the actual minimum cost will be different).

2. Subtract the smallest number in each column from the other elements in that column. Note now that the preference order of each task by each operator is preserved (Observation 1): The best task for an operator remains the best and its worst remains worst (see Tableau 5.3).

	Operator			
Task	a	b	c	d
A	0	0	2	0
B	1	4	0	5
C	0	0	0	6
D	1	2	0	2

Tableau 5.3

The assignment that minimizes the cost of Tableau 5.3 also minimizes the cost of Tableau 5.2.

3. Test to see if an allocation can be made that utilizes only the zero elements of the tableau; that is, in which

$$\sum_{i=1}^{N} \sum_{j=1}^{N} C_{ij} X_{ij} = 0$$

A convenient way to do this is to check the *minimum* number of lines taken to cross out *all* the zeroes (diagonals not allowed; see Tableau 5.4).

Tableau 5.4

		Operator		
Task	a	b	c	d
A	0	0	2	0
B	1	4	0	5
C	0	0	0	6
D	1	2	0	2

If this minimum number of lines is equal to the size of the matrix, then an assignment satisfying Observation 2 is possible. Here the minimum number of lines is three and the matrix size is four, so an assignment is not possible. The reason for this step is to discover the number of independent zeros in the matrix, for if any one of the zeroes in a line or column is allocated, the other zeroes are no longer available for allocation.

We must discover additional independent zeros by adding and subtracting from rows and columns. Any further subtraction from a row or column will result in negative numbers, and a solution among zeros would no longer be optimal. However, we could remove these negative entries by additions to columns or rows, again restoring all nonnegative entries. An easy way of achieving this is as follows:

4. Select the smallest element not covered by a line (in this case 1).

5. Subtract this number from all the uncovered elements and add it to all elements covered with *two* lines (see Tableau 5.5).

Tableau 5.5

Task	a	b	c	d
A	0	0	3	0
B	0	3	0	4
C	0	0	1	6
D	0	1	0	1

6. Again check the minimum number of lines required to cover all the zeros (Tableau 5.6).

Task	a	b	c	d
A	0	0	3	0
B	0	3	0	4
C	0	0	1	6
D	0	1	0	1

Tableau 5.6

Now the number of lines required equals the matrix size (4), so observation 2 is satisfied.

In fact there are two alternative solutions that allow for zeros to be allocated as solutions. These are

Ad   Ba   Cb   Dc

or

Ad   Bc   Cb   Da

Difficulties encountered in the method of operation are (1) drawing the minimum number of lines and (2) making the final allocation. These difficulties can usually be dealt with by proceeding as follows:

1. Draw the lines in such a way that each successive line crosses out as many zeros as possible.

2. Look for single zeros in a row or a column. These *must* be assigned. Then cross out that row and column from further consideration. The assignment usually "falls out" after that. However, there may be more than one solution, as in this example. Following the consequences of the alternative permutations will produce all the solutions.

One final point: Should an assignment be required that maximizes, rather than minimizes, subtract all the C_{ij} elements from the maximum C_{ij} element (or a larger value) and then minimize the resulting matrix.

COMPUTER SOLUTION

It is debatable whether it is worthwhile using a computer over the longhand method to solve an assignment problem. However, once logged on the system, the formulation of the assignment in general LP terms is easily achieved. Consider the problem just solved. It is formulated as a LINDO file thus:

```
MIN
8X11+7X12+9X13+8X14
+4X21+6X22+2X23+8X24
+2X31+1X32+1X33+8X34
+7X41+7X42+5X43+8X44
ST
X11+X12+X13+X14=1
X21+X22+X23+X24=1
X31+X32+X33+X34=1
X41+X42+X43+X44=1
X11+X21+X31+X41=1
X12+X22+X32+X42=1
X13+X23+X33+X43=1
X14+X24+X34+X44=1
END
```

The computer solution is as follows:

```
LP OPTIMUM FOUND AT STEP 9
        OBJECTIVE FUNCTION VALUE
1)  18.000000

VARIABLE        VALUE           REDUCED COST
X11            .000000             .000000
X12            .000000             .000000
X13            .000000            2.000000
X14           1.000000             .000000
X21            .000000            1.000000
X22            .000000            4.000000
X23           1.000000             .000000
X24            .000000            5.000000
X31            .000000             .000000
X32           1.000000             .000000
X33            .000000             .000000
X34            .000000            6.000000
X41           1.000000             .000000
X42            .000000            1.000000
X43            .000000            1.000000
X44            .000000            1.000000
```

ROW	SLACK OR SURPLUS	DUAL PRICES
2)	.000000	-8.000000
3)	.000000	-3.000000
4)	.000000	-2.000000
5)	.000000	-7.000000
6)	.000000	.000000
7)	.000000	1.000000
8)	.000000	1.000000
9)	.000000	.000000

Compare this solution with the output from the transportation problem in Chapter 4. Similar remarks apply. Parametric analysis is also similar.

MULTI-DIMENSIONAL ASSIGNMENT

The assignment problem just illustrated is that normally covered in texts. However, the solution method is actually a special case of the more general multi-dimensional problem. Instead of pairings (two dimensions), one may wish to discover triple groupings (three dimensions) or multiple groupings (N dimensions) that minimize a particular cost function. The same procedure is applicable. For example, consider the following three-dimensional case:

	W				X				Y				Z			
	P	Q	R	S	P	Q	R	S	P	Q	R	S	P	Q	R	S
A	8	5	7	3	4	3	3	6	9	4	4	3	8	6	3	6
B	4	2	5	9	8	4	5	3	8	2	2	8	6	4	4	9
C	5	3	6	2	9	9	6	4	9	4	8	3	9	8	3	2
D	4	8	5	6	8	8	7	6	4	4	6	2	4	3	8	2

Tableau 5.7

In the two-dimensional Hungarian method, solution proceeds by addition and subtraction of equal values from lines and columns. Because there are now three variables, the procedure must be carried out in planes.

Taking the smallest element equally from the respective planes W (top), X, Y, and Z (bottom), we obtain

Tableau 5.8		W				X				Y				Z			
		P	Q	R	S	P	Q	R	S	P	Q	R	S	P	Q	R	S
	A	6	3	5	1	1	0	0	3	7	2	2	1	6	4	1	4
	B	2	0	3	7	5	1	2	0	6	0	0	6	4	2	2	7
	C	3	1	4	0	6	6	3	1	7	2	6	1	7	6	1	0
	D	2	6	3	4	5	5	4	3	2	2	4	0	2	1	6	0

All planes now contain at least one zero, with the exception of the P plane.

Tableau 5.9		W				X				Y				Z			
		P	Q	R	S	P	Q	R	S	P	Q	R	S	P	Q	R	S
	A	6	3	5	1	1	0	0	3	7	2	2	1	6	4	1	4
	B	2	0	3	7	5	1	2	0	6	0	0	6	4	2	2	7
	C	3	1	4	0	6	6	3	1	7	2	6	1	7	6	1	0
	D	2	6	3	4	5	5	4	3	2	2	4	0	2	1	6	0

Taking the smallest element from this plane gives us

Tableau 5.10		W				X				Y				Z			
		P	Q	R	S	P	Q	R	S	P	Q	R	S	P	Q	R	S
	A	5	3	5	1	0	0	0	3	6	2	2	1	5	4	1	4
	B	1	0	3	7	4	1	2	0	5	0	0	6	3	2	2	7
	C	2	1	4	0	5	6	3	1	6	2	6	1	6	6	1	0
	D	1	6	3	4	4	5	4	3	1	2	4	0	1	1	6	0

Now, does the cube contain sufficient independent zeros? This is checked by considering the minimum number of planes needed to cross out the zeros. Three are needed, A, B, and S:

Tableau 5.11

	W				X				Y				Z			
	P	Q	R	S	P	Q	R	S	P	Q	R	S	P	Q	R	S
A	a5	a3	a5	a1s	a0	a0	a0	a3s	a6	a2	a2	a1s	a5	a4	a1	a4s
B	b1	b0	b3	b7s	b4	b1	b2	b0s	b5	b0	b0	b6s	b3	b2	b2	b7s
C	2	1	4	0s	5	6	3	1s	6	2	6	1s	6	6	1	0s
D	1	6	3	4s	4	5	4	3s	1	2	4	0s	1	1	6	0s

Therefore there are only three independent zeros from which a three-parameter allocation might be made. Because 1 is the smallest uncovered element, further independent zeros might be discovered by subtracting 1 from all uncovered elements, and adding 1 to those double-covered elements. (For those who wonder at the validity of this last statement, the same result is obtained by subtracting and adding in planes, as previously).

Tableau 5.12

	W				X				Y				Z			
	P	Q	R	S	P	Q	R	S	P	Q	R	S	P	Q	R	S
A	5	3	5	2	0	0	0	4	6	2	2	2	5	4	1	5
B	1	0	3	8	4	1	2	1	5	0	0	7	3	2	2	8
C	1	0	3	0	4	5	2	1	5	1	5	1	5	5	0	0
D	0	5	2	4	3	4	3	3	0	1	3	0	0	0	5	0

Now four planes are required to cross out the zero elements. An allocation is now possible using zero entries:

Tableau 5.13

	W				X				Y				Z			
	P	Q	R	S	P	Q	R	S	P	Q	R	S	P	Q	R	S
A	5	3	5	2	**0**	0	0	4	6	2	2	2	5	4	1	5
B	1	**0**	3	8	4	1	2	1	5	0	0	7	3	2	2	8
C	1	0	3	0	4	5	2	1	5	1	5	1	5	5	**0**	0
D	0	5	2	4	3	4	3	3	0	1	3	**0**	0	0	5	0

In terms of the original matrix, the allocation is

Tableau 5.14		W				X				Y				Z			
		P	Q	R	S	P	Q	R	S	P	Q	R	S	P	Q	R	S
	A	8	5	7	3	**4**	3	3	6	9	4	4	3	8	6	3	6
	B	4	**2**	5	9	8	4	5	3	8	2	2	8	6	4	4	9
	C	5	3	6	2	9	9	6	4	9	4	8	3	9	8	**3**	2
	D	4	8	5	6	8	8	7	6	4	4	6	**2**	4	3	8	2

The three-dimensional allocation is thus the four points

W B Q
X A P
Y D S
Z C R

with a cost of 11.

MORE THAN THREE VARIABLES

The same method is applicable to higher dimensions. With four variables a hypercube of four dimensions is used, and elements are subtracted equally from the various three-dimensional spaces contained therein. Although interesting, the method becomes tedious for these higher-dimension problems. Further, building a LINDO file to solve problems in more than two dimensions is time consuming. It would be preferable to solve such problems via specialized software packages, but none have been developed yet.

As an illustration of three-dimensional spaces contained within a four-dimensional hypercube, which is mapped into two dimensions, the representations of the X cube and the L cube, contained in the four-dimensional hypercube, are shown outlined. Similar cubes for each of the parameters, such as the Q cube and the C cube, can be delineated.

		W				X				Y				Z				
		P	Q	R	S	P	Q	R	S	P	Q	R	S	P	Q	R	S	**Tableau 5.15**
K	A	8	5	7	3	4	3	3	6	9	4	4	3	8	6	3	6	
	B	4	2	5	9	8	4	5	3	8	2	2	8	6	4	4	9	
	C	5	3	6	2	9	9	6	4	9	4	8	3	9	8	3	2	
	D	4	8	5	6	8	8	7	6	4	4	6	2	4	3	8	2	
L	A	6	4	7	3	4	8	3	9	2	1	3	3	5	6	7	6	
	B	1	5	2	9	8	4	5	2	3	4	3	3	4	5	4	9	
	C	3	6	6	5	9	2	4	7	7	3	7	5	7	3	7	1	
	D	4	5	8	8	4	3	8	5	8	1	1	6	2	8	9	8	
M	A	7	4	5	3	4	3	3	3	9	2	3	4	1	4	1	4	
	B	4	8	8	8	3	3	4	8	4	6	4	3	7	5	7	9	
	C	5	9	6	9	6	7	2	9	3	4	1	3	9	4	5	3	
	D	1	3	3	6	8	9	7	2	6	9	8	3	4	6	3	2	
N	A	2	5	3	6	3	3	4	2	1	2	2	5	4	4	5	4	
	B	5	6	7	8	3	4	7	7	6	6	4	8	7	8	4	8	
	C	8	9	6	2	8	7	3	5	4	4	3	2	9	9	9	2	
	D	4	2	5	6	4	9	4	1	9	7	8	2	3	2	8	2	

CHAPTER 5 PROBLEMS

The solutions to Problems 1 through 10 are found at the end of the chapter.

       1. The credit control manager of a large retail company supervises six accounting data operatives, who post entries onto sales accounts. Sales are nationwide and for control purposes are analyzed into six geographical regions designated A through F. Each region's sales accounts are kept in separate sales ledgers. Each machine operator takes a different length of time to post the sales ledgers because the number of postings differ from ledger to ledger, and each operator has a different level of skill and efficiency. The following table is the expected time, in hours, that each operator will take to complete any of the sales ledgers:

	Machine Operators					
Ledger	a	b	c	d	e	f
A	28	26	30	32	28	30
B	32	34	32	36	33	34
C	30	32	31	33	34	35
D	34	32	32	38	35	33
E	30	31	31	37	35	36
F	32	30	34	35	33	32

What is the total minimum number of hours in which the work can be completed? Indicate which ledger should be assigned to which operator.

What is the minimum cost, assuming that the operators earn $5 per hour, except operator a, who is a supervisor and earns $7 per hour, and operator b, the assistant supervisor, who earns $6 per hour?

What would be the minimum cost of "idle time" assuming all operators are paid for a 40-hour basic week?

2. The vice president of Ma Annie Vise tool company is faced with an unexpected increase in demand for five of its products. The main plant is fully loaded, but overtime labor is available. There is a small branch available with spare capacity. The excess may be subcontracted. The costs of each decision are as follows:

Product	Overtime	Branch	Subcontract
A	4,200	3,800	4,000
B	4,000	4,000	5,000
C	3,000	3,200	2,800
D	1,500	1,400	1,200
E	5,000	5,200	4,800

The company and the branch are able to cope with the demand for only one product each. Which product would you send to the branch and which would you subcontract?

3. Doobits is a small jobbing company that produces custom microchip devices for the electronics industry. There are currently six contracts in the shop and past quality-control records indicate that the six employees working in the shop have had the following percentage of their work returned by their supervisor on these contracts:

	Contract Worker					
Worker	a	b	c	d	e	f
1	1.4%	2.2%	1.8%	2.3%	2.1%	3.0%
2	3.0	2.4	1.6	2.6	3.0	2.2
3	2.5	1.8	1.4	1.6	1.6	2.8
4	3.2	1.4	1.0	1.4	1.8	2.0
5	2.2	2.8	1.4	3.0	2.0	1.3
6	1.8	1.6	2.5	1.4	1.2	2.2

How would you advise Doobits to organize its schedule of contract work?

4. The supervisor of the probation department, Reece Idivest, is considering the allocation of five new cases to his six probation officers. Two of his officers have only qualified for four of the five types of cases currently due for scheduling. The success rate, as defined by the state guidelines, of the officers for the case types are as follows:

	Case				
Officer	A	B	C	D	E
1	90%	95%	91%	90%	88%
2	86	97		94	93
3	87	94	93	86	80
4	80	91	84	92	86
5	91		91	90	92
6	89	90	90	88	90

How should the caseload be allocated to maximize departmental effectiveness? If officer 1 would prefer not to take the assignment allocated, can Idivest offer another assignment?

5. Formulate Problem 4 as a LINDO input file.

6. The van pool at Democratic Republic has initiated a system in which each of their six van drivers can submit choices for the six routes covered by the vans. The proposals for the six have been opened and the transportation manager must now make the route allocation to optimize the preferences of the van drivers.

Driver	Van Route					
	A	B	C	D	E	F
1	4	3	2	1	5	6
2	6	5	3	2	1	4
3	5	3	4	1	2	6
4	3	4	2	1	5	6
5	3	5	2	1	4	6
6	5	6	3	2	1	4

7. Three equal partners, Danielle, Jo, and Mack, operate a consulting business. They currently have four projects in progress, which they are considering completing. The partners estimate their profit from completing the projects as follows:

	Project			
	A	B	C	D
Danielle	$40,000	$48,000	$42,000	$44,000
Jo	38,000	43,000	36,000	36,000
Mack	47,000	49,000	46,000	48,000

How should they schedule the projects such that the profit to the partnership is maximized?

8. Danielle proposes to Jo and Mack that they should bring into the partnership a friend, Fred Rater, in order that they can complete the four projects. Fred has estimated that he could complete the projects in the portfolio at a profit of

A	$41,000
B	$45,000
C	$39,000
D	$41,000

What effect does this have on the partners' plans?

9. Prepare the LINDO input file for the assignment problem with the following cost matrix:

	a	b	c	d	e	f	g	h	i	j	k	l	m	n	o	p	q	r	s	t
A	6	9	6	4	6	9	2	3	1	2	7	8	6	1	2	1	6	8	2	3
B	3	8	9	9	1	2	6	4	7	8	2	3	3	8	6	7	3	6	8	3
C	4	5	3	7	2	5	6	2	8	9	4	6	3	5	2	5	6	3	2	4
D	2	5	8	2	4	2	4	3	5	7	2	9	9	6	7	4	5	3	5	2
E	2	3	5	3	4	5	6	3	4	2	6	7	8	5	4	2	3	5	2	7
F	3	5	1	3	5	7	3	8	6	9	4	3	6	7	8	4	1	2	1	3
G	3	5	3	2	5	4	5	3	8	7	6	9	6	7	3	6	4	6	2	3
H	7	6	8	7	9	6	7	7	8	6	5	9	4	7	3	6	7	9	6	3
I	5	6	3	4	5	3	7	8	9	4	5	3	2	6	7	5	8	9	3	5
J	6	7	8	9	5	4	3	5	7	5	8	9	4	3	5	7	8	9	6	4
K	5	6	4	7	8	9	5	4	8	6	7	7	4	9	9	3	3	3	10	5
L	6	7	9	12	3	4	8	7	9	6	7	4	3	6	7	3	6	4	2	1
M	5	6	7	4	3	7	8	9	4	5	3	5	2	1	2	8	9	5	4	5
N	4	5	6	3	7	3	4	7	8	9	3	4	2	3	12	5	6	7	8	3
O	6	7	4	3	5	6	4	7	8	5	6	7	3	4	5	2	7	8	4	5
P	2	3	4	2	5	7	8	4	3	2	4	3	5	6	6	7	8	4	5	3
Q	5	6	7	4	5	3	4	7	2	8	9	4	2	3	1	4	6	7	4	7
R	5	6	8	9	4	3	2	5	6	7	8	5	6	3	2	3	2	1	4	5
S	4	5	6	7	4	5	6	3	4	8	6	5	4	3	1	3	4	2	5	6
T	5	6	3	7	8	9	5	7	4	8	9	3	2	1	5	4	5	2	4	6
U	5	3	6	7	8	9	4	3	2	5	6	3	5	2	7	8	9	3	2	1
V	5	6	7	4	3	2	5	6	7	5	4	3	2	6	7	8	9	4	5	3
W	4	5	7	8	5	4	6	3	5	4	2	1	4	3	2	6	7	8	5	4
X	4	3	6	7	3	5	7	8	9	4	3	2	1	5	4	7	8	6	5	3

10. Psi-Clone Computers is evaluating software. It wishes to optimize a mix of hardware, operating system, and applications software and has determined the following data:

Machine	Applications package	Operating System			
		A	B	C	D
Alpha	a	4	7	6	3
	b	5	2	4	3
	c	8	6	7	2
	d	9	1	6	3
Beta	a	8	2	4	6
	b	5	4	6	3
	c	9	1	7	2
	d	8	4	5	3
Gamma	a	6	1	7	4
	b	8	3	9	2
	c	8	6	7	3
	d	9	1	2	4
Delta	a	6	1	6	2
	b	8	4	9	2
	c	4	1	1	8
	d	6	2	3	6

The data within the three-dimensional matrix refer to a preference ordering scheme where 1 is most preferred.

Which hardware-application-operating system trios would you advise Psi-Clone to package?

Next, formulate the above problem as a LINDO input file.

HOMEWORK PROBLEMS

11. Dr. Highroller is head of the Management Department at Golden Hills University. The accreditation board will conduct a review visit following the next semester, and Dr. Highroller is anxious to give a good impression as to the department's standing. Dr. Highroller has four professors to assign to courses and wishes to maximize the students' performance evaluation of the department over five major courses. She will have to assign the course not allocated to the professors to a part-time instructor, and this will not be counted in the professorial evaluation. On checking departmental records, she discovers that the student evaluations received by these professors in these courses were as follows (4.0 being perfect):

	Course				
	A	B	C	D	E
Professor Smiley	3.8	3.4	3.2	3.9	3.1
Professor Einstein	1.9	1.8	2.4	2.6	1.7
Professor Nobel	2.3	2.0	2.3	1.8	2.1
Professor Bland	4.0	3.8	3.9	3.9	3.7

Advise Dr. Highroller on her optimal allocation.

12. M. A. Barracudas owns the L.A. Cab Company. There are currently seven cabs out on the streets, and six customers have called in for service. On checking the locator board, Barracudas discovers the following estimated times from each location to each prospective customer (in minutes):

Cab	Customer					
	A	B	C	D	E	F
1	10	4	8	9	12	15
2	2	7	12	7	10	8
3	8	5	8	4	11	8
4	9	10	9	8	12	9
5	8	5	2	3	11	8
6	6	4	11	8	9	10
7	9	6	9	7	10	9

How should the cabs be dispatched?

13. The Makepiece jobbing shop has five machines, that can make each of several items for which there are a backlog of orders. The time taken for each machine to make each item is as follows:

			Machine		
Item	1	2	3	4	5
A	5	8	9	10	8
B	4	7	8	3	9
C	7	3	8	2	12
D	9	5	9	3	11
E	6	9	11	6	9
F	8	6	9	3	12
G	9	6	8	5	8
H	8	5	3	8	9

How should Makepiece schedule the machines?

14. Interpret the matrix in Problem 13 as the contribution to profit per minute generated by allocating a machine to fabricate a specific part.

15. Formulate Problem 14 as a LINDO input file.

16. Suppose that in Problem 13, Makepiece discovers that partial breakdowns leave some machines unable to make some parts. The new matrix now becomes:

			Machine		
Item	1	2	3	4	5
A	5	8	9	10	8
B		7	8		9
C	7		8		12
D	9	5	9	3	11
E		9	11	6	9
F	8	6	9		12
G	9	6	8	5	8
H	8			8	9

What is Makepiece's best strategy now?

17. Suppose that customer F in Problem 12 insists that only cab number 1 is suitable. What should be Barracudas's response?

18. In Problem 11, Professor Einstein insists on teaching course E. How does this affect Dr. Highroller's plans?

19. Solve the allocation problem with the following cost matrix:

	a	b	c	d	e	f	g	h	i	j	k	l	m	n	o	p	q	r	s	t
A	16	9	6	4	6	9	2	3	1	2	7	8	6	1	2	1	6	8	2	3
B	3	8	9	9	1	12	6	4	7	8	2	3	3	8	6	7	3	6	8	3
C	4	5	3	17	2	5	6	2	8	9	4	16	3	5	2	5	6	3	2	4
D	2	5	8	2	4	2	4	3	5	7	2	9	9	6	7	14	5	3	5	2
E	2	3	5	3	4	5	6	3	4	2	6	7	8	5	4	2	3	5	2	7
F	3	5	1	3	5	7	3	8	6	9	4	3	6	7	8	4	1	2	1	3
G	3	5	3	2	5	4	5	3	8	7	6	9	6	7	3	6	4	6	2	3
H	7	6	8	7	9	6	7	7	8	6	5	9	4	7	3	6	7	9	6	3
I	5	6	3	4	5	13	7	8	9	4	5	3	2	6	7	5	8	9	3	5
J	6	7	8	9	5	4	3	5	7	5	8	9	4	3	5	7	8	9	6	4
K	5	6	4	7	8	9	5	4	8	6	7	17	4	9	9	3	3	3	10	5
L	6	7	9	12	3	4	8	7	9	6	7	4	3	6	7	3	6	4	2	1
M	5	6	7	4	3	7	8	9	4	5	3	5	2	1	2	8	9	5	4	5
N	4	5	6	3	7	3	4	7	8	9	3	4	2	3	12	5	6	7	8	3
O	6	7	4	3	5	6	4	7	8	5	6	7	3	4	5	2	7	8	4	5
P	2	3	4	2	5	7	8	4	3	2	4	3	5	6	6	7	8	4	5	3
Q	5	6	7	4	5	3	4	7	2	8	9	4	2	3	1	4	6	7	4	7
R	5	6	8	9	4	3	2	5	6	7	8	5	6	3	2	3	2	1	4	5
S	4	5	6	7	4	5	6	3	4	8	6	5	4	3	1	3	4	2	5	6
T	5	6	3	7	8	9	5	7	4	8	9	3	2	1	5	4	5	2	4	6
U	5	3	6	7	8	9	4	3	2	5	6	3	5	2	7	8	9	3	2	1
V	5	6	7	4	3	2	5	6	7	5	4	3	2	6	7	8	9	4	5	3
W	4	5	7	8	5	4	6	3	5	4	2	1	4	3	2	6	7	8	5	4
X	4	3	6	7	3	5	7	8	9	4	3	2	1	5	4	7	8	6	5	3

20. Formulate the LINDO input file for Problem 19.

CHAPTER 5 SOLUTIONS

1. Minimum hours to complete the work:

	Machine Operators					
Ledger	a	b	c	d	e	f
A	28	26	30	32	28	30
B	32	34	32	36	33	34
C	30	32	31	33	34	35
D	34	32	32	38	35	33
E	30	31	31	37	35	36
F	32	30	34	35	33	32

Step 1. Subtract the smallest entry in each row from the other elements in that row.

	Machine Operators					
Ledger	a	b	c	d	e	f
A	2	0	4	6	2	4
B	0	2	0	4	1	2
C	0	2	1	3	4	5
D	2	0	0	6	3	1
E	0	1	1	7	5	6
F	2	0	4	5	3	2

Step 2. Subtract the smallest entry in each column from the other elements in that column.

	Machine Operators					
Ledger	a	b	c	d	e	f
A	2	0	4	3	1	3
B	0	2	0	1	0	1
C	0	2	1	0	3	4
D	2	0	0	3	2	0
E	0	1	1	4	4	5
F	2	0	4	2	2	1

Step 3. Test for the number of independent zeros.

		Machine Operators				
Ledger	a	b	c	d	e	f
A	2	0	4	3	1	3
B	0	2	0	1	0	1
C	0	2	1	0	3	4
D	2	0	0	3	2	0
E	0	1	1	4	4	5
F	2	0	4	2	2	1

Step 4. As the minimum number of lines needed to cover all the zeros is 5, which is one less than the size of the matrix (6), the smallest uncovered element is located (1), and subtracted from each uncovered element and added to each double-covered element.

		Machine Operators				
Ledger	a	b	c	d	e	f
A	2	0	3	3	0	2
B	1	3	0	2	0	1
C	0	2	0	0	2	3
D	3	1	0	4	2	0
E	0	1	0	4	3	4
F	2	0	3	2	1	0

Step 5. As the minimum number of lines is equal to the matrix size, the solution is obtained.

The total minimum number of hours is obtained by the following assignment:

a	E	30 hours
b	A	26 hours
c	D	32 hours
d	C	33 hours
e	B	33 hours
f	F	32 hours
		Total 186 hours

The minimum cost is achieved with the same assignment. The difference in pay scales applies equally across rows and will be removed at step 1 of the procedure.

a	E	Cost = 30 × $7.00 = $ 210
b	A	Cost = 26 × $6.00 = $ 150
c	D	Cost = 32 × $5.00 = $ 160
d	C	Cost = 33 × $5.00 = $ 165
e	B	Cost = 33 × $5.00 = $ 165
f	F	Cost = 32 × $5.00 = $ 160

Total Cost = $1,016

Cost of "idle time" is:

a	Cost = 10 × $7.00 = $ 70
b	Cost = 14 × $6.00 = $ 84
c	Cost = 8 × $5.00 = $ 40
d	Cost = 7 × $5.00 = $ 35
e	Cost = 7 × $5.00 = $ 35
f	Cost = 8 × $5.00 = $ 40

Total Idle Time Cost = $304

2. The costs associated with each product are:

Product	O	B	S1	S2	S3
A	4,200	3,800	4,000	4,000	4,000
B	4,000	4,000	5,000	5,000	5,000
C	3,000	3,200	2,800	2,800	2,800
D	1,500	1,400	1,200	1,200	1,200
E	5,000	5,200	4,800	4,800	4,800

Solution:

Produce	By
B	Overtime work
A	Branch working
C	Subcontract work
D	Subcontract work
E	Subcontract work

Solution: The total cost of this decision is: $16,600.

3. Optimal allocation for Doobits:

Worker	Contract	Defects
1	a	1.4%
2	c	1.6%
3	b	1.8%
4	d	1.4%
5	f	1.3%
6	e	1.2%

Solution: Average defect rate (equivalent workrate) = $8.7 / 6 = 1.45\%$

4. Reece Idivest's optimal assignments are:

Officer		Case		Officer		Case
1	to	B	and	1	to	A
2		E		2		B
3		C		3		C
4		D		4		D
5		A		5		E
6		Free		6		Free

He may therefore use either of the alternative solutions to satisfy officer 1.

5. The LINDO file for problem 4 is:

```
MAX
        90 A1 +  95 B1 +  91 C1 +  90 D1 +  88 E1
     +  86 A2 +  97 B2            +  94 D2 +  93 E2
     +  87 A3 +  94 B3 +  93 C3 +  86 D3 +  80 E3
     +  80 A4 +  91 B4 +  84 C4 +  92 D4 +  86 E4
     +  91 A5 +            +  91 C5 +  90 D5 +  92 E5
     +  89 A6 +  90 B6 +  90 C6 +  88 D6 +  90 E6
     ST
     A1 + A2 + A3 + A4 + A5 + A6 = 1
     B1 + B2 + B3 + B4 + B5 + B6 = 1
     C1 + C2 + C3 + C4 + C5 + C6 = 1
     D1 + D2 + D3 + D4 + D5 + D6 = 1
     E1 + E2 + E3 + E4 + E5 + E6 = 1
     A1 + B1 + C1 + D1 + E1 < 1
     A2 + B2 + C2 + D2 + E2 < 1
     A3 + B3 + C3 + D3 + E3 < 1
     A4 + B4 + C4 + D4 + E4 < 1
     A5 + B5 + C5 + D5 + E5 < 1
     A6 + B6 + C6 + D6 + E6 < 1
END
```

6. Democratic Republic has a choice of 12 equally desirable
alternatives:

Route	Driver
A	4 4 4 4 4 4 4 4 5 5 5 5
B	3 3 5 5 1 1 1 1 3 3 1 1
C	5 5 1 1 4 4 5 5 1 1 4 4
D	1 1 3 3 5 5 4 4 4 4 3 3
E	2 6 2 6 2 6 2 6 2 6 2 6
F	6 2 6 2 6 2 6 2 6 2 6 2

7. The optimal assignments are:

	Project		Project	
Danielle	D	(44) or	B	(48)
Jo	B	(43)	A	(38)
Mack	A	(47)	D	(48)
Partnership profit =	$134,000		$134,000	
($44,667 each)				

8. Adding Fred, the optimal assignments would be:

	Project		Project	
Danielle	C	(42) or	D	(44)
Jo	B	(43)	B	(43)
Mack	D	(48)	C	(46)
Fred	A	(41)	A	(41)
Partnership profit =	$174,000		$174,000	
($43,500 each)				

9. LINDO file for assignment problem:

```
MIN
6Aa + 9Ab + 6Ac + 4Ad + 6Ae + 9Af + 2Ag + 3Ah + 1Ai + 2Aj + 7Ak +
8Al + 6Am + 1An + 2Ao + 1Ap + 6Aq + 8Ar + 2As + 3At + 3Ba + 8Bb +
9Bc + 9Bd + 1Be + 2Bf + 6Bg + 4Bh + 7Bi + 8Bj + 2Bk + 3Bl + 3Bm +
8Bn + 6Bo + 7Bp + 3Bq + 6Br + 8Bs + 3Bt + 4Ca + 5Cb + 3Cc + 7Cd +
2Ce + 5Cf + 6Cg + 2Ch + 8Ci + 9Cj + 4Ck + 6Cl + 3Cm + 5Cn + 2Co +
5Cp + 6Cq + 3Cr + 2Cs + 4Ct + 2Da + 5Db + 8Dc + 2Dd + 4De + 2Df +
4Dg + 3Dh + 5Di + 7Dj + 2Dk + 9Dl + 9Dm + 6Dn + 7Do + 4Dp + 5Dq +
3Dr + 5Ds + 2Dt + 2Ea + 3Eb + 5Ec + 3Ed + 4Ee + 5Ef + 6Eg + 3Eh +
4Ei + 2Ej + 6Ek + 7El + 8Em + 5En + 4Eo + 2Ep + 3Eq + 5Er + 2Es +
7Et + 3Fa + 5Fb + 1Fc + 3Fd + 5Fe + 7Ff + 3Fg + 8Fh + 6Fi + 9Fj +
4Fk + 3Fl + 6Fm + 7Fn + 8Fo + 4Fp + 1Fq + 2Fr + 1Fs + 3Ft + 3Ga +
5Gb + 3Gc + 2Gd + 5Ge + 4Gf + 5Gg + 3Gh + 8Gi + 7Gj + 6Gk + 9Gl +
6Gm + 7Gn + 3Go + 6Gp + 4Gq + 6Gr + 2Gs + 3Gt + 7Ha + 6Hb + 8Hc +
7Hd + 9He + 6Hf + 7Hg + 7Hh + 8Hi + 6Hj + 5Hk + 9Hl + 4Hm + 7Hn +
3Ho + 6Hp + 7Hq + 9Hr + 6Hs + 3Ht + 5Ia + 6Ib + 3Ic + 4Id + 5Ie +
3If + 7Ig + 8Ih + 9Ii + 4Ij + 5Ik + 3Il + 2Im + 6In + 7Io + 5Ip +
```

(continued)

```
8Iq + 9Ir + 3Is + 5It + 6Ja + 7Jb + 8Jc + 9Jd + 5Je + 4Jf + 3Jg +
5Jh + 7Ji + 5Jj + 8Jk + 9J1 + 4Jm + 3Jn + 5Jo + 7Jp + 8Jq + 9Jr +
6Js + 4Jt + 5Ka + 6Kb + 4Kc + 7Kd + 8Ke + 9Kf + 5Kg + 4Kh + 8Ki +
6Kj + 7KK + 7K1 + 4Km + 9Kn + 9Ko + 3Kp + 3Kq + 3Kr + 10Ks + 5Kt +
6La + 7Lb + 9Lc + 12Ld + 3Le + 4Lf + 8Lg + 7Lh + 9Li + 6Lj + 7LK +
4L1 + 3Lm + 6Ln + 7Lo + 3Lp + 6Lq + 4Lr + 2Ls + 1Lt + 5Ma + 6Mb +
7Mc + 4Md + 3Me + 7Mf + 8Mg + 9Mh + 4Mi + 5Mj + 3MK + 5M1 + 2Mm +
1Mn + 2Mo + 8Mp + 9Mq + 5Mr + 4Ms + 5Mt + 4Na + 5Nb + 6Nc + 3Nd +
7Ne + 3Nf + 4Ng + 7Nh + 8Ni + 9Nj + 3NK + 4N1 + 2Nm + 3Nn + 12No +
5Np + 6Nq + 7Nr + 8Ns + 3Nt + 6Oa + 7Ob + 4Oc + 3Od + 5Oe + 6Of +
4Og + 7Oh + 8Oi + 5Oj + 6Ok + 7O1 + 3Om + 4On + 5Oo + 2Op + 7Oq +
8Or + 4Os + 5Ot + 2Pa + 3Pb + 4Pc + 2Pd + 5Pe + 7Pf + 8Pg + 4Ph +
3Pi + 2Pj + 4PK + 3P1 + 5Pm + 6Pn + 6Po + 7Pp + 8Pq + 4Pr + 5Ps +
3Pt + 5Qa + 6Qb + 7Qc + 4Qd + 5Qe + 3Qf + 4Qg + 7Qh + 2Qi + 8Qj +
9Qk + 4Q1 + 2Qm + 3Qn + 1Qo + 4Qp + 6Qq + 7Qr + 4Qs + 7Qt + 5Ra +
6Rb + 8Rc + 9Rd + 4Re + 3Rf + 2Rg + 5Rh + 6Ri + 7Rj + 8RK + 5R1 +
6Rm + 3Rn + 2Ro + 3Rp + 2Rq + 1Rr + 4Rs + 5Rt + 4Sa + 5Sb + 6Sc +
7Sd + 4Se + 5Sf + 6Sg + 3Sh + 4Si + 8Sj + 6SK + 5S1 + 4Sm + 3Sn +
1So + 3Sp + 4Sq + 2Sr + 5Ss + 6St + 5Ta + 6Tb + 3Tc + 7Td + 8Te +
9Tf + 5Tg + 7Th + 4Ti + 8Tj + 9TK + 3T1 + 2Tm + 1Tn + 5To + 4Tp +
5Tq + 2Tr + 4Ts + 6Tt + 5Ua + 3Ub + 6Uc + 7Ud + 8Ue + 9Uf + 4Ug +
3Uh + 2Ui + 5Uj + 6UK + 3U1 + 5Um + 2Un + 7Uo + 8Up + 9Uq + 3Ur +
2Us + 1Ut + 5Va + 6Vb + 7Vc + 4Vd + 3Ve + 2Vf + 5Vg + 6Vh + 7Vi +
5Vj + 4VK + 3V1 + 2Vm + 6Vn + 7Vo + 8Vp + 9Vq + 4Vr + 5Vs + 3Vt +
4Wa + 5Wb + 7Wc + 8Wd + 5We + 4Wf + 6Wg + 3Wh + 5Wi + 4Wj + 2WK +
1W1 + 4Wm + 3Wn + 2Wo + 6Wp + 7Wq + 8Wr + 5Ws + 4Wt + 4Xa + 3Xb +
6Xc + 7Xd + 3Xe + 5Xf + 7Xg + 8Xh + 9Xi + 4Xj + 3XK + 2X1 + 1Xm +
5Xn + 4Xo + 7Xp + 8Xq + 6Xr + 5Xs + 3Xt
ST
AA + AB + AC + AD + AE + AF + AG + AH + AI + AJ + AK + AL + AM +
AN + AO + AP + AQ + AR + AS + AT < 1
BA + BB + BC + BD + BE + BF + BG + BH + BI + BJ + BK + BL + BM +
BN + BO + BP + BQ + BR + BS + BT < 1
CA + CB + CC + CD + CE + CF + CG + CH + CI + CJ + CK + CL + CM +
CN + CO + CP + CQ + CR + CS + CT < 1
DA + DB + DC + DD + DE + DF + DG + DH + DI + DJ + DK + DL + DM +
DN + DO + DP + DQ + DR + DS + DT < 1
EA + EB + EC + ED + EE + EF + EG + EH + EI + EJ + EK + EL + EM +
EN + EO + EP + EQ + ER + ES + ET < 1
FA + FB + FC + FD + FE + FF + FG + FH + FI + FJ + FK + FL + FM +
FN + FO + FP + FQ + FR + FS + FT < 1
GA + GB + GC + GD + GE + GF + GG + GH + GI + GJ + GK + GL + GM +
GN + GO + GP + GQ + GR + GS + GT < 1
HA + HB + HC + HD + HE + HF + HG + HH + HI + HJ + HK + HL + HM +
HN + HO + HP + HQ + HR + HS + HT < 1
IA + IB + IC + ID + IE + IF + IG + IH + II + IJ + IK + IL + IM +
IN + IO + IP + IQ + IR + IS + IT < 1
JA + JB + JC + JD + JE + JF + JG + JH + JI + JJ + JK + JL + JM +
JN + JO + JP + JQ + JR + JS + JT < 1
```

```
KA + KB + KC + KD + KE + KF + KG + KH + KI + KJ + KK + KL + KM +
KN + KO + KP + KQ + KR + KS + KT < 1
LA + LB + LC + LD + LE + LF + LG + LH + LI + LJ + LK + LL + LM +
LN + LO + LP + LQ + LR + LS + LT < 1
MA + MB + MC + MD + ME + MF + MG + MH + MI + MJ + MK + ML + MM +
MN + MO + MP + MQ + MR + MS + MT < 1
NA + NB + NC + ND + NE + NF + NG + NH + NI + NJ + NK + NL + NM +
NN + NO + NP + NQ + NR + NS + NT < 1
OA + OB + OC + OD + OE + OF + OG + OH + OI + OJ + OK + OL + OM +
ON + OO + OP + OQ + OR + OS + OT < 1
PA + PB + PC + PD + PE + PF + PG + PH + PI + PJ + PK + PL + PM +
PN + PO + PP + PQ + PR + PS + PT < 1
QA + QB + QC + QD + QE + QF + QG + QH + QI + QJ + QK + QL + QM +
QN + QO + QP + QQ + QR + QS + QT < 1
RA + RB + RC + RD + RE + RF + RG + RH + RI + RJ + RK + RL + RM +
RN + RO + RP + RQ + RR + RS + RT < 1
SA + SB + SC + SD + SE + SF + SG + SH + SI + SJ + SK + SL + SM +
SN + SO + SP + SQ + SR + SS + ST < 1
TA + TB + TC + TD + TE + TF + TG + TH + TI + TJ + TK + TL + TM +
TN + TO + TP + TQ + TR + TS + TT < 1
UA + UB + UC + UD + UE + UF + UG + UH + UI + UJ + UK + UL + UM +
UN + UO + UP + UQ + UR + US + UT < 1
VA + VB + VC + VD + VE + VF + VG + VH + VI + VJ + VK + VL + VM +
VN + VO + VP + VQ + VR + VS + VT < 1
WA + WB + WC + WD + WE + WF + WG + WH + WI + WJ + WK + WL + WM +
WN + WO + WP + WQ + WR + WS + WT < 1
XA + XB + XC + XD + XE + XF + XG + XH + XI + XJ + XK + XL + XM +
XN + XO + XP + XQ + XR + XS + XT < 1
AA + BA + CA + DA + EA + FA + GA + HA + IA + JA + KA + LA + MA +
NA + OA + PA + QA + RA + SA + TA + UA + VA + WA + XA = 1
AB + BB + CB + DB + EB + FB + GB + HB + IB + JB + KB + LB + MB +
NB + OB + PB + QB + RB + SB + TB + UB + VB + WB + XB = 1
AC + BC + CC + DC + EC + FC + GC + HC + IC + JC + KC + LC + MC +
NC + OC + PC + QC + RC + SC + TC + UC + VC + WC + XC = 1
AD + BD + CD + DD + ED + FD + GD + HD + ID + JD + KD + LD + MD +
ND + OD + PD + QD + RD + SD + TD + UD + VD + WD + XD = 1
AE + BE + CE + DE + EE + FE + GE + HE + IE + JE + KE + LE + ME +
NE + OE + PE + QE + RE + SE + TE + UE + VE + WE + XE = 1
AF + BF + CF + DF + EF + FF + GF + HF + IF + JF + KF + LF + MF +
NF + OF + PF + QF + RF + SF + TF + UF + VF + WF + XF = 1
AG + BG + CG + DG + EG + FG + GG + HG + IG + JG + KG + LG + MG +
NG + OG + PG + QG + RG + SG + TG + UG + VG + WG + XG = 1
AH + BH + CH + DH + EH + FH + GH + HH + IH + JH + KH + LH + MH +
NH + OH + PH + QH + RH + SH + TH + UH + VH + WH + XH = 1
AI + BI + CI + DI + EI + FI + GI + HI + II + JI + KI + LI + MI +
NI + OI + PI + QI + RI + SI + TI + UI + VI + WI + XI = 1
AJ + BJ + CJ + DJ + EJ + FJ + GJ + HJ + IJ + JJ + KJ + LJ + MJ +
NJ + OJ + PJ + QJ + RJ + SJ + TJ + UJ + VJ + WJ + XJ = 1
```

(continued)

```
AK + BK + CK + DK + EK + FK + GK + HK + IK + JK + KK + LK + MK +
NK + OK + PK + QK + RK + SK + TK + UK + VK + WK + XK = 1
AL + BL + CL + DL + EL + FL + GL + HL + IL + JL + KL + LL + ML +
NL + OL + PL + QL + RL + SL + TL + UL + VL + WL + XL = 1
AM + BM + CM + DM + EM + FM + GM + HM + IM + JM + KM + LM + MM +
NM + OM + PM + QM + RM + SM + TM + UM + VM + WM + XM = 1
AN + BN + CN + DN + EN + FN + GN + HN + IN + JN + KN + LN + MN +
NN + ON + PN + QN + RN + SN + TN + UN + VN + WN + XN = 1
AO + BO + CO + DO + EO + FO + GO + HO + IO + JO + KO + LO + MO +
NO + OO + PO + QO + RO + SO + TO + UO + VO + WO + XO = 1
AP + BP + CP + DP + EP + FP + GP + HP + IP + JP + KP + LP + MP +
NP + OP + PP + QP + RP + SP + TP + UP + VP + WP + XP = 1
AQ + BQ + CQ + DQ + EQ + FQ + GQ + HQ + IQ + JQ + KQ + LQ + MQ +
NQ + OQ + PQ + QQ + RQ + SQ + TQ + UQ + VQ + WQ + XQ = 1
AR + BR + CR + DR + ER + FR + GR + HR + IR + JR + KR + LR + MR +
NR + OR + PR + QR + RR + SR + TR + UR + VR + WR + XR = 1
AS + BS + CS + DS + ES + FS + GS + HS + IS + JS + KS + LS + MS +
NS + OS + PS + QS + RS + SS + TS + US + VS + WS + XS = 1
AT + BT + CT + DT + ET + FT + GT + HT + IT + JT + KT + LT + MT +
NT + OT + PT + QT + RT + ST + TT + UT + VT + WT + XT = 1
END
```

10. Clone's optimum configuration is:

Alpha	A	a
Beta	D	b
Gamma	B	d
Delta	C	c

The LINDO input file is:

```
MIN
      4 X111 + 7 X112 + 6 X113 + 3 X114
    + 5 X121 + 2 X122 + 4 X123 + 3 X124
    + 8 X131 + 6 X132 + 7 X133 + 2 X134
    + 9 X141 + 1 X142 + 6 X143 + 3 X144

    + 8 X211 + 2 X212 + 4 X213 + 6 X214
    + 5 X221 + 4 X222 + 6 X223 + 3 X224
    + 9 X231 + 1 X232 + 7 X233 + 2 X234
    + 8 X241 + 4 X242 + 5 X243 + 3 X244
```

```
+  6  X311  +  1  X312  +  7  X313  +  4  X314
+  8  X321  +  3  X322  +  9  X323  +  2  X324
+  8  X331  +  6  X332  +  7  X333  +  3  X334
+  9  X341  +  1  X342  +  2  X343  +  4  X344

+  6  X411  +  1  X412  +  6  X413  +  2  X414
+  8  X421  +  4  X422  +  9  X423  +  2  X424
+  4  X431  +  1  X432  +  1  X433  +  8  X434
+  6  X441  +  2  X442  +  3  X443  +  6  X444

ST
     X111  +     X112  +     X113  +     X114
+    X121  +     X122  +     X123  +     X124
+    X131  +     X132  +     X133  +     X134
+    X141  +     X142  +     X143  +     X144  =  1

     X211  +     X212  +     X213  +     X214
+    X221  +     X222  +     X223  +     X224
+    X231  +     X232  +     X233  +     X234
+    X241  +     X242  +     X243  +     X244  =  1

     X311  +     X312  +     X313  +     X314
+    X321  +     X322  +     X323  +     X324
+    X331  +     X332  +     X333  +     X334
+    X341  +     X342  +     X343  +     X344  =  1

     X411  +     X412  +     X413  +     X414
+    X421  +     X422  +     X423  +     X424
+    X431  +     X432  +     X433  +     X434
+    X441  +     X442  +     X443  +     X444  =  1

     X111  +     X121  +     X131  +     X141
+    X211  +     X221  +     X231  +     X241
+    X311  +     X321  +     X331  +     X341
+    X411  +     X421  +     X431  +     X441  =  1

     X112  +     X122  +     X132  +     X142
+    X212  +     X222  +     X232  +     X242
+    X312  +     X322  +     X332  +     X342
+    X412  +     X422  +     X432  +     X442  =  1

     X113  +     X123  +     X133  +     X143
+    X213  +     X223  +     X233  +     X243
+    X313  +     X323  +     X333  +     X343
+    X413  +     X423  +     X433  +     X443  =  1
```

(continued)

```
     X114 +    X124 +    X134 +    X144
+    X214 +    X224 +    X234 +    X244
+    X314 +    X324 +    X334 +    X344
+    X414 +    X424 +    X434 +    X444 =  1


     X111 +    X112 +    X113 +    X114
+    X211 +    X212 +    X213 +    X214
+    X311 +    X312 +    X313 +    X314
+    X411 +    X412 +    X413 +    X414 =  1


     X121 +    X122 +    X123 +    X124
+    X221 +    X222 +    X223 +    X224
+    X321 +    X322 +    X323 +    X324
+    X421 +    X422 +    X423 +    X424 =  1


     X131 +    X132 +    X133 +    X134
+    X231 +    X232 +    X233 +    X234
+    X331 +    X332 +    X333 +    X334
+    X431 +    X432 +    X433 +    X434 =  1


     X141 +    X142 +    X143 +    X144
+    X241 +    X242 +    X243 +    X244
+    X341 +    X342 +    X343 +    X344
+    X441 +    X442 +    X443 +    X444 =  1
```

6

Decomposition

We have noted that the feasible region in a linear programming problem may be viewed as a geometrical space bounded by the constraint boundaries. We have also noted that, in a convex feasible region, straight lines drawn between any two feasible points will never go outside the feasible region. If we therefore take the intersecting points of the constraints that make up the extremities of the feasible region, and then take weighted averages of these points, it is possible to define all the points within the feasible region. This observation is used in the principle of decomposition of linear programs.

Decomposition is a technique that allows very large and somewhat "lumpy" programs to be solved as a series of smaller problems. By "lumpy" we mean problems that arise from combining smaller problems with linking equations and ending up with many zero coefficients associated with the variables. This may be clearer with an example.

Consider the case of a corporation that has taken over a new firm. Let us alter our original example to the case where the original company makes margarine and cooking oil using the resource supplies as depicted in our first example. (Note that the actual coefficients have been altered.)

Firm A

$$X1a + X2a \leqslant 8 \qquad \text{Sunflower oil}$$
$$2\,X1a + X2a \leqslant 6 \qquad \text{Plant capacity (A)}$$

(Note that the a appendage for X1 signifies X1 made by Firm A.) Firm A's processed profit per ton is \$4 for each X1 and \$5 for each X2.

Firm A now takes over another company, Firm B; a manufacturer of cooking oil and cooking fat, again using sunflower oil as a raw material. There is thus some but not complete overlap between the two firms' activities. The constraints for Firm B are as follows:

Firm B

$$X2b + 4\,X3b \leqslant 4 \qquad \text{Sunflower oil}$$
$$X2b + X3b \leqslant 8 \qquad \text{Plant capacity (B)}$$

Firm B's processed profit per ton is \$2 on cooking oil and \$3 for cooking fat.

The new corporation may then pool its total resources to make a new linear programming problem:

Maximize $\qquad 4\,X1a + 5\,X2a + 2\,X2b + 3\,X3b$

subject to

$$
\begin{aligned}
X1a + X2a + X2b + 4\,X3b &\leqslant 12 \\
2\,X1a + X2a &\leqslant 6 \\
X2b + X3b &\leqslant 8
\end{aligned}
$$

Notice that this LP formulation has the following schematic:

Figure 6.1

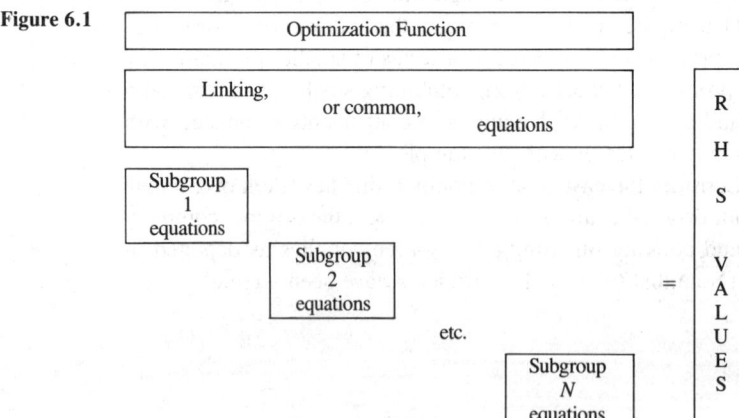

This is a common structure observed in corporate LP formulations. The sparse nature of the matrix, many elements being zero, gives rise to the description "lumpy" used earlier.

The corporate example is solved directly, using the condensed tableaux in which Si refers to subgroup i:

	X1a	X2a	X2b	X3b		
S1	1	1	1	4	12	
S2	2	1	0	0	6	*
S3	0	0	1	1	8	
	− 4	− 5	− 2	− 3	0	

D = 1   **

Tableau 6.1

	X1a	S2	X2b	X3b		
S1	− 1	− 1	1	4	6	*
X2a	2	1	0	0	6	
S3	0	0	1	1	8	
	6	5	− 2	− 3	30	

D = 1   **

Tableau 6.2

	X1a	S2	X2b	S1		
X3b	− 1	− 1	1	1	6	*
X2a	8	4	0	0	24	
S3	1	1	3	− 1	26	
	21	17	− 5	3	138	

D = 4   **

Tableau 6.3

	X1a	S2	X3b	S1	
X2b	− 1	− 1	4	1	6
X2a	2	1	0	0	6
S3	1	1	− 3	− 1	2
	4	3	5	2	42

D = 1

Tableau 6.4

This indicates a solution for the new corporate structure to manufacture 6 tons of cooking oil at Plant A and 6 tons of cooking oil at Plant B. Notice that this is an improvement over the two separate company LP optimizations: Firm A's was $X1a = 0$, $X2a = 6$, profit $= \$30$; Firm B's was $X2b = 4$, $X3b = 0$, profit $= \$8$, for a combined total profit of $38.

Suppose that this problem is too large to be solved directly. One way around the difficulty is to allow each company management to ignore the other company, and also to ignore the common constraints. Furthermore, we can split up the objective function into two parts. This gives us the two problems

For A,

Maximize	$4\,X1a + 5\,X2a$
subject to	$2\,X1a + X2a \le 6$

For B,

Maximize	$2\,X2b + 3\,X3b$
	$X2b + X3b \le 8$

leaving the common sunflower oil constraint to be put aside by the corporate management until the subsystems A and B place their proposals.

Manager A considers the problem for Firm A and proposes being given enough sunflower oil to manufacture 6 tons of cooking oil. The division's profit arising from this proposed strategy would be $30.

Manager B considers the problem for Firm B and proposes being given sufficient sunflower oil to manufacture 8 tons of cooking fat. B's divisional profit arising from this proposal would generate $24.

Corporate management, reviewing these proposals, notes that they have a healthy total of $54. Unfortunately, the two plans call for the consumption of 38 tons of sunflower oil, 6 tons by division A and 32 tons by division B. There are only 12 tons available to the corporation. How should it be rationed?

One way is to observe that if the two plans are scaled by amounts Ka1 for Firm A, and Kb1 for Firm B, where $0 \le K \le 1$, this will maintain a feasible demand on the resource by the divisions, although they will no longer be optimal, either as independent subsystems or as part of the total corporate system. However, as a first step, the corporate management forms the system:

					Requested
X1a	becomes	$X1a \cdot Ka1$	$=$	0	
X2a	becomes	$X2a \cdot Ka1$	$=$	$6 \cdot Ka1$	6 tons
X2b	becomes	$X2b \cdot Kb1$	$=$	0	
X3b	becomes	$4\,X3b \cdot Kb1$	$=$	$4 \cdot 8\,Kb1$	32 tons

The total profit being given by

$$P = 30 \text{ Ka1} + 24 \text{ Kb1}$$

(P indicates positive profit, while the Z used in simplex solutions can designate either positive or negative profit.) In order to discover the best choice of scaling factors to apply to these proposals to optimize corporate profits, corporate management forms the linear program

Maximize $30 \text{ Ka1} + 24 \text{ Kb1}$
subject to $0 + 6 \text{ Ka1} + 32 \text{ Kb1} + 0 \le 12$
$$\text{Ka1} \qquad\qquad \le\ 1$$
$$\text{Kb1} \quad \le\ 1$$

The solution of which is

$$\text{Ka1} = 1$$
$$\text{Kb1} = 0.1875$$
$$\text{Corporate profit} = \$34.50$$

Hence, the more efficient plant gets all that it requires while the less efficient has to go short.

With this small example, it is easy to see that the best plan has not been considered. The principle of decomposing large linear problems now results in the corporate management requesting the subsystem management that, in light of their allocation of the scarce resource, they supply alternative proposals that would lead to a more profitable corporate plan. In more complex problems, the corporate management would aid the subsystem managements in their search by using the sensitivity coefficients generated by the coordinating problem.

Following through the coordinating problem, we see that

Tableau 6.5

	Ka1	Kb1		
S1	6	32	12	
S2	1	0	1	*
S3	0	1	1	
	−30	−24	0	D = 1
	**			

Tableau 6.6

	S2	Kb1		
S1	−6	32	6	*
Ka1	1	0	1	
S3	0	1	1	
	30	−24	30	D = 1
		**		

Tableau 6.7

	S2	S1		
Kb1	−6	1	6	
Ka1	32	0	32	
S3	6	−1	26	
	816	24	1104	D = 32

Thus

$$Ka1 = 1, Kb1 = 0.1875, S3 = 0.8125$$
$$\text{Profit function} = 34.50 + 0.75\ S1 + 25.5\ S2$$

The shadow values of the linking equations (here S1, the only linking equation) may now be utilized to provide an indicator connecting the coordinating system to the subsystems, which in turn provide alternative proposals for solution. Since the shadow value for a constraint reflects the extra improvement possible in the objective function, it is possible to write new optimization functions for the component subsystems based upon the previous optimization function merged with the shadow values multiplied by the constraint coefficients. For example, for system A, a new optimization function would be

Maximize
$$(4 - 1\ S1)\ X1a + (5 - 1\ S1)\ X2a$$
$$= (4 - 0.75)\ X1a + (5 - 0.75)\ X2a$$
$$= \qquad 3.25\ X1a + \qquad 4.25\ X2a$$

This is to be optimized over the subsystem constraints

$$2 X1a + X2a \leqslant 6$$

For system B it would be

Maximize
$$(2 - 1 S1) X2b + (3 - 4 S1) X3b$$
$$= (2 - 0.75) X2b + (3 - 4 \times 0.75) X3b$$
$$= \quad 1.25 \quad X2b$$

subject to $X2b + X3b \leqslant 8$

The second solution proposed by system A, after optimizing the new LP problem, is the same basis as previously, namely

$$X1a2 = 0$$
$$X2a2 = 6$$

For system B, however, the coordinating corporate system has effectively reduced the profit on cooking fat to zero. The new proposal now becomes

$$X2b2 = 8$$
$$X3b2 = 0$$

Corporate management has before it now three proposals, one from system A and two from system B: the one made in the first round, (the preferred proposal) which assumes unlimited supply of sunflower oil, and the other a result of the modified function supplied by corporate management. As before, corporate management wishes to choose the optimum weights to apply to these proposals.

X1a2	becomes	X1a2 · Ka2	=	0
X2a2	becomes	X2a2 · Ka2	=	6Ka2
X2b2	becomes	X2b2 · Kb2	=	8Kb2
X3b2	becomes	X3b2 · Kb1	=	4 × 8Kb11

Ka2 is thus the weight to be applied to A's proposal, Kb2 is the weight to be applied to B's modified proposal, and Kb11 is the weight to be applied to B's initial proposal in the light of the new proposals.

Corporate management then devises its LP problem to choose the best values for these weights to optimize profit:

Maximize	$30\ Ka2 + 16\ Kb2 + 24\ Kb11$
subject to	$6\ Ka2 + \ 8\ Kb2 + 32\ Kb11 \leq 12$
	$Ka2 \qquad\qquad\qquad\qquad\ \ \leq\ 1$
	$Kb2 \qquad\qquad\qquad \leq\ 1$
	$Kb11 \leq\ 1$
	$Kb2 + \quad Kb11 \leq\ 1$

$$Ka2,\ Kb2,\ Kb11 \geq 0$$

The solution is

$$Ka2 = 1,\ Kb2 = 0.75,\ Kb11 = 0$$
$$\text{Total profit} = (30 \times 1) + (0.75 \times 16) = \$42$$

We have now achieved the same solution as the single overall simplex optimization. A check on the shadow values in the profit function is now made to see whether further proposals are required for the subsystems. They are not, and the problem is completely solved.

You may wonder whether the extensive iterations of problem formulation and solution are an advantage over the direct single-stage solution. They are not in the pure mathematical optimization case, where all the information is available to the problem solver and the hardware capable of single-stage solution is available. However, the technique was developed when the demand for large-scale LP solutions was greater than the capacity of the computer hardware to handle in one stage. This is unlikely to be the case with current hardware, but corporate problems may still occur where, although total information is available within a corporation, it is held in decentralized locations. A decomposition analysis may be used to supply specific sensitive information effectively to corporate headquarters without disturbing decentralized autonomy, providing information overload at corporate headquarters, or treating a diversified corporation as a single mega-firm. (This topic is examined by the author in *A Proposal for a Corporate Control System*, Management International Review, 16(2), 1976.)

In addition, the approach used to handle decomposition of linear programming is closely allied to the technique of separable programming and the general programming problem. These advanced techniques are used to allow a simplex-type approach to nonlinear programming problems. These will be discussed in later chapters.

7

Solution of Two-Person Zero-Sum Games by Linear Programming

In Von Neumann's classical games theory formulation for two-person zero-sum games, the decision situation is normally expressed as shown in Tableau 7.1. The numerical elements of the matrix are payoffs, expressed in terms of payment to Player A, when both players simultaneously choose their strategies, which intersect at that element.

Tableau 7.1

		Player B		
		(i)	(ii)	(iii)
Player A	(i)	1	3	5
	(ii)	−4	0	−1

In this example, Player A chooses to adopt strategy A(i) while Player B adopts strategy B(ii), resulting in a payoff of 3 to Player A. Choice of the strategies A(ii) and B(iii) results in a payoff of −1, meaning that A pays B one unit.

The assumptions for this analysis are:

   1. Both players are fully informed, intelligent opponents whose aims directly conflict. All information is available to both opponents, and any analysis that might be undertaken by one player is also available to the opponent.

2. No external payments into or out of the game matrix are possible. What one player gains, the other player loses. There is no benefit gained by the two opponents colluding to their mutual benefit at the expense of others.

3. Both players adopt strategies that attempt to minimize their maximum loss. In fact, since the game matrix is completed in terms of payments to A, then B will choose strategies to minimize B's maximum loss, the *minimax criterion*, while A will choose strategies to maximize A's minimum gain, the *maximin criterion*.

Certain of these conflict situations may be solved very quickly. This is because the payoff matrix exhibits some special characteristics. The most important of these is called a *saddle point*. In this case, the maximizing player can see a strategy that yields no worse than a certain payoff, while the minimizing player can also perceive a like strategy. Thus both players choose their strategies and end up with this maximum minimum payoff.

The payoff matrix in Tableau 7.1 exhibits such a saddle point because, in this case, Player A perceives that strategy A(i) means doing no worse than to receive 1 unit of payment from Player B. Simultaneously, Player B perceives that playing strategy B(i) means doing no worse than to give player A 1 unit. Thus, under the minimax theory, they would play strategies A(i) and B(i).

Other game matrixes, such as Tableau 7.2, may be reduced to smaller matrixes by observing that some strategies would never be used by minimax players. For example, consider strategies A(i) and A(ii) in Tableau 7.2.

Tableau 7.2

		Player B		
		(i)	(ii)	(iii)
	(i)	1	2	3
Player A	(ii)	2	3	4
	(iii)	5	2	1

Player A perceives that strategy A(ii) will in all cases provide a better result than strategy A(i), whatever strategy player B might adopt. Thus strategy A(i) is dominated by strategy A(ii), and A(i) would never be used.

However, in some cases it is not possible for the players to perceive which strategy they should adopt. Consider the example in Tableau 7.3:

Tableau 7.3

		Player B	
		(i)	(ii)
Player A	(i)	1	3
	(ii)	4	0

Here there is no saddle point. (Test the minima of each row, which in no case equals the maxima of any column.) Neither is there any dominance on either player's strategies. Here both players will select strategies in a way that will make the expectation of each of the opponent's strategies equal, thus removing any influence from the opponent.

Consider Player A. Let A allocate the probabilities of playing A's strategies as X to A(i) and $(1 - X)$ to A(ii). Player B's strategy B(i) will then have an expectation (E):

$$E(B(i)) = 1\,X + 4\,(1 - X)$$

and for B(ii)

$$E(B(ii)) = 3\,X + 0\,(1 - X)$$

These are linear relationships in X. In order to maximize A's minimum expectation, and simultaneously minimize B's expectation, a solution is found where these expressions are equal (see Figure 7.1). Equating these gives

$$-3\,X + 4 = 3\,X$$
$$X = 0.6667$$
$$1 - X = 0.3333$$

Strategy A(i) will be chosen according to a probability weight 0.6667; A(ii) is half as likely to be played, with a probability weight of 0.3333. This is depicted graphically in Figure 7.2.

The value of the game is the expectation of Player A at the solution; for example

$$Value = (1 \times 2/3) + (4 \times 1/3)$$
$$= (3 \times 2/3) + (0 \times 1/3)$$
$$= 2$$

Figure 7.1

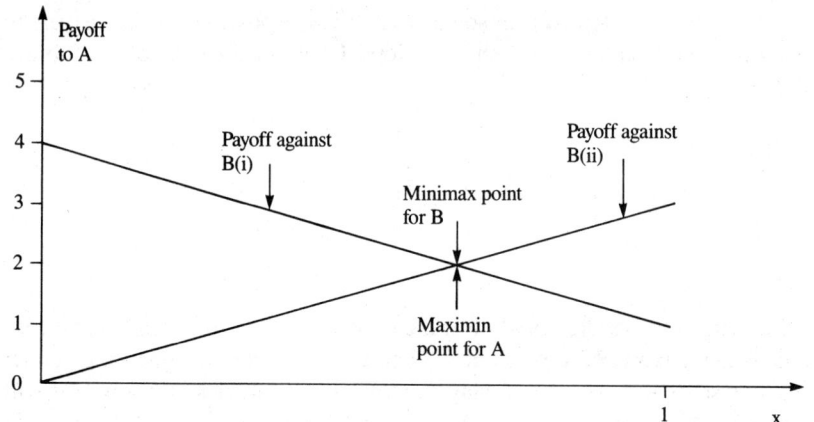

Figure 7.2

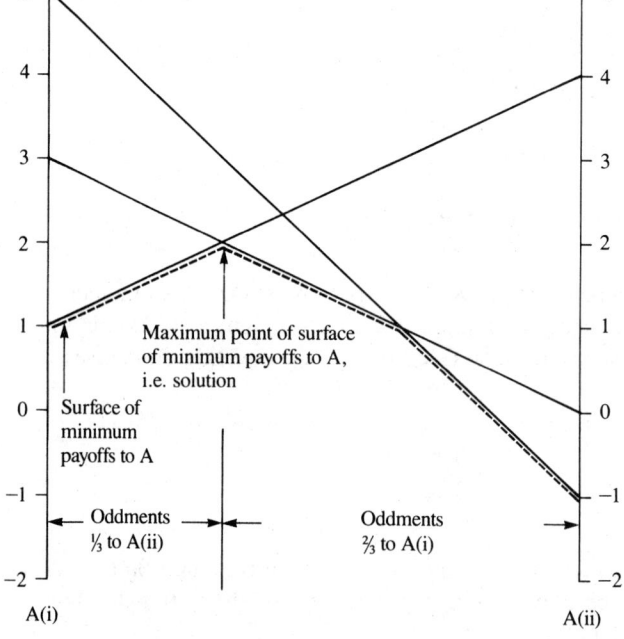

A similar exercise may be conducted for Player B, but an arithmetic shortcut known as the *method of oddments* may always be used for a 2 by 2 game matrix as shown in Tableau 7.4. Player B would work downward, Player A to the right.

Tableau 7.4

Player B

		(i)	(ii)	
Player A	(i)	1	3	$(4 - 0) = 4$
	(ii)	4	0	$(3 - 1) = 2$
				Sum = 6

$(3 - 0)$ $(4 - 1)$
= 3   = 3

Sum = 6

The probability for A's strategies are thus

$$A(i) = 4/6 = 0.6667$$
$$A(ii) = 2/6 = 0.3333$$

For B,

$$B(i) = 3/6 = 0.5$$
$$B(ii) = 3/6 = 0.5$$

The method of oddments is extendable to larger matrixes, but does not always work. The reader who would like to pursue this further is referred to an article by the author, *A Note on the Method of Oddments*, Operational Research Quarterly, 31 (1980): 349–352.

LINEAR PROGRAMMING METHOD

Consider the example illustrated in the game matrix of Tableau 7.5:

Tableau 7.5

Player B

		(i)	(ii)	(iii)
	(i)	1	3	5
Player A	(ii)	4	0	-1

If we consider the matrix algebraically and let player A's strategies have the probability weights X and Y, so that

$$X \geqslant 0; Y \geqslant 0; \text{ and } X + Y = 1$$

then the expected payoffs against each of B's strategies are

$$1\,X + 4\,Y \text{ against B(i)}$$
$$3\,X + 0\,Y \text{ against B(ii)}$$
$$5\,X - 1\,Y \text{ against B(iii)}$$

Now in general, X and Y will not be the solution of the game; thus the expectations computed above will, in general, be different. Let g' be the smallest of these expectations, so that

$$1\,X + 4\,Y \geqslant g'$$
$$3\,X + 0\,Y \geqslant g'$$
$$5\,X - 1\,Y \geqslant g'$$

As we have an inequality sign present, it is advantageous to ensure that all entries are nonnegative. By adding 1 to all the coefficient entries, we ensure that, because X and Y are nonnegative, the smallest expectation will also be nonnegative. This does not destroy the relative probability weights of the solution but merely alters the value of the game solution. The entries for this problem are then rewritten as

$$2\,X + 5\,Y \geqslant G$$
$$4\,X + 1\,Y \geqslant G$$
$$6\,X + 0\,Y \geqslant G$$

where G is the new minimum expectation of this set of relations.

Now that we have a positive G value, we may write a new set of relations:

$$\frac{X}{G} \geqslant 0 \qquad \frac{Y}{G} \geqslant 0$$

while

$$\frac{X}{G} + \frac{Y}{G} = \frac{1}{G} = M$$

$$2\frac{X}{G} + 5\frac{Y}{G} \geq 1$$

$$4\frac{X}{G} + 1\frac{Y}{G} \geq 1$$

$$6\frac{X}{G} + 0\frac{Y}{G} \geq 1$$

Player A wishes to choose X and Y so that G, A's minimum expectation, is maximized. A therefore wishes to maximize 1/G. Written more neatly, and substituting V for X/G and W for Y/G, we obtain:

$$V \geq 0 \qquad W \geq 0$$

$$2\,V + 5\,W \geq 1$$

$$4\,V + 1\,W \geq 1$$

$$6\,V + 0\,W \geq 1$$

with the optimization function to minimize

$$\text{Minimize } V + W = M$$

A simplex linear programming problem results. A similar treatment of B's strategies produces the dual of this linear program. Using R, S, and T for the dual variables, this generates

Maximize $R + S + T$
subject to $2\,R + 4\,S + 6\,T \leq 1$
              $5\,R + 1\,S + 0\,T \leq 1$
              $R \geq 0, S \geq 0, T \geq 0$

The solution of these formulations is conveniently and elegantly handled by the condensed simplex tableau. The equation representation,

Tableau 7.6

			R	S	T
X0	=	0	-1	-1	-1
V	=	1	-2	-4	-6
W	=	1	-5	-1	0

becomes in the condensed tableau format

Tableau 7.7

			Player B				
			(i)	(ii)	(iii)		
		(i)	2	4	6	1	
Player A	(ii)		5	1	0	1	
			-1	-1	-1	0	D = 1

(D is initially given the value 1. It is used in the condensed simplex tableau format as a multiplier to avoid the occurrence of fractions as the updating proceeds.) Here we have a tie for a possible pivot column. A useful tie-breaking rule used in these decision theory matrixes is as follows: For each positive possible pivot element, calculate the value of

$$- \frac{\text{(Outer column element} \times \text{Outer row element)}}{\text{(Corresponding positive matrix element)}}$$

(note the minus sign), which is

0.5	0.25	0.1667
0.2	1.0	—

Choose the smallest entry in each column:

—	0.25	0.1667
0.2	—	—

Choose the largest of these as the position of the pivot element:

$$
\begin{array}{ccc}
- & 0.25 & - \\
- & - & -
\end{array}
$$

The pivot element is therefore A(i), B(ii), and the tableau is updated to

	B (i)	A (ii)	B (iii)		
B(ii)	2	1	6	1	
A(ii)	18	−1	−6	3	
	−2	1	2	1	D = 4

Tableau 7.8

which is optimized as

	A (ii)	A (i)	B (iii)		
B(ii)	−2	5	30	3	
B(i)	4	−1	−6	3	
	2	4	6	6	D = 18

Tableau 7.9

The interpretation of this solution is as follows:

Probability weights of A's strategies:
A(i):A(ii)   (0.667 : 0.333)

Probability weights of B's strategies:
B(i):B(ii):B(iii)   (0.5 : 0.5 : 0)

Value of the objective function $= 1/G = 6/18$

Value of the Game solved $= G = 18/6 = 3$

Value of the original Game $= g' = 3 - 1 = 2$

A refinement of the LP method used in the analysis of decision problems where the decision maker is unsure if the opponent is malevolent and fully informed, together with the use of further information, which could be false, is presented by the author in *A Criterion for Decision Making When Supplied with Extra Imperfect Information as to the Opponent's Intentions*, Operational Research Quarterly, 24 (1973): 117–120.

CHAPTER 7 PROBLEMS

The solutions to Problems 1 through 5 are found at the end of the chapter.

1. Adam Upp has just obtained his M.B.A. degree after graduating as a mathematician. Having perceived a market for educational toys for aspiring mathematicians, he has invented three products:

> Infinite plastic sets
> Boxes of rubber tensors
> Four-dimensional jigsaws

Unfortunately, the market is also being entered by an established firm, Mathemagics, which has recently obtained the patent rights to the following products:

> Electric abacus
> Clockwork integrators
> Laplace transformers

In addition, Mathemagics is currently marketing pocket analog computers widely used by engineers, and this product is likely to be packaged for the perceived market.

Both Upp and Mathemagics have only enough resources to market one product effectively for the coming Christmas selling season. Upp has received infallible information that Mathemagics will be going all out for the market with one of their products, but does not know which specific product will be marketed by them. Upp wishes to capture as much market share as possible and has analyzed the market shares resulting from each decision as follows:

Market share to Adam Upp:

	Mathemagics			
Adam Upp	Abacus	Integrator	Laplace Transformer	Analog Computer
Do nothing	0	0	0	0
Plastic sets	100	50	70	100
Tensors	20	40	80	90
4-D jigsaws	80	70	60	80

Look at your watch, note the exact time recorded, and advise Adam of which strategy he should adopt.

How does the time affect the decision? What is the expected market share for Adam Upp?

2. Richard Vigorous is an ultra-conservative businessman who makes use of the latest management information technology in order to operate his business activities. He would like to undertake one of four mutually exclusive projects.

Making use of the firm's impressive computer installations, he has calculated the NPVs (net present values) of each project using a simulation package under four possible scenarios. The results are summarized in the table below:

Project	Scenario			
	a	b	c	d
A	-10	12	15	0
B	-14	14	17	0
C	0	3	5	0
D	7	-2	0	9

Knowing that Vigorous makes decisions on the most conservative criterion, how should he select which project to undertake?

3. Ann Gola is president of a developing country. Her economic advisers have produced plans for seven possible economic strategies that might be adopted. The outcomes of adopting any of these strategies are affected by the activities of the superpower that has influence in the area (U.S.S.A.). The superpower may adopt one of three possible postures, which Gola has code-named Red, White, and Blue. The president has called for a full analysis of the situation and has received the following matrix, which summarizes the effect of adopting each of the strategies under each possible posture. The entries are supplied in units of socioeconomic utiles.

Gola's Strategy	U.S.S.A. Posture		
	Red	White	Blue
A	16	16	57
B	98	63	89
C	1	3	9
D	29	7	16
E	72	61	80
F	71	11	41
G	61	5	66

President Gola wishes to make the most conservative decision possible. How would you advise her? The advisers, to their horror, discover that the figures supplied to the president for strategy B against White and Blue postures have been transposed. How does this affect the president's decision?

4. Inspector "Clue" Seaut of the Paris Interpol has gained intelligence through an informer that the international terrorist "Kartos" has decided to attack one of four major airports: Los Angeles, London, Amsterdam, or Paris. The inspector, who has made a specialist study of international terrorism, and Kartos in particular, is confident that he can adequately forestall any attempt by being present at the scene. Unfortunately, should the press discover that he was defending a different airport than the one attacked, the results, in terms of a hostility scale of 1 to 5, will be as shown below.

Seaut	Kartos			
	LAX	LHR	ASPL	PORLY
LAX	0	3	5	5
LHR	2	0	2	3
ASPL	4	3	0	2
PORLY	4	3	1	0

Where should the inspector proceed?

5. Solve the following games theory matrix:

		B			
		i	ii	iii	iv
	i	3	7	8	4
A	ii	4	8	9	5
	iii	6	2	3	7
	iv	5	1	2	6

HOMEWORK PROBLEMS

6. Eric Dickerperson is currently negotiating his football playing contract with the Los Angeles Devils. After months of negotiating and dickering, Eric must now make a decision. He has two possible strategies available to him. He also notes that the managers of the Devils have three possible strategies available to them. The value, to Eric, of each of the possibilities is calculated in terms of $10,000 units in the table below.

Eric's	Management Option		
Option	A	B	C
a	98	45	40
b	85	70	80

How should Eric decide his position?

7. General Quatars is facing an enemy in a mock battle and has three possible strategies to adopt. An analysis of the situation presented by his staff shows the estimated relative loss of troops under each alternative:

Quatars's	Enemy Strategy		
Strategy	A	B	C
1	3,000	7,000	4,000
2	8,000	−3,000	6,000
3	−5,000	9,000	−3,000

How would you advise the general, who wishes to maintain the most effective position in terms of troops?

8. Solve the following games theory matrix:

		B				
		i	ii	iii	iv	v
A	i	20	13	10	−11	−16
	ii	−15	−1	−13	−5	15

Use the minimax criterion, then the maximin criterion.

9. On visiting Cal Worthyman's used car lot, you are approached by the owner, Cal, who proposes that, for fun, you should play a game of matching quarters. To play this game, each of you takes a quarter from your pocket and if they match, heads or tails, you lose your quarter. If they do not match, then, if yours is tails, you will get 50 cents, otherwise you each keep your quarter. Do you think that this is a good game to play?

10. Formulate the LP solution for the following games theory matrix:

Player A   Player B

	a	b	c	d	e	f	g	h	i	j
i	5	−4	5	8	−4	9	2	−6	8	6
ii	−3	3	7	1	5	−6	−2	7	−4	2
iii	1	−6	2	8	2	8	5	3	8	−10
iv	10	3	−5	−2	12	−6	2	−5	−11	12
v	−4	2	−4	6	−1	4	−12	4	−5	−8
vi	8	−7	11	−6	−14	1	−6	−8	15	5
vii	2	11	5	−12	−6	18	−3	12	−4	11

CHAPTER 7 SOLUTIONS

1. Check first for saddle point or dominance:

Adam Upp	Mathemagics				Row Minima
	Abacus	Integrator	Laplace Transformer	Analog Computer	
Do nothing	0	0	0	0	0
Plastic sets	100	50	70	100	50
Tensors	20	40	80	90	20
4-D jigsaws	80	70	60	80	60
Column Maxima	100	70	80	100	

The row minima and column maxima do not coincide; therefore there is no saddle point. The first row is dominated and so does not need to enter the remainder of the solution. Continue using the condensed simplex tableaux:

	X1	X2	X3	
Y1	10	5	7	1
Y2	2	4	8	1
Y3	8	7	6	1	*
	−1	−1	−1	0

D = 1

*

	X1	Y3	X3	
Y1	30	−5	19	2
Y2	−18	−4	32	3	*
X2	8	1	6	1
	1	1	−1	1

D = 7

*

	X1	Y3	Y2	
Y1	186	−9	−19	1
X3	−18	−4	7	3
X2	52	8	−6	2
	2	4	1	5

D = 32

Thus, the solution is

Mathemagics' strategy	(0:2:3)
Upp's strategy	(0:1:4)

The value of the LP is 5/32; thus the value of the games matrix solved $= 32/5 = 6.4$.

The matrix used in the solution was obtained by dividing each element by 10. Therefore, the value of the expected market share captured is 64 percent.

The watch is used to produce the sampling probability for the choice of strategy. For example, divide the face of the watch into two segments, 0–12 and 12–60 seconds. If the second hand (or if digital, the second indicator) is in the first segment, then Adam Upp should adopt strategy 2 (market the tensors). If the indicator lies in the second segment, then Upp should adopt strategy 3 and market the four-dimensional jigsaws.

2. The LP formulation of Richard's problem is:

Maximize
$$a + b + c + d$$
subject to
$$-10a + 12b + 15c \leq 1$$
$$-14a + 14b + 17c \leq 1$$
$$+ 3b + 5c \leq 1$$
$$7 - 2b + 9d \leq 1$$

Primal solution: $a = 0.21875$
$b = 0.265625$

Dual solution: $A = 0.140625$
$B = 0.34375$

Probability oddment for Richard to choose project A:
$$= \frac{0.140625}{0.140625 + 0.343750} = 0.2903225$$

Probability oddment for Richard to choose project B:
$$= \frac{0.343750}{0.140625 + 0.343750} = 0.7096774$$

Richard would therefore choose from among the mutually exclusive projects in order to apply the ultra-conservative maximin criterion by the employment of a random-number-generating device such that the projects would be selected with these probabilities:

$$(A:B:C:D) = (0.29:0.71:0:0)$$

3. Checking first for a saddle point:

Gola's Strategy	U.S.S.A. Posture			Row Minima
	Red	White	Blue	
A	16	16	57	16
B	98	63	89	63 *
C	1	3	9	3
D	29	7	16	7
E	72	61	80	61
F	71	11	41	11
G	61	5	66	5
Column Maxima	98	63	89	
		*		

A saddle point exists between President Gola's economic strategy B and the U.S.S.A.'s posture "White." Gola should therefore choose to adopt the pure strategy B.

The advisers have reason to be concerned. The problem now is to solve the LP formulation:

Maximize $\quad R + W + B$

subject to
$$16R + 16W + 57B \leq 1$$
$$98R + 89W + 63B \leq 1$$
$$1R + 3W + 9B \leq 1$$
$$29R + 7W + 16B \leq 1$$
$$72R + 61W + 80B \leq 1$$
$$71R + 11W + 41B \leq 1$$
$$61R + 5W + 66B \leq 1$$

Two iterations produce the solution

White = 0.005188
Blue = 0.008544

Dual solution:

B = 0.005798
E = 0.007934

Probability weights for Ann Gola's government are

(A : B : C : D : E : F : G)
(0 : 0.422 : 0 : 0 : 0.577 : 0 : 0)

4. The LP to be solved by Inspector "Clue" Seaut is

Maximize	KLAX + KLHR + KASPL + KPORLY
subject to	$-$ 3 KLHR $-$ 5 KASPL $-$ 5 KPORLY $\leq$ 1
	$-$ 2 KLAX $-$ 2 KASPL $-$ 3 KPORLY $\leq$ 1
	$-$ 4 KLAX $-$ 3 KLHR $-$ 2 KPORLY $\leq$ 1
	$-$ 4 KLAX $-$ 3 KLHR $-$ 1 KASPL $\leq$ 1

This formulation contains negative entries. An attempt at direct solution of this formulation will result in an unbounded solution being reported. The problem is solved by making all entries in the formulation take non-negative values; in this case a constant value, 6 has been added to each coefficient of the payoff matrix.

(Remember that in the analysis for the LP solution of games theory problems, the maximum of the minimum payoffs expected for each strategy must be non-negative, otherwise the signs of the inequalities will cause an error.)

The LINDO solution follows:

```
LOOK ALL
MAX KLAX + KLHR + KASPL + KPORLY
SUBJECT TO
2)   KLAX + 3 KLHR +   KASPL +   KPORLY <= 1
3) 4 KLAX + 6 KLHR + 4 KASPL + 3 KPORLY <= 1
4) 2 KLAX + 3 KLHR + 6 KASPL + 4 KPORLY <= 1
5) 2 KLAX + 3 KLHR + 5 KASPL + 6 KPORLY <= 1
END

: GO
LP OPTIMUM FOUND AT STEP 4
OBJECTIVE FUNCTION VALUE
 1)  .270833333
```

```
      VARIABLE            VALUE          REDUCED COST
       KLAX              .145833           .000000
       KLHR              .000000           .125000
       KASPL             .041667           .000000
       KPORLY            .083333           .000000
        ROW         SLACK OR SURPLUS     DUAL PRICES
        2)              .000000           .062500
        3)              .000000           .104167
        4)              .125000           .000000
        5)              .000000           .104167
NO. ITERATIONS= 4
```

The solution indicates that, for the greatest impact on public opinion, "Kartos" would plan his attack on the airports with probabilities

(Los Angeles	:	London	:	Amsterdam	:	Paris)
ratios (0.145833	:	0	:	0.041667	:	0.0833)
probabilities (0.538	:	0	:	0.1538	:	0.307)

The inspector, in order to counteract Kartos, should plan to be present at the airports with probabilities

(Los Angeles	:	London	:	Amsterdam	:	Paris)
ratios (0.0625	:	0.104167	:	0	:	0.104167)
probabilities (0.231	:	0.3846	:	0	:	0.3846)

5. Solution by LINDO:

```
: LOOK ALL
 MAX B1 + B2 + B3 + B4
 SUBJECT TO
 2) 3 B1 + 7 B2 + 8 B3 + 4 B4 <= 1
 3) 4 B1 + 8 B2 + 9 B3 + 5 B4 <= 1
 4) 6 B1 + 2 B2 + 3 B3 + 7 B4 <= 1
 5) 5 B1 +   B2 + 2 B3 + 6 B4 <= 1
 END
```

(continued)

```
: GO
LP OPTIMUM FOUND AT STEP 2
OBJECTIVE FUNCTION VALUE
1) .200000000
      VARIABLE              VALUE          REDUCED COST
          B1               .150000           .000000
          B2               .050000           .000000
          B3               .000000           .200000
          B4               .000000           .200000
         ROW        SLACK OR SURPLUS      DUAL PRICES
          2)               .200000           .000000
          3)               .000000           .100000
          4)               .000000           .100000
          5)               .200000           .000000
NO. ITERATIONS= 2
```

8

Optimizing Critical-Path Formulations

One type of large constrained problem frequently encountered in business and engineering production involves the scheduling of many separate tasks; the tasks may themselves be complex and numerous. In learning to handle such complex situations, two approaches were developed: PERT (Program Evaluation and Review Technique) and CPM (Critical Path Method). PERT was used in the late 1950s to control complex projects that had never been attempted previously. It was thus designed for situations where uncertainty exists in the data. The system was used to manage the Polaris Missile Project with such success that it became a standard tool for project control in industry. CPM, in contrast, was designed for the control of industrial projects and maintenance schedules where the activities and data are considered to be well known. Both techniques use essentially the same concepts and have evolved to the point that no difference exists in commercial packages. In this chapter, the method will first be described for the case where the data are well known and therefore deterministic (the CPM approach), and the modifications that allow uncertainty to be considered (the PERT approach) will be addressed at the end of the chapter.

Consider the following project:

Task Label	Task Description	Completion Time	Immediate Predecessors
A	Feasibility study	8 weeks	—
B	Raise extra financing	6 weeks	—
C	Develop market	8 weeks	B
D	R&D	4 weeks	A
E	Training	5 weeks	A
F	Test production	3 weeks	D
G	Market launch	3 weeks	C
H	Product plan	4 weeks	E, F
I	On-line system trials	2 weeks	G, H

This project can be depicted in the form of a directed graph or network. The branches of the network represent the activities, and the nodes of the graph (Xi) are the points at which activities start and are completed, as shown in Figure 8.1.

Figure 8.1

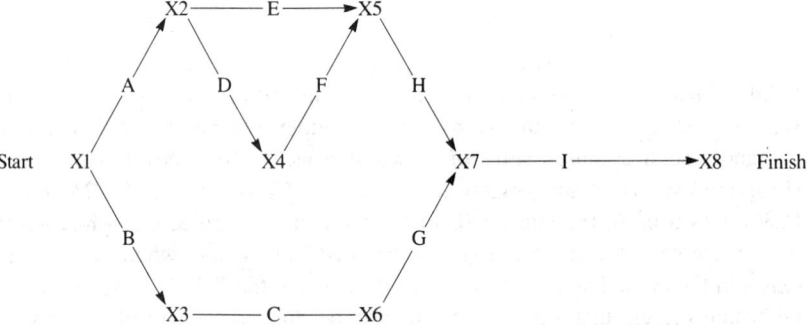

Note that in this example no two activities have the same starting and ending node. Although such a situation causes no particular problem in practice, it is convenient to separate the tasks with the use of dummy activities; thus is replaced by Figure 8.2.

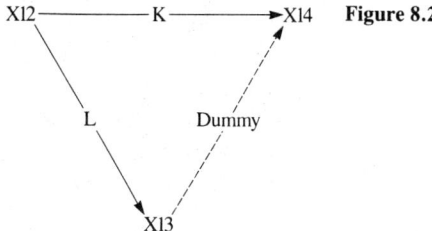

Figure 8.2

Using this graphical representation, we are now able to prepare an activity schedule for the project:

Activity	Earliest Start ES	Latest Start LS	Earliest Finish EF	Latest Finish LF	Slack (LS − ES) (LF − EF)
A	0	0	8	8	0
B	0	0	6	8	2
C	6	8	14	16	2
D	8	8	12	12	0
E	8	10	13	15	2
F	12	12	15	15	0
G	14	16	17	19	2
H	15	15	19	19	0
I	19	19	21	21	0

The activities that have a zero slack are those that are critical to the project and thus should have most managerial attention focused upon them. In the case of activity E, the 2 weeks of slack, or allowable slippage, is applied to E alone. However, on the lower arm of the graph B − C − G, the 2 weeks slack time is shared by all the activities of the arm. If one activity slips by 2 weeks, the others have no slack remaining. In some representations of the technique a distinction is made between slack that is applicable to a single activity and slack shared between activities. The upper path, A − D − F − H − I, is the critical path for this project. Considering this critical path, the minimum time to complete this project is 21 weeks.

Suppose now that this time is considered too long. A technique might be adopted to reduce this overall time. First consider the cheapest alternatives to speed up, or crash, activities on the critical path. Then, as these are reduced, consider the interactions with currently noncritical activities that might become critical.

Normally, when activities are undertaken in a reduced time, an increase in cost is expected. It is reasonable to assume (1) that there are practical limits to the amount of time reduction possible, and (2) that the reduction in time is proportionate to the increase in costs incurred. These assumptions are implicit in Figure 8.3.

Figure 8.3

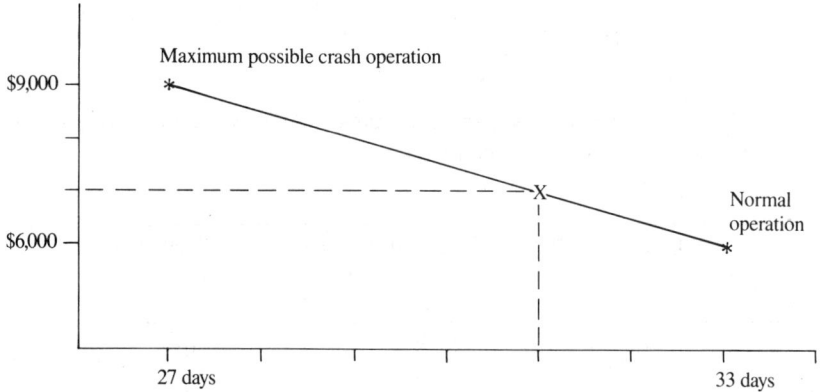

The illustrated case has a normal operating time of 33 days at a normal cost of $6,000, and a maximum possible crash operating condition of 27 days at a total cost of $9,000. The proportionality condition therefore predicts that this task can be completed in 31 days at a total cost of $7,000. Thus shortening the schedule by 2 days will cost $1,000. We can formalize these points as follows:

$$
\begin{aligned}
t \;&=\; \text{Normal operation time, weeks} \\
T \;&=\; \text{Maximum crash operation time, weeks} \\
C_n \;&=\; \text{Normal cost of operation} \\
C_c \;&=\; \text{Cost of operation at maximum crash condition} \\
M \;&=\; \text{Maximum time reduction, } t - T, \text{ weeks} \\
K \;&=\; \text{Crash cost per unit time, } (C_c - C_n)/M, \text{ \$/week}
\end{aligned}
$$

For the example project, we obtain the following data:

Activity	Normal Time t	Crash Time T	Normal Cost Cn	Crash Cost Cc	M	K
A	8	4	500	900	4	100
B	6	4	400	600	2	100
C	8	5	200	300	3	33.33
D	4	2	300	400	2	50
E	5	4	200	300	1	100
F	3	2	600	1000	1	400
G	3	2	400	800	1	400
H	4	2	800	900	2	50
I	2	1	300	600	1	300

We now define one more variable,

$$Y_i = \text{Amount of crash time used for activity } i$$
$$(\text{where } i = A, B, C, \text{ etc.})$$

We now wish to choose which activities must be crashed, and by how much, in order to meet a defined project deadline and at the same time incur the minimum extra cost over the normal cost in achieving this objective. Our optimization function then becomes

$$\text{Min} \sum_i K_i Y_i$$

or

Minimize $100\ YA + 100\ YB + 33.33\ YC + 50\ YD + 100\ YE + 400\ YF$
$+ 400\ YG + 50\ YH + 300\quad YI$

The constraints upon this optimization function are provided by a description of the network:

1. The time of an event (the node Xj) must be greater than or equal to the time to complete all activities leading to that node.

2. An activity start time is equal to the event time of the node preceding that activity.

3. An activity time is the normal time of the event, less the time by which it is crashed. Thus for event 2 (node X2),

Occurrence time for event 2	$\geqslant$	Actual time for activity	$+$	Start time for activity
		A		A
X2	$\geqslant$	tA $-$ YA	$+$	X1
X2	$\geqslant$	8 $-$ YA	$+$	0

or

$$X2 + YA \geqslant 8$$

For event 3,

$$X3 \geqslant 6 - YB \text{ or } X3 + YB \geqslant 6$$

For event 4,

$$X4 \geqslant 4 - YD + X2 \text{ or } X4 - X2 + YD \geqslant 4$$

For event 5,

$$X5 \geqslant 5 - YE + X2 \text{ and } X5 \geqslant 3 - YF + X4$$

because two activities enter the same node at this point, or

$$X5 - X2 + YE \geqslant 5$$
$$X5 - X4 + YF \geqslant 3$$

For event 6,

$$X6 \geqslant 8 - YC + X3 \text{ or } X6 - X3 + YC \geqslant 8$$

For event 7,

$$X7 \geqslant 4 - YH + X5 \text{ or } X7 - X5 + YH \geqslant 4$$
$$X7 \geqslant 3 - YG + X6 \text{ or } X7 - X6 + YG \geqslant 3$$

For event 8,

$$X8 \geqslant 2 - YI + X7 \text{ or } X8 - X7 + YI \geqslant 2$$

And finally, for event 8, the constraint desired on the completion time of the project. Let us assume that it must be done in 15 weeks:

$$X8 \leqslant 15$$

In addition, there are constraints on the maximum allowable crash times, the Yi. The optimization problem therefore becomes:

	YA	YB	YC	YD	YE	YF	YG	YH	YI	X2	X3	X4	X5	X6	X7	X8	
Minimize	100	100	33.3	50	100	400	400	50	300	0	0	0	0	0	0	0	

subject to

(Network constraints)

YA	YB	YC	YD	YE	YF	YG	YH	YI	X2	X3	X4	X5	X6	X7	X8	
1									1							$\geqslant 8$
	1									1						$\geqslant 6$
		1								−1	1					$\geqslant 4$
			1							−1		1				$\geqslant 5$
				1							−1	1				$\geqslant 3$
					1							−1	1			$\geqslant 8$
						1							−1	1		$\geqslant 4$
							1						−1	1		$\geqslant 3$
								1						−1	1	$\geqslant 2$

(Maximum allowable crash constraints)

YA	YB	YC	YD	YE	YF	YG	YH	YI	
1									$\leqslant 4$
	1								$\leqslant 2$
		1							$\leqslant 3$
			1						$\leqslant 2$
				1					$\leqslant 1$
					1				$\leqslant 1$
						1			$\leqslant 1$
							1		$\leqslant 2$
								1	$\leqslant 1$

(Project completion constraint)

$$1 \leqslant 15$$

This linear programming approach will not only take into account the activities on the critical path, but will also consider the noncritical path activities, which in their turn become critical as the project time shrinks. As a LINDO input file, this will be

```
MIN
    100 YA + 100 YB + 33.333 YC + 50 YD
+ 100 YE + 400 YF + 400 YG + 50 YH
+ 300 YI
ST
YA + X2 > 8
YB + X3 > 6
YC - X3 + X6 > 8
YD - X2 + X4 > 4
YE - X2 + X5 > 5
YF - X4 + X5 > 3
YG - X6 + X7 > 3
YH - X5 + X7 > 4
YI - X7 + X8 > 2
YA < 4
YB < 2
YC < 3
YD < 2
YE < 1
YF < 1
YG < 1
YH < 2
YI < 1
X8 < 15
END
```

This produces the output

```
LP OPTIMUM FOUND AT STEP 12
OBJECTIVE FUNCTION VALUE
1) 599.99999
     VARIABLE              VALUE           REDUCED COST
        YA             2.000000              .000000
        YB             1.000000              .000000
        YC             3.000000              .000000
        YD             2.000000              .000000
        YE              .000000            50.000000
        YF              .000000           350.000000
        YG              .000000           300.000000
        YH             2.000000              .000000
        YI              .000000           100.000000
        X2             6.000000              .000000
        X3             5.000000              .000000
        X6            10.000000              .000000
        X4             8.000000              .000000
        X5            11.000000              .000000
        X7            13.000000              .000000
        X8            15.000000              .000000

      ROW         SLACK OR SURPLUS      DUAL PRICES
       2)              .000000          -100.000000
       3)              .000000          -100.000000
       4)              .000000          -100.000000
       5)              .000000           -50.000000
       6)              .000000           -50.000000
       7)              .000000           -50.000000
       8)              .000000          -100.000000
       9)              .000000          -100.000000
      10)              .000000          -200.000000
      11)             2.000000              .000000
      12)             1.000000              .000000
      13)              .000000            66.666666
      14)              .000000              .000000
      15)             1.000000              .000000
      16)             1.000000              .000000
      17)             1.000000              .000000
      18)              .000000            50.000000
      19)             1.000000              .000000
      20)              .000000           200.000000
NO. ITERATIONS= 12
DO RANGE (SENSITIVITY) ANALYSIS? > Y
```

(continued)

```
RANGES IN WHICH THE BASIS IS UNCHANGED
                OBJ COEFFICIENT RANGES
VARIABLE        CURRENT       ALLOWABLE       ALLOWABLE
                 COEF          INCREASE        DECREASE
    YA          100.            50.             50.
    YB          100.           100.             66.66
    YC           33.33          66.66        INFINITY
    YD           50.            50.             50.
    YE          100.         INFINITY           50.
    YF          400.         INFINITY          350.
    YG          400.         INFINITY          300.
    YH           50.            50.          INFINITY
    YI          300.         INFINITY          100.
    X2            0.            50.             50.
    X3            0.            66.66          100.
    X6            0.           100.            100.
    X4            0.            50.            100.
    X5            0.            50.            100.
    X7            0.           200.            100.
    X8            0.           200.         INFINITY


                RIGHTHAND SIDE RANGES
   ROW          CURRENT       ALLOWABLE       ALLOWABLE
                 RHS           INCREASE        DECREASE
    2            8.             2.              2.
    3            6.             1.              1.
    4            8.             1.              1.
    5            4.             0.              2.
    6            5.             2.              0.
    7            3.             0.              2.
    8            3.             1.              1.
    9            4.             2.              2.
   10            2.             1.              1.
   11            4.          INFINITY           2.
   12            2.          INFINITY           1.
   13            3.             1.              1.
   14            2.          INFINITY           0.
   15            1.          ·INFINITY          1.
   16            1.          INFINITY           1.
   17            1.          INFINITY           1.
   18            2.             2.              2.
   19            1.          INFINITY           1.
   20           15.             1.              1.
```

This solution shows that we are able to complete the project within the 15-week deadline by spending an extra $600. Subtracting the crash-time amounts from the normal completion times, we see that activity A should be completed in 6 weeks, B in 5 weeks, C in 5 weeks, D in 2 weeks, and H in 2 weeks. The nodes of the network, or milestones of the project, are indicated by X2 through X8 in the basis, X2 being achieved by week 6, X5 by week 11, etc.

Further interesting results appear in the computer output. The "reduced cost" figures indicate the net increase in cost by investing in activities that were not crashed by this solution; these are alternatives that achieve the same project deadline.

On the row report, the "slack or surplus" column indicates the amount of available reduction used, while the dual prices indicate the net cost of changing the solution by one unit in each constraint. For example, row 20 shows that there will be a cost increase of $200 to shorten the overall project time (X8) by 1 week from the 15-week deadline.

Finally, the sensitivity analysis indicates the range of the coefficient values over which the solution will remain optimal.

NONDETERMINISTIC ACTIVITY TIMES

The analysis of this chapter has assumed that the activity times used in calculating the critical path are known with certainty. Earlier, it was stated that the PERT procedure was developed for situations where the activity time could be only imperfectly known or estimated. In practice, under these conditions, it has been discovered that a good estimate of an activity time can be made by obtaining three estimates for each activity:

a, the optimistic time—the time the project would take if everything progressed in an ideal manner.

m, the most probable time—the time the activity will take under normal conditions.

b, the pessimistic time—the time if significant breakdowns and reverses occur.

These estimates are then combined, using the assumption that a beta distribution is an appropriate model of the process. That is a suitable assumption for most practical cases and may be tested using statistical techniques. The beta distribution allows us to estimate the expected value, t, of the normal time that a project activity will take by use of the formula

$$t = \frac{(a + 4m + b)}{6}$$

This expected time is not the most probable time. The most probable time, m, is the modal value of the statistical distribution of times for the project activity, while the expected time t is the mean of this distribution. In addition, we can also estimate the spread of the time that the activity will take by calculating the variance of this beta distribution:

$$v = \left(\frac{b - a}{6}\right)^2$$

The standard deviation is the square root of this value,

$$s = \left(\frac{b - a}{6}\right)$$

We may obtain these estimates for the example project:

Activity	a	m	b	t	v	Critical Path
A	5	7	15	8	2.777	*
B	4	5	12	6	1.777	
C	4	6	20	8	7.111	
D	2	4	6	4	0.444	*
E	2	5	8	5	1	
F	2	3	4	3	0.111	*
G	1	3	5	3	0.444	
H	1	4	7	4	1	*
I	1	2	3	2	0.111	*

We see that, under normal conditions, the project will take 21 weeks to complete as constrained by the activities on the critical path, A, D, F, H, and I. Statistically, times taken from the distributions representing these activities will be combined in the distribution representing the sum of these activities, which will have an expected value, or mean, equal to the sum of the means of the constituent distributions, and a variance equal to the sum of the variances of each constituent distribution. Moreover, by the central limit theorem, the distribution of overall project times will tend to be represented by a normal distribution. In the example here, the variance of this normal distribution of project completion times (Vp) will be

$$Vp = vA \quad + vD \quad + vF \quad + vH + vI$$
$$= 2.777 + 0.444 + 0.111 + 1 \quad + 0.111$$
$$= 4.444$$

The standard deviation of this distribution is the square root of the variance,

$$sp = 2.1 \text{ weeks}$$

Under these conditions, we may say that the project has a 50 percent chance of being completed within the 21-week period. Using normal distribution tables, we can also work out the probability that the project will be completed within any other time limit. A 15-week limit is 6 weeks under the mean value, or 2.85 standard deviations below the mean.

Normal distribution tables indicate that approximately 0.002 of the normal distribution lies more than 2.85 standard deviations below the mean. Thus there is only a 0.2 percent chance that the project will be completed within 15 weeks under normal conditions.

The linear programming analysis carried out to determine the optimal policy of investing in extra resources in order to meet a deadline is applicable to the expected or mean values of a stochastic variable. The optimal investment may be determined, using this method, to meet a deadline with a desired probability. By investing in extra resources to reduce the expected time for task completion, the variance of the distribution associated with that task will also change. This will result in a change in the total project variance. There is, however, some empirical evidence that, particularly for large complex projects, the coefficient of variation—the ratio of the standard deviation to the mean—tends to remain constant with a change in the mean. For the example shown,

$$\text{Coefficient of variation} = \frac{s}{X} = \frac{2.1}{21}$$

Therefore, for a total project time of X, the standard deviation s expected will be

$$s = 0.1 X$$

If we decide that we want to be 95 percent certain of completing the project within the 15-week deadline, we can calculate the extra costs that would be incurred by first finding the value on the normal curve that exceeds 95 percent of its area. This is found from the table in Appendix 2 to be 1.645 standard deviations above the mean of the distribution. Hence our 15-week deadline should be 1.645 standard deviations above the project mean value:

$$15 - X = 1.645 \, s = 1.645 \ (0.1 \, X)$$
$$15 - X = 0.1645 \, X$$
$$X = 12.88$$

It is easy to find the schedule that optimizes the project to be completed within this expected value. The LINDO input file can be modified using an editor, such as XEDIT, or, if you are using a microcomputer, by use of the microcomputer's editing capabilities. It is also possible to use the ALTER command within LINDO to simply modify the last line (line 20) to the desired project completion time.

Should the completion time of 12.88 be inserted for node X8, then an optimal schedule showing a cost increase of $1,160 would result. However, it is more likely that a practical decision maker will consider options in terms of whole weeks. In this case, the following analysis can quickly be produced with an alteration to constraint row 20 for each run:

	Project Completion Deadline of 15 Weeks			
Option	Expected Completion (weeks)	Probability of 15-week Completion	Extra Costs ($)	Total Project Cost ($)
A	21	0.002	0	3700
B	15	0.5	600	4300
(C)	(12.88)	(0.95)	(1160)	(4860)
D	13	0.94	1100	4800
E	12	0.99	1600	5300

Suppose that management will be satisfied with a 94 percent chance of having the project completed by the 15-week deadline. Option D, with an expected completion date of 13 weeks, is chosen:

Activity	Amount Crashed (weeks)	Scheduled Completion (weeks)	Cost ($)
A	3	5	800
B	2	4	600
C	3	5	300
D	2	2	400
E	—	5	200
F	—	3	600
G	—	3	400
H	2	2	900
I	1	1	600
			$4,800

The project milestones given by the nodes of the graph are

Milestone	Week
X1	0
X2	5
X3	4
X4	7
X5	10
X6	9
X7	12
X8	13

CHAPTER 8 PROBLEMS

The solutions to Problems 1 through 5 will be found at the end of the chapter.

1. Ben Gunn, a big-shot businessman, hired advisers of the highest caliber to assist in project control. He has the following data on a project:

Project Label	Completion Times (days)		Costs ($)		Immediate Predecessor
	Normal	Crash	Normal	Crash	
A	9	6	200	500	—
B	3	1	600	900	—
C	2	1	800	1500	B
D	4	2	400	800	A
E	7	3	300	700	C,D
F	6	3	500	950	C,D
G	9	3	200	900	E

Gunn asks the advisers to calculate the minimum cost of completing the project in 25 days.

2. Ben Gunn has decided to fire the hot shots and has now asked you to work on the more complex project management problem summarized below.

Project Label	Estimated Normal Completion Time (weeks)			Immediate Predecessor
	Optimistic (a)	Most Likely (m)	Pessimistic (b)	
A	1	2	3	—
B	4	5	12	—
C	1	4	7	B
D	2	4	6	A
E	2	5	8	C
F	2	3	4	C
G	1	3	5	E
H	5	7	15	F
I	4	6	20	D
J	1	1	1	G,H,I

Project Label	Max Crash Time	Costs ($)	
		Normal	Crash
A	1	100	200
B	4	500	800
C	3	100	300
D	2	300	700
E	3	500	600
F	1	100	400
G	2	100	300
H	5	900	1500
I	5	300	1200
J	1	300	300

What is the cost of completing the project in 18 weeks with a 50:50 chance?

3. You are assigned the following project:

Project Label	Estimated Normal Completion Time (weeks)			Immediate Predecessor
	Optimistic (a)	Most Likely (m)	Pessimistic (b)	
A	7	8	9	—
B	8	10	12	—
C	5	6	13	A
D	3	5	7	A
E	5	6	7	B
F	5	7	21	B
G	2	5	8	D
H	1	2	3	E
I	2	5	8	H
J	6	8	16	F,I
K	3	4	5	C,G
L	1	3	5	K,J

Project Label	Max Crash Time	Costs ($)	
		Normal	Crash
A	6	500	700
B	7	300	750
C	6	150	300
D	3	400	800
E	2	400	600
F	6	300	600
G	3	400	1000
H	1	100	200
I	2	300	750
J	5	800	1600
K	1	50	200
L	2	160	200

What is the minimum cost of completing the project in 30 weeks with at least a 95 percent chance of completion? (Work to integer whole weeks.)

4. You are assigned the following project:

Project Label	Estimated Normal Completion Time (weeks)			Immediate Predecessor
	Optimistic (a)	Most Likely (b)	Pessimistic (c)	
A	7	9	11	—
B	7	8	9	A
C	3	4	5	A
D	1	3	5	B
E	4	5	12	B
F	5	7	21	D
G	3	4	11	B
H	3	4	5	G
I	2	4	6	C
J	1	2	3	C
K	4	5	12	C
L	4	5	6	K
M	1	4	7	L
N	6	8	16	I
O	4	6	20	J,N,M
P	6	7	8	E,F,H
Q	4	5	12	P

Project Label	Max Crash Time	Costs ($)	
		Normal	Crash
A	7	200	400
B	4	200	1000
C	2	300	600
D	2	300	600
E	2	1000	2000
F	2	250	1000
G	4	50	100
H	2	200	400
I	3	100	200
J	1	200	400
K	3	250	1000
L	3	400	1000
M	2	300	600
N	3	400	1000
O	2	900	1800
P	5	400	800
Q	4	600	1000

What is the minimum cost of completing the project within 35 weeks with at least a 90 percent chance of completion? (Work to integer whole weeks.)

5. You are assigned the following project:

Project Label	Estimated Normal Completion Time (weeks)			Immediate Predecessor
	Optimistic (a)	Most Likely (m)	Pessimistic (b)	
A	8	9	10	—
B	6	8	10	—
C	4	5	12	—
D	3	5	7	A
E	3	4	5	A
F	4	5	12	B
G	2	3	4	B
H	1	2	3	C
I	1	1	1	D,E
J	4	6	14	D,E
K	8	12	28	F,G
L	2	3	4	H

Project Label	Max Crash Time	Costs ($)	
		Normal	Crash
A	5	600	1000
B	5	450	900
C	3	400	600
D	3	500	1000
E	2	400	800
F	3	300	600
G	2	100	200
H	1	100	150
I	—	—	—
J	2	2000	3500
K	6	1200	2400
L	2	100	200

What is the minimum cost of completing the project within 25 weeks with a probability of 99 percent? (Work to partial weeks.)

HOMEWORK PROBLEMS

6. What is the minimum cost of completing the following project in 26 days?

Project Label	Completion Times (days)		Costs ($)		Immediate Predecessor
	Normal	Crash	Normal	Crash	
A	6	4	200	400	—
B	5	4	600	750	A
C	8	4	200	400	A
D	3	2	450	570	B
E	4	2	500	1000	C
F	6	3	400	700	D,E
G	9	5	400	800	D,E
H	2	1	650	950	G

7. Can the following project be accomplished in 30 days? If so, what is its minimum cost?

Project Label	Completion Times (days)		Costs ($)		Immediate Predecessor
	Normal	Crash	Normal	Crash	
A	8	4	500	700	—
B	4	3	600	900	—
C	6	3	200	500	A
D	9	5	400	1000	A
E	8	3	500	1000	C
F	5	3	700	800	D
G	4	2	300	900	F
H	10	8	400	500	B
I	12	6	900	1800	E,G
J	11	8	400	700	H
K	8	5	300	450	H
L	6	4	200	800	J
M	5	2	500	600	K
N	2	1	20	30	L,M

8. You are given the following project data:

Project Label	Estimated Normal Completion Time (weeks)			Immediate Predecessor
	Optimistic (a)	Most Likely (m)	Pessimistic (b)	
A	7	8	9	—
B	3	4	11	—
C	5	7	9	A
D	5	6	7	A
E	2	3	14	A
F	1	2	23	B
G	2	5	18	F
H	1	1	1	E
I	4	5	12	C,D,H

Project Label	Max Crash Time	Costs ($)	
		Normal	Crash
A	4	400	800
B	2	400	1000
C	5	500	1000
D	3	450	900
E	2	100	200
F	1	50	100
G	3	600	1000
H	—	—	—
I	5	500	900

What is the minimum cost of completing the project within 18 days with a 50:50 chance of completion?

9. You are given the following project data:

Project Label	Estimated Normal Completion Time (weeks)			Immediate Predecessor
	Optimistic (a)	Most Likely (m)	Pessimistic (b)	
A	8	9	10	—
B	6	8	10	A
C	4	5	12	A
D	3	5	7	A
E	3	4	5	B
F	4	5	12	B
G	2	3	4	C
H	1	2	3	D
I	1	1	1	E,F
J	4	6	14	H
K	8	12	28	H
L	2	3	4	I,G,J,K

Project Label	Max Crash Time	Costs ($)	
		Normal	Crash
A	5	600	1000
B	5	450	900
C	3	400	600
D	3	500	1000
E	2	400	800
F	3	300	600
G	2	100	200
H	1	100	150
I	—	—	—
J	2	2000	3500
K	6	1200	2400
L	2	100	200

What is the minimum cost of completion for this project within 30 days with a 95 percent chance of completion by the deadline? (Work to whole weeks.)

10. You are given the following project data:

Project Label	Estimated Normal Completion Time (weeks)			Immediate Predecessor
	Optimistic (a)	Most Likely (m)	Pessimistic (b)	
A	7	8	9	—
B	8	10	12	A
C	5	6	13	A
D	3	5	7	A
E	5	6	7	B
F	5	7	21	B
G	2	5	8	C
H	1	2	3	D
I	2	5	8	E,F
J	6	8	16	D
K	3	4	5	E,F
L	1	3	5	C

Project Label	Max Crash Time	Costs ($)	
		Normal	Crash
A	6	500	700
B	7	300	750
C	6	150	300
D	3	400	800
E	2	400	600
F	6	300	600
G	3	400	1000
H	1	100	200
I	2	300	750
J	5	800	1600
K	1	50	200
L	2	160	200

Can this project be completed within 28 weeks with a 99 percent probability of sucessful completion by the deadline?

CHAPTER 8 SOLUTIONS

1. The network for this problem is shown in Figure 8.4.

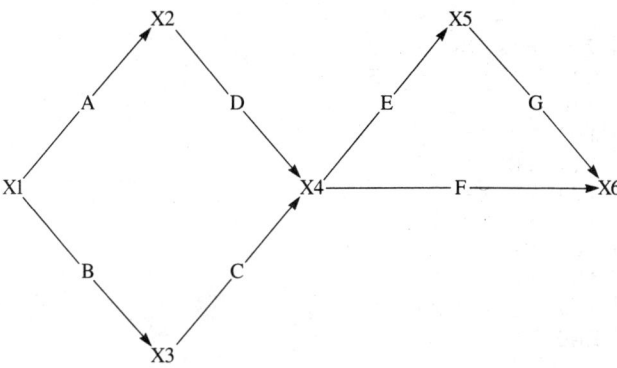

Figure 8.4

Activity	Units of Crash	Cost per Unit
A	3	100
B	2	150
C	1	700
D	2	200
E	4	100
F	3	150
G	6	116.667

The LINDO file would be

```
look all
MIN
100 YA + 150 YB + 700 YC + 200 YD + 100 YE
   + 150 YF + 116.6667 YG
```

(continued)

```
SUBJECT TO
2)  YA + X2 >= 9
3)  YB + X3 >= 3
4)  YC - X3 + X4 >= 2
5)  YD - X2 + X4 >= 4
6)  YE - X4 + X5 >= 7
7)  YF - X4 + X6 >= 6
8)  YG - X5 + X6 >= 9
9)  YA <= 3
10) YB <= 2
11) YC <= 1
12) YD <= 2
13) YE <= 4
14) YF <= 3
15) YG <= 6
16) X6 <= 25
END


: go
LP OPTIMUM FOUND AT STEP 8
OBJECTIVE FUNCTION VALUE
1)  400.000000
       VARIABLE           VALUE           REDUCED COST
          YA             .000000             .000000
          YB             .000000          150.000000
          YC             .000000          700.000000
          YD             .000000          100.000000
          YE            4.000000             .000000
          YF             .000000          150.000000
          YG             .000000           16.666700
          X2            9.000000             .000000
          X3           11.000000             .000000
          X4           13.000000             .000000
          X5           16.000000             .000000
          X6           25.000000             .000000


        ROW        SLACK OR SURPLUS     DUAL PRICES
         2)             .000000         -100.000000
         3)            8.000000             .000000
         4)             .000000             .000000
         5)             .000000         -100.000000
         6)             .000000         -100.000000
         7)            6.000000             .000000
         8)             .000000         -100.000000
         9)            3.000000             .000000
        10)            2.000000             .000000
        11)            1.000000             .000000
        12)            2.000000             .000000
```

```
       13)              .000000           .000000
       14)             3.000000           .000000
       15)             6.000000           .000000
       16)              .000000        100.000000
NO. ITERATIONS= 8
DO RANGE (SENSITIVITY) ANALYSIS? > y

RANGES IN WHICH THE BASIS IS UNCHANGED
OBJ COEFFICIENT RANGES
                    OBJ COEFFICIENT RANGES
VARIABLE        CURRENT        ALLOWABLE       ALLOWABLE
                 COEF          INCREASE        DECREASE
   YA         100.000000       INFINITY          .000000
   YB         150.000000       INFINITY       150.000000
   YC         700.000000       INFINITY       700.000000
   YD         200.000000       INFINITY       100.000000
   YE         100.000000        .000000       100.000000
   YF         150.000000       INFINITY       150.000000
   YG         116.666700       INFINITY        16.666700
   X2           .000000         .000000       100.000000
   X3           .000000         .000000       100.000000
   X4           .000000         .000000       100.000000
   X5           .000000       100.000000       16.666700
   X6           .000000       100.000000        INFINITY


                    RIGHTHAND SIDE RANGES
   ROW          CURRENT        ALLOWABLE       ALLOWABLE
                 RHS           INCREASE        DECREASE
    2           9.000000        .000000         4.000000
    3           3.000000       8.000000         INFINITY
    4           2.000000       8.000000         INFINITY
    5           4.000000        .000000         4.000000
    6           7.000000        .000000         4.000000
    7           6.000000       6.000000         INFINITY
    8           9.000000        .000000         4.000000
    9           3.000000       INFINITY         3.000000
   10           2.000000       INFINITY         2.000000
   11           1.000000       INFINITY         1.000000
   12           2.000000       INFINITY         2.000000
   13           4.000000       INFINITY          .000000
   14           3.000000       INFINITY         3.000000
   15           6.000000       INFINITY         6.000000
   16          25.000000       4.000000          .000000
```

The solution is found by crashing activity E by 4 days at a cost of $400.

2. The network for this project is shown in Figure 8.5.

Figure 8.5

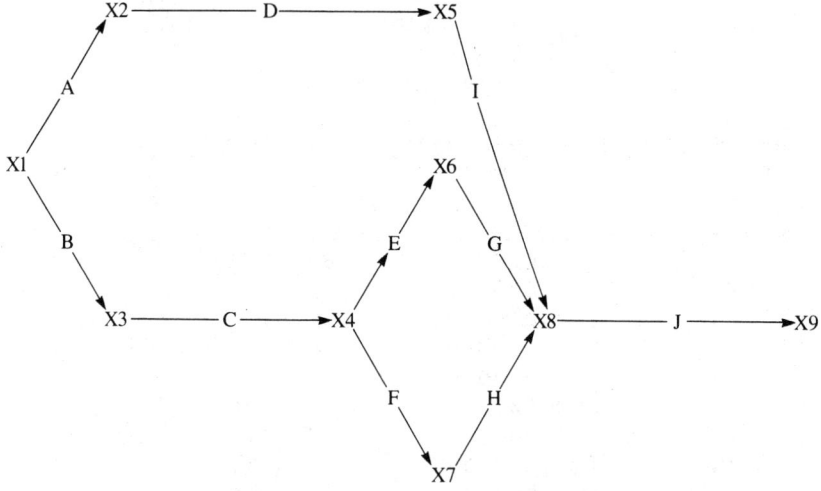

Activity	Mean Time	Variance	Max Crash	Cost/Unit Crash
A	2	0.11	1	100
B*	6	1.77*	2	150
C*	4	1 *	1	200
D	4	0.44	2	200
E	5	1	2	50
F*	3	0.11*	2	150
G	3	0.44	1	200
H*	8	2.77*	3	200
I	8	7.11	3	300
J*	1	—*	—	—

The LINDO file would be

```
: look all
 MIN 100 YA +  150 YB +  200 YC +  200 YD +  50 YE +
     150 YF +  200 YG +  200 YH +  300 YI
```

```
SUBJECT TO
2)  YA + X2 >= 2
3)  YB + X3 >= 6
4)  YC - X3 + X4 >= 4
5)  YD - X2 + X5 >= 4
6)  YE - X4 + X6 >= 5
7)  YF - X4 + X7 >= 3
8)  YG - X6 + X7 >= 3
9)  YH - X7 + X8 >= 8
10)  YI - X5 + X8 >= 8
11)  - X8 + X9 >= 1
12)  YA <= 1
13)  YB <= 2
14)  YC <= 1
15)  YD <= 2
16)  YE <= 2
17)  YF <= 2
18)  YG <= 1
19)  YH <= 3
20)  YI <= 3
21)  X9 <= 18
END


: go
LP OPTIMUM FOUND AT STEP 15
OBJECTIVE FUNCTION VALUE
1)  1400.00000

       VARIABLE          VALUE          REDUCED COST
           YA           .000000        100.000000
           YB          2.000000           .000000
           YC          1.000000           .000000
           YD           .000000        200.000000
           YE          2.000000           .000000
           YF           .000000        150.000000
           YG          1.000000           .000000
           YH          3.000000           .000000
           YI           .000000        300.000000
           X2          2.000000           .000000
           X3          4.000000           .000000
           X4          7.000000           .000000
           X5          6.000000           .000000
           X6         10.000000           .000000
           X7         12.000000           .000000
           X8         17.000000           .000000
           X9         18.000000           .000000
```

(continued)

```
        ROW        SLACK OR SURPLUS      DUAL PRICES
         2)              .000000            .000000
         3)              .000000        -200.000000
         4)              .000000        -200.000000
         5)              .000000            .000000
         6)              .000000        -200.000000
         7)             2.000000            .000000
         8)              .000000        -200.000000
         9)              .000000        -200.000000
        10)             3.000000            .000000
        11)              .000000        -200.000000
        12)             1.000000            .000000
        13)              .000000          50.000000
        14)              .000000            .000000
        15)             2.000000            .000000
        16)              .000000         150.000000
        17)             2.000000            .000000
        18)              .000000            .000000
        19)              .000000            .000000
        20)             3.000000            .000000
        21)              .000000         200.000000
```

```
NO. ITERATIONS= 15
DO RANGE (SENSITIVITY) ANALYSIS? > y


RANGES IN WHICH THE BASIS IS UNCHANGED
                  OBJ COEFFICIENT RANGES
VARIABLE        CURRENT         ALLOWABLE         ALLOWABLE
                  COEF          INCREASE          DECREASE
    YA        100.000000        INFINITY        100.000000
    YB        150.000000       50.000000         INFINITY
    YC        200.000000        INFINITY          .000000
    YD        200.000000        INFINITY        200.000000
    YE         50.000000      150.000000         INFINITY
    YF        150.000000        INFINITY        150.000000
    YG        200.000000         .000000         INFINITY
    YH        200.000000         .000000         INFINITY
    YI        300.000000        INFINITY        300.000000
    X2          .000000       100.000000          .000000
    X3          .000000        INFINITY         50.000000
    X4          .000000         .000000          INFINITY
    X5          .000000       100.000000          .000000
    X6          .000000         .000000          INFINITY
    X7          .000000         .000000          INFINITY
    X8          .000000       200.000000         INFINITY
    X9          .000000       200.000000         INFINITY
```

ROW	CURRENT RHS	RIGHTHAND SIDE RANGES ALLOWABLE INCREASE	ALLOWABLE DECREASE
2	2.000000	3.000000	2.000000
3	6.000000	.000000	1.000000
4	4.000000	.000000	1.000000
5	4.000000	3.000000	6.000000
6	5.000000	.000000	1.000000
7	3.000000	2.000000	INFINITY
8	3.000000	.000000	1.000000
9	8.000000	.000000	1.000000
10	8.000000	3.000000	INFINITY
11	1.000000	.000000	1.000000
12	1.000000	INFINITY	1.000000
13	2.000000	1.000000	.000000
14	1.000000	INFINITY	.000000
15	2.000000	INFINITY	2.000000
16	2.000000	1.000000	.000000
17	2.000000	INFINITY	2.000000
18	1.000000	1.000000	.000000
19	3.000000	1.000000	.000000
20	3.000000	INFINITY	3.000000
21	18.000000	1.000000	.000000

The solution with the minimum cost, in order to have a mean completion time of 18 weeks, is

Crash Activity	by	Weeks
B		2
C		1
E		2
G		1
H		3

The minimum cost is $1,400.

3. The network representing this project is shown in Figure 8.6.

Figure 8.6

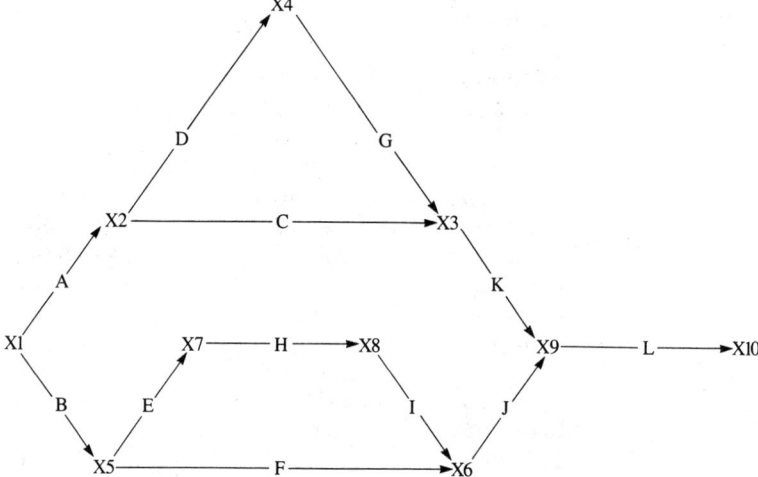

Activity	Mean Time	Variance	Max Crash	Cost/Unit Crash
A	8	0.11	2	100
B*	10	0.44*	3	150
C	7	1.78	1	150
D	5	0.44	2	200
E*	6	0.11*	4	50
F	9	7.11	3	100
G	5	1	2	300
H*	2	0.11*	1	100
I*	5	1 *	3	150
J*	9	2.78*	4	200
K	4	0.11	3	50
L*	3	0.44*	1	40

Variance of total project is the sum of the variance of the critical path activities:

$$\text{Project variance} = 0.44 + 0.11 + 0.11 + 1 + 2.78 + 0.44$$
$$= 4.88$$

Project standard deviation = 2.209

The project expected time, under normal conditions, is 35 weeks. To reduce the time to 30 weeks, then, the variance may be estimated to be reduced by the same ratio as the mean of the distribution is reduced.

For a 95 percent chance of project completion we require 1.645 standard deviations from the mean, thus

$$30 - X = \frac{1.645 \times 2.209}{35} X$$

or

$$X = 27.18$$

Working to integer weeks, we require X = 27. The LINDO input file is therefore:

```
MIN
100 YA + 150 YB + 150 YC + 200 YD + 50 YE + 100 YF
  + 300 YG + 100 YH + 150 YI + 200 YJ + 50 YK
  + 40 YL
ST
YA + X2 > 8
YB + X5 > 10
YC - X2 + X3 > 7
YD - X2 + X4 > 5
YE - X5 + X7 > 6
YF - X5 + X6 > 9
YG - X4 + X3 > 5
YH - X7 + X8 > 2
YI - X8 + X6 > 5
YJ - X6 + X9 > 9
YK - X3 + X9 > 4
YL - X9 + X10 > 3
YA < 2
YB < 3
YC < 1
YD < 2
YE < 4
YF < 3
YG < 2
YH < 1
YI < 3
YJ < 4
YK < 3
YL < 1
X10 < 27
END
```

4. This project network is shown in Figure 8.7.

Figure 8.7

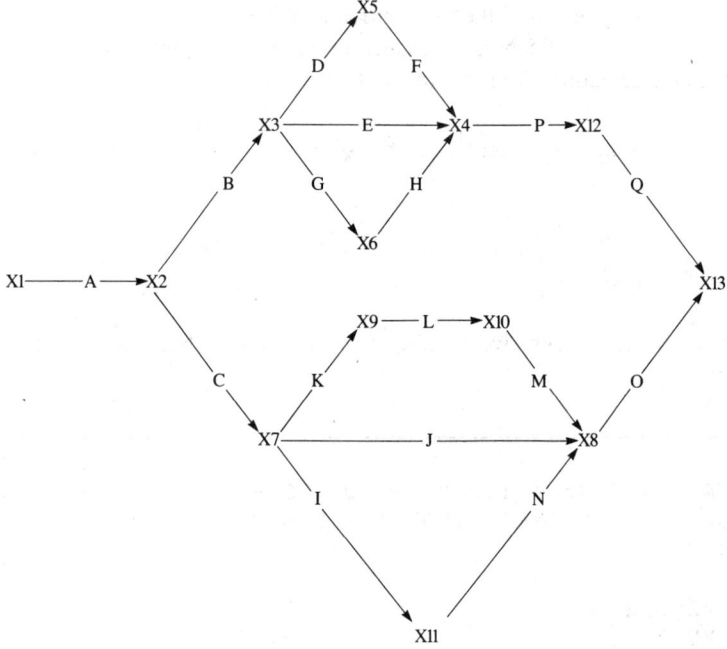

Activity	Mean Time	Variance	Max Crash	Cost/Unit Crash
A	9	0.44*	2	100
B	8	0.11*	4	200
C	4	0.11	2	150
D	3	0.44*	1	300
E	6	1.78	4	250
F	7	7.11*	5	150
G	5	1.77	1	50
H	4	0.11	2	100
I	4	0.44	1	100
J	2	0.11	1	200
K	6	1.77	3	250
L	5	0.11	2	300
M	4	1	2	150
N	9	2.77	6	100
O	8	7.11	6	150
P	7	0.11*	2	200
Q	6	1.77*	2	200

Project total variance = 9.998
Standard deviation = 3.159

For a 90 percent chance of completion of the project, we require 1.285 standard deviations:

$$35 - X = \frac{1.285 \times 3.159}{40} X$$

or

$$X = 31.775 \text{ weeks.}$$

The LINDO file is therefore:

```
MIN
100 YA + 200 YB + 150 YC + 300 YD + 250 YE
  + 150 YF + 50 YG + 100 YH + 100 YI + 200 YJ
  + 250 YK + 330 YL + 150 YM + 100 YN + 150 YO
  + 200 YP + 200 YQ
ST
YA + X2 > 9
YB - X2 + X3 > 8
YC - X2 + X7 > 4
YD - X3 + X5 > 3
YE - X3 + X4 > 6
YF - X5 + X4 > 7
YG - X3 + X6 > 5
YH - X6 + X4 > 4
YI - X7 + X11 > 4
YJ - X7 + X8 > 2
YK - X7 + X9 > 6
YL - X9 + X10 > 5
YM - X10 + X8 > 4
YN - X11 + X8 > 9
YO - X8 + X13 > 8
YP - X4 + X12 > 7
YQ - X12 + X13 > 6
```

(continued)

```
YA < 2
YB < 4
YC < 2
YD < 1
YE < 4
YF < 5
YG < 1
YH < 2
YI < 1
YJ < 1
YK < 3
YL < 2
YM < 2
YN < 6
YO < 6
YP < 2
YQ < 2
X13 < .31
END
```

5. This project is represented by Figure 8.8.

Figure 8.8

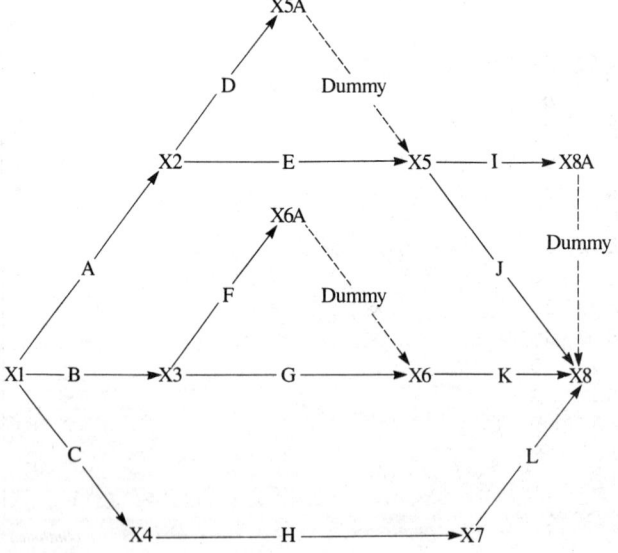

Activity	Mean Time	Variance	Max Crash	Cost/Unit Crash
A	9	0.11	4	100
B	8	0.44*	3	150
C	6	1.77	3	200
D	5	0.44	2	250
E	4	0.11	2	200
F	6	1.77*	3	100
G	3	0.11	1	100
H	2	0.11	1	50
I	1	—	—	—
J	7	2.77	5	300
K	14	2.77*	8	150
L	3	0.11	1	100

Project total variance = 4.98

Standard deviation = 2.23

For a 99 percent chance of project completion, 2.32 standard deviations are required:

$$X = 21.088$$

The LINDO input file is therefore:

```
MIN
100 YA + 150 YB + 200 YC + 250 YD + 200 YE
 + 100 YF + 100 YG + 50 YH + 300 YJ + 150 YK
 + 100 YL
ST
YA + X2 > 9
YB + X3 > 8
YC + X4 > 6
YD - X2 + X5A > 5
D1 - X5A + X5 = 0
YE - X2 + X5 > 4
YF - X3 + X6A > 6
D2 - X6A + X6 = 0
YG - X3 + X6 > 3
YH - X4 + X7 > 2
D3 - X8A + X8 = 0
YI - X5 + X8A > 1
YJ - X5 + X8 > 7
YK - X6 + X8 > 14
YL - X7 + X8 > 3
YA < 4
YB < 3
YC < 3
YD < 2
YE < 2
YF < 3
YG < 1
YH < 1
YI = 0
YJ < 5
YK < 8
YL < 1
X8 < 21,088
END
```

9

Integer
Programming

In the calculations carried out in Chapters 1 through 8, there were no requirements that solutions should be integers or whole number values. The structure of some special cases, such as the formulations used for the transportation and assignment problems, were such that integer-valued solutions resulted. Also, the formulation used to optimize the CPM project network was structured in such a way that integer crash times resulted. In most cases, however, a formulation for a linear program will result in a noninteger solution. The reason for this becomes clear on considering that the intersection of constraint boundaries will not, in general, occur neatly at whole numbers. While little difficulty occurs with variables that are continuous, such as tons of product to be shoveled or poured, variables such as ships, bridges, aircraft, and the like cannot be divided. It was stressed in Chapter 1 that the requirements for the applicability of linear programming were that the variables be continuous and there be no problem in manufacturing or supplying fractional quantities of materials. However, even in situations where these restrictions are not met, linear programming may be used to some advantage, for example when optimizing the deployment of factories or the design of a bridge. A noninteger solution may then indicate to management the possibility of the existence of a more optimal, previously unconsidered option.

Unfortunately, in many cases this is not satisfactory and a true optimal integer solution must be obtained. Mathematically stated,

$$\sum_{i=1}^{n} C_i X_i$$

over

$$\sum_{i=1}^{n} A_{ij} X_i = b_j \qquad j = 1 \text{ to } m$$

with X_i being nonnegative integers.

Let us return to the example of the edible fats manufacturer. In the original problem, the solution obtained did in fact contain integer values of the products. If, however, the profit margins on the two products were different, say $1 per ton and $3 per ton, the objective function would then be represented as

$$\text{Max } Z = X1 + 3 X2$$

The solution of this problem would then be discovered at the intersection of the contraints

$$1 X1 + 4 X2 = 12 \text{ and } 1 X1 + 2 X2 = 7$$

This will be depicted by the simplex equation formulation

Equation Set 9.1

			X4	X5	
X0	=	−9.5	0.5	0.5	(i)
X3	=	1.5	1.5	−0.5	(ii)
X1	=	2	−2	1	(iii)
X2	=	2.5	0.5	−0.5	(iv)

This solution to the problem indicates a noninteger profit and calls for noninteger amounts of X2 (tons of margarine) to be produced. If this had not been tons of margarine, but number of ships, then the solution as indicated would be unacceptable and a true integer solution would be sought. Consider Equation 9.1(iv), which states that

$$X2 = 2.5 + 0.5\ X4 - 0.5\ X5$$

Because we now insist on integer values, we can rewrite it as two equations, one having only integer values, the other collecting all the fractional parts.

X2 =	2	+ 1	X4	− 1	X6		(i)	**Equation Set 9.2**
X6 =	−0.5	+ 0.5	X4	+ 0.5	X5		(ii)	

Writing this integer equation for X2, together with the fractional equation, into the solution set of equations results in

X0 =	−9.5 + 0.5 X4 + 0.5 X5	(i)	**Equation Set 9.3**
X3 =	1.5 + 1.5 X4 − 0.5 X5	(ii)	
X1 =	2 − 2 X4 + 1 X5	(iii)	
X2 =	2 + 1 X4 − X6	(iv)	
X6 =	−0.5 + 0.5 X4 + 0.5 X5	(v)	

This representation of a solution point is, by the rules of the regular simplex procedure, an infeasible solution, for it contains a negative variable in the current basis. Equation Set 9.3 may be pivoted to restore feasibility by removing variable X6 from the basis and replacing it with variable X4.

		X5	X6		**Equation Set 9.4**
X0	=	−9	0	1	(i)
X3	=	3	−2	3	(ii)
X1	=	0	3	−4	(iii)
X2	=	3	−1	1	(iv)
X4	=	1	−1	2	(v)

This restores feasibility to the equation set and has resulted in a solution equivalent to point E on Figure 1.1. In addition to being a feasible solution, it is also optimal. Moreover, it is also integer. Note that the cost of forcing an integer solution has resulted in a loss of profit from the previously obtained $9.50 to a new profit of $9.00.

This approach is a version of the cutting-plane algorithm. The new restriction imposed by rewriting Equation 9.1(iv) reduces the feasible region by cutting off a small part of the feasible region that does not contain any integer point values (Figure 9.1). The cutting plane (a line in the two-dimensional problem representation) is

$$X6 = -0.5 + 0.5 X4 + 0.5 X5$$

Substituting for X4 and X5 in terms of X1 and X2, the expression

$$X6 = 9 - X1 - 3 X2$$

Figure 9.1

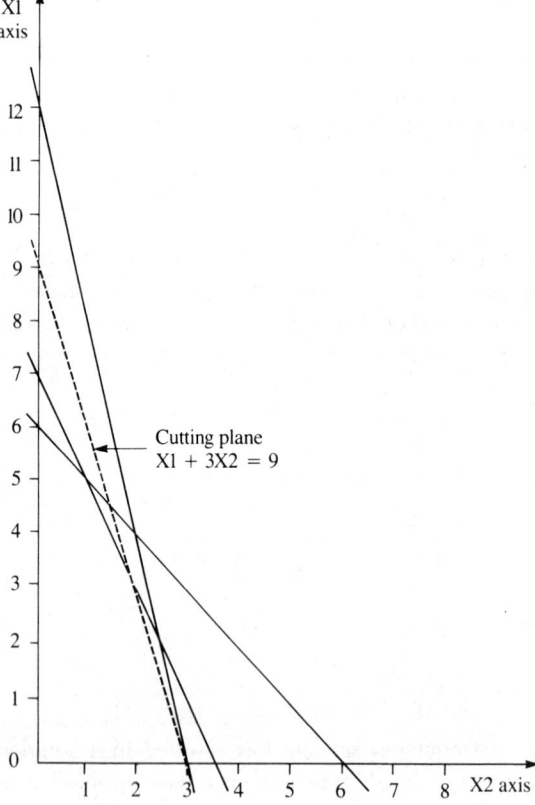

is obtained, as shown in Figure 9.1. Incidentally, the figure also indicates an alternative integer solution to this problem at the point X1 = 3, X2 = 2, with the same profit of $9.00. This is obtained by substituting X5 for X6 instead of X4 for X6 when reducing the infeasibility of the Equation Set 9.3. The simplex equation set that results is

			X4	X5	
X0	=	−9	0	1	(i)
X3	=	1	2	−1	(ii)
X1	=	3	−3	2	(iii)
X2	=	2	1	−1	(iv)
X5	=	1	−1	2	(v)

Equation Set 9.5

TREE TECHNIQUES

The cutting-plane method is by no means the only method of discovering integer solutions to constrained linear optimization problems. Indeed, mathematicians spend much time attempting to discover integer solutions to equations. In the context of discovering integer solutions within a feasible region, useful techniques for practical situations have been developed based on a *tree approach*. The most successful methods are those of "branch and bounds" and "partial enumeration."

Branch and Bounds

The practical decision maker may, at this stage, wonder if the benefit gained by finding a true integer solution merits the considerable effort expended in extra computation over merely rounding off the standard LP solution. Where the solution variables take large values and the gradient of the objective function is not too steep, rounding off will give a solution not far removed from the true optimal value. Where the variables take small values, and thus relatively few alternatives need to be considered, then rounding down the LP solution can be far from the true optimum, and a technique such as the cutting plane or tree becomes attractive.

We have previously considered the CPM/PERT technique, which employs a graph or chart to depict a project as a set of nodes and branches connected together. That chart allowed branches to diverge from a node, and also allowed several branches to converge at a node. With a tree technique, a similar approach of nodes and branches is used, but the tree is an open graph that allows branches to diverge from nodes, but not to converge, as shown in Figure 9.2.

Figure 9.2

Start

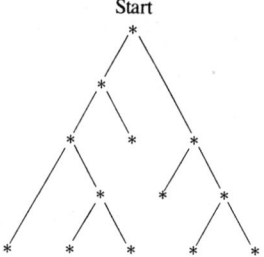

In the above tree structure, not all the branches are the same length. Some of the branches end in nodes and are not developed further. This structure is used in the branch and bounds approach. Consider the problem solved by the cutting-plane technique:

Maximize $X1 + 3 X2$
subject to $X1 + X2 \leqslant 6$
                 $X1 + 2 X2 \leqslant 7$
                 $X1 + 4 X2 \leqslant 12$
                 $X1, X2 \geqslant 0$ $X1, X2$ integer

The LP solution to this problem is not integer valued; its simplex equation format is

Equation Set 9.6

			X4	X5
X0	=	−9.5	0.5	0.5
X3	=	1.5	1.5	−0.5
X1	=	2	−2	1
X2	=	2.5	0.5	−0.5

The solution value of variable X1 is already an integer. We wish to discover a solution value for X2 that is also an integer. We should specify bounds for the two variables that encompass the feasible region. However, for clarity of illustration, only the single bound on the X2 variable will be stated:

$$0 \leqslant X2 \leqslant 3$$

Equation Set 9.6 depicts a value of 2.5 for X2. In seeking an integer value, we can split up the feasible region by removing the section that lies between $X2 = 2$ and $X2 = 3$, because no integer values of X2 will

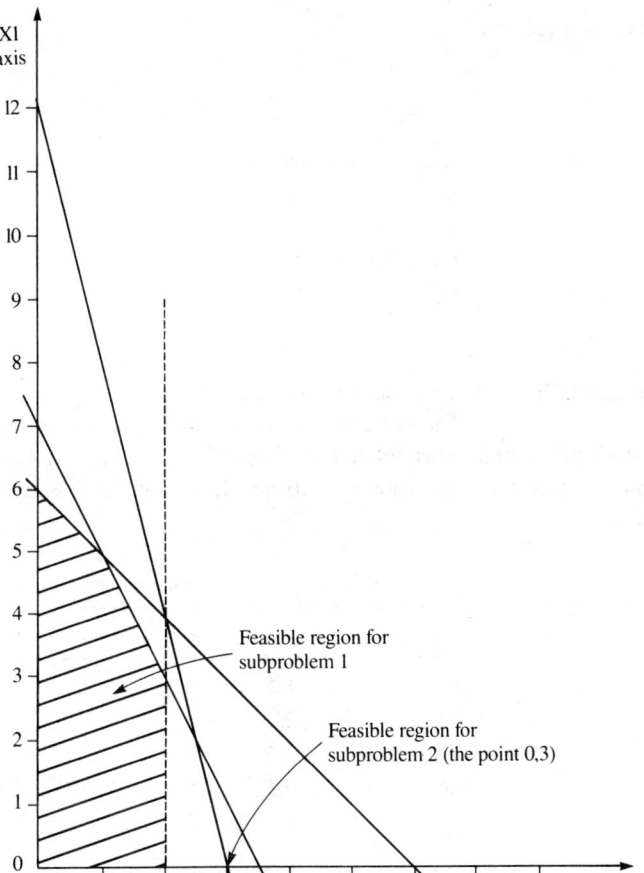

Figure 9.3

Feasible region for
subproblem 1

Feasible region for
subproblem 2 (the point 0,3)

be found inside this band. We can therefore force the method to seek the best solution that lies below $X2 = 3$ and within the bound on X2, and also within the separated region that lies above $X2 = 3$ and within the bound on X2. These two regions are depicted in Figure 9.3. The equations for the two subproblems covering the two regions are as follows. For subproblem 1,

Equation Set 9.7

			X1	X2
X0	=	0	-1	-3
X3	=	6	-1	-1
X4	=	7	-1	-2
X5	=	12	-1	-4
X6	=	2		-1

For subproblem 2,

Equation Set 9.8

			X1	X2
X0	=	0	− 1	− 3
X3	=	6	− 1	− 1
X4	=	7	− 1	− 2
X5	=	12	− 1	− 4
**X2	=	3		**

Equation Set 9.8 represents the single point $X2 = 3$.

The structure of the solution thus far is that of splitting a single problem into two related problems. Consider each of these subproblems. Substituting the expression for X6 into Equation Set 9.6 gives, for subproblem 1,

Equation Set 9.9

			X4	X5
X0	=	− 9.5	0.5	0.5
X3	=	1.5	1.5	− 0.5
X1	=	2	− 2	1
X2	=	2.5	0.5	− 0.5
X6	=	− 0.5	− 0.5	0.5

As we would expect, this results in an infeasible formulation, because this section has been excluded from the feasible region. Pivoting X6 and X5 restores feasibility and results in the optimal solution, which is also integer:

Equation Set 9.10

			X4	X6
X0	=	− 9	1	1
X3	=	1	1	− 1
X1	=	3	− 1	2
X2	=	2		− 1
X5	=	1	1	2

Subproblem 1 has resulted in a solution that is an integer and has an objective function value of 9. Because the noninteger solution to the LP problem has an objective function value of 9.5, we know that no integer solution can improve on 9. (This is only because we also have integer values in the coefficients of the objective function.) We have not, however, explored the other arm of the tree. A further solution might be found there. In general, arms of the tree are investigated until an integer solution is discovered or until a noninteger solution is discovered with a worse value of the objective function than a previously discovered integer solution. At that point the search is terminated and the best integer solution is optimal.

Consider subproblem 2. This rewrites as

			X1	**
X0	=	0	−1	−3(3)
X3	=	6	−1	−1(3)
X4	=	7	−1	−2(3)
X5	=	12	−1	−4(3)
X2	=	3		

Equation Set 9.11

or

			X1
X0	=	−9	−1
X3	=	3	−1
X4	=	1	−1
X5	=	0	−1
X2	=	3	

Equation Set 9.12

This point solution also has the objective function value of 9, so that there are two equal-valued solutions. This tree is illustrated in Figure 9.4.

Figure 9.4

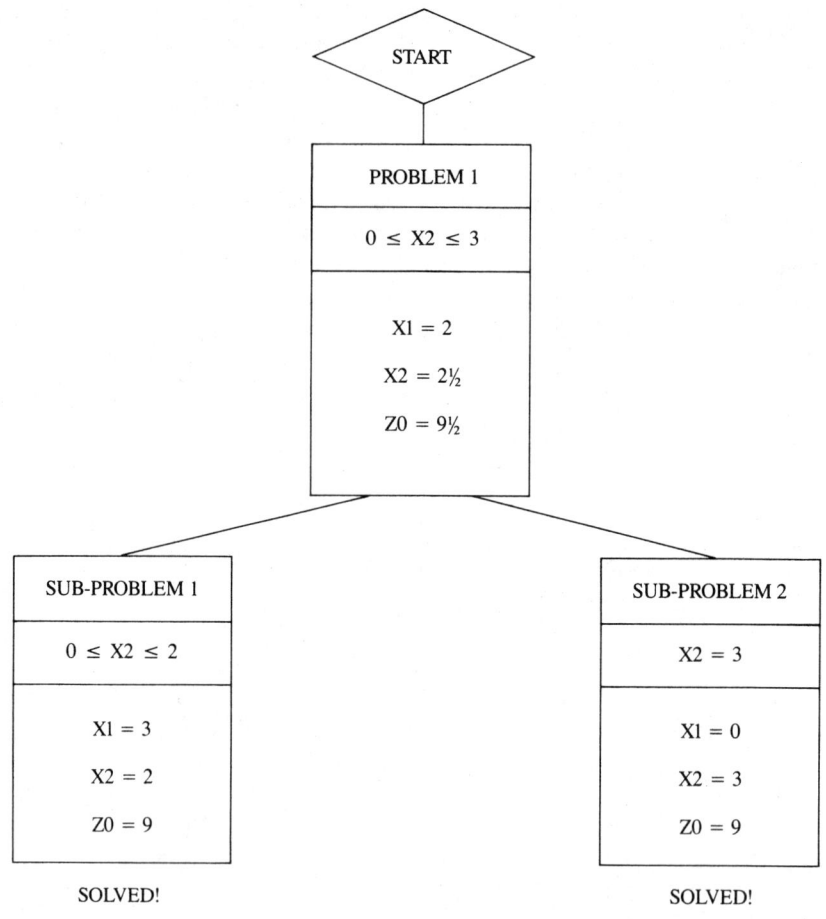

In a larger problem than the one illustrated, the best integer solution to a subproblem is kept for reference as the "best solution to date." It is subject to review as the calculation proceeds; any solution worse than it is rejected and further examination of the subproblem is terminated. Any noninteger solution better than the best integer solution to date is split into further subproblems, like the original problem. The procedure ends when no further splits can be made and the best integer solution to date is thus optimal.

The computation involved with the branch and bounds method in larger problems varies considerably with the choice of subproblems. In cases where the tree is large, the number of subproblems to be investigated becomes too large for this to be a practical proposition.

0 1 Partial Enumeration

A variation on the branch and bounds method is that of 0 1 partial enumeration. This is useful when the integer requirements are of a 0 or 1 variety. This is a particularly attractive technique for financial management, in cases where the goal is to compute the optimum allocation of funds to a series of projects, where projects are either accepted ($Xi = 1$) or rejected ($Xi = 0$). For capital budgeting decisions, this method drastically reduces the computation time in comparison with other methods. In fact, it is possible to transform other integer problems into this form to take advantage of its computational savings.

In order to use the technique, a nonnegative integer variable X, having an upper bound U, can be replaced by a set of variables, Yi ($i = 1$ to k), each of which may take the values 0 or 1:

$$X = 1\ Y1 + 2\ Y2 + 4\ Y3 + \ldots + 2^{k-1}\ Yk$$

where k is chosen such that $2^k - 1 \geqslant U$. For our example,

$$0 \leqslant X1 \leqslant 7 \qquad X1 = 1\ Y1 + 2\ Y2 + 4\ Y3$$
$$0 \leqslant X2 \leqslant 3 \qquad X2 = 1\ Y4 + 2\ Y5$$

The integer problem faced by the edible fats manufacturer then becomes, in terms of 0,1 variables,

Maximize	$1\ Y1 + 2\ Y2 + 4\ Y3 + 3\ Y4 + 6\ Y5$
subject to	$1\ Y1 + 2\ Y2 + 4\ Y3 + 1\ Y4 + 2\ Y5 \leqslant 6$
	$1\ Y1 + 2\ Y2 + 4\ Y3 + 2\ Y4 + 4\ Y5 \leqslant 7$
	$1\ Y1 + 2\ Y2 + 4\ Y3 + 4\ Y4 + 8\ Y5 \leqslant 12$

This method, for reasons which will become clearer later, is normally applied as a minimization problem, rather than maximization. Therefore, for convention

Maximize $1\ Y1 + 2\ Y2 + 4\ Y3 + 3\ Y4 + 6\ Y5$

becomes

Minimize $-1\ Y1 - 2\ Y2 - 4\ Y3 - 3\ Y4 - 6\ Y5$

This optimization function now contains negative coefficients. Unfortunately, we require a function that is monotonically decreasing (has no negative coefficients). We therefore adopt the transformation

$$Yi = 1 - Zi$$

which results in

Minimize $1\,Z1 + 2\,Z2 + 4\,Z3 + 3\,Z4 + 6\,Z5 - 16$
subject to $1\,Z1 + 2\,Z2 + 4\,Z3 + 1\,Z4 + 2\,Z5 - 10 \geqslant 6$
              $1\,Z1 + 2\,Z2 + 4\,Z3 + 2\,Z4 + 4\,Z5 - 13 \geqslant 7$
              $1\,Z1 + 2\,Z2 + 4\,Z3 + 4\,Z4 + 8\,Z5 - 19 \geqslant 12$
              $Zi = 0$ or $Zi = 1$

or

Minimize $1\,Z1 + 2\,Z2 + 4\,Z3 + 3\,Z4 + 6\,Z5\,(-16)$
subject to $1\,Z1 + 2\,Z2 + 4\,Z3 + 1\,Z4 + 2\,Z5 - 4 \geqslant 0$
              $1\,Z1 + 2\,Z2 + 4\,Z3 + 2\,Z4 + 4\,Z5 - 6 \geqslant 0$
              $1\,Z1 + 2\,Z2 + 4\,Z3 + 4\,Z4 + 8\,Z5 - 7 \geqslant 0$
              $Zi = 0$ or $Zi = 1$

To solve this problem, the best value of the objective function discovered to date is set at infinity. We could use an exhaustive enumeration technique where all the possible combinations of 0,1 variables are calculated, with the value of the optimization function evaluated for each combination. For this case there are $2^5 = 32$ solutions to the problem. However, not all of the possible combinations need to be evaluated, because use may be made of partial vector ordering. For a graphical representation see Figure 9.5. The set of 32 possible solutions has only 18 solutions within the feasible region. Figure 9.5 also shows that the vector ordering of the possible integer solutions in the form Z1, Z2, Z3, Z4, Z5 does not show a monotonically increasing objective function (in other words, the pattern of the vector values is not a simple one). Rewriting the vector ordering in the form Z5, Z4, Z3, Z2, Z1, however, produces Figure 9.6. (Note that the matrix of elements of Z is the transpose of the elements of Y.) Examination of Figure 9.6 shows the point of the apparently devious mathematical manipulation and the reason for rewriting the problem.

The point 0,0,0,0,0 in the binary Z5,Z1 coordinates is the same as the point 7,3 in the X1,X2 coordinates. Furthermore, counting down

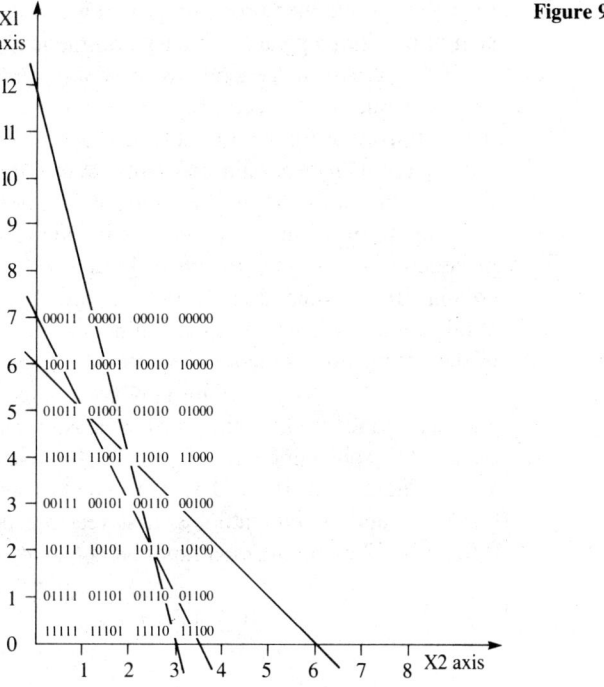

Figure 9.5

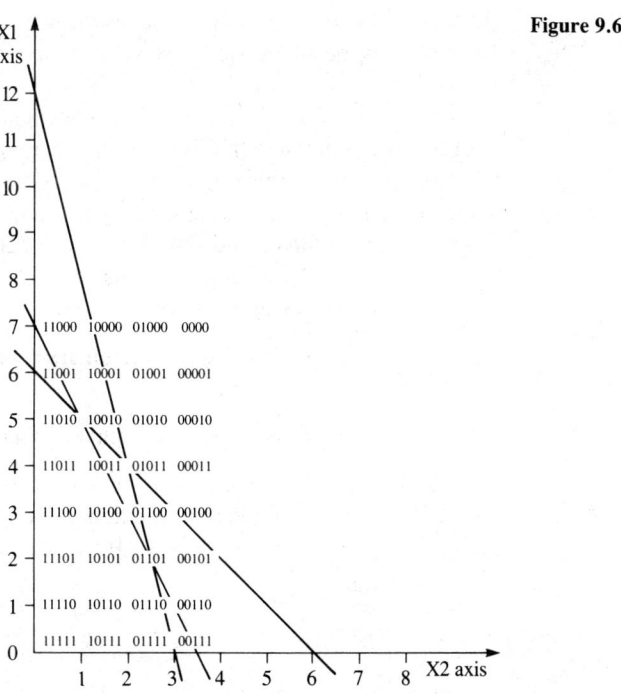

Figure 9.6

the columns and then proceeding by rows from 0,0,0,0,0 indicates a progression of counting upward in binary arithmetic (0,1,10,11,100,101, etc.).

In general, we have ensured that by increasing the binary number value of the vector representing the coordinates of any integer point, the optimization function is not increased. (We cannot say that it would decrease, for it may remain constant.) While this binary number value of an integer solution will not increase the optimization function, there is the problem of changing from one vector progression to the next; in other words, at the end of counting down one column of Figure 9.6, one starts at the top of a new column. In the general case, this is a new dimension or variable. Obviously, at this point, there will be a discontinuous "jump" in the downward evaluation of the optimization function, and it could increase.

A procedure may be adopted as follows: Start at 0,0,0,0,0 and determine whether the point is feasible. In our example, this point is outside the constraint set and so it is eliminated from further consideration. Try the next point 0,0,0,0,1: This point is again infeasible. Try the next, 0,0,0,1,0 and so on until the first feasible point is reached, in this case 0,0,1,1,1. The optimization function is then evaluated as -9. That is,

$$1\,Z1 + 2\,Z2 + 4\,Z3 + 3\,Z4 + 6\,Z5 - 16$$
$$= 1\,(1) + 2\,(1) + 4\,(1) + 3\,(0) + 6\,(0) - 16$$
$$= -9$$

As this is the first feasible point discovered, the value of the optimization function now becomes the "best value found to date." The search can then proceed.

It is important to decide when a new point might signify the start of a new column. (In this case, the next point does; see Figure 9.6). In a more complex case, this might not be so easy to spot, but some reflection will show the reader that when the next highest power of 2 in the binary representation is reached, this could signify a new direction.

An algebraic means of discovering this possible new point is to adopt the following method:

1. Write down the binary representation, in this case 00111.

2. Subtract 1 via binary arithmetic: $00111 - 1 = 00110$.

3. Compare this number with the original number, and write a third number that has a zero where *both* other numbers had a 0 and a 1 everywhere else:

```
00111    Original
00110    Original − 1
─────
00111    New number
```

4. Add 1 to this new number: $00111 + 1 = 01000$

This is a possible new column header number, so check its feasibility. In our example, this is in fact a new column header, and is checked against the constraint set and discovered to be infeasible. Thus far, we have discovered only one feasible point and evaluated it. That is our optimum to date.

We proceed as before, testing points until a feasible point is reached. In this example, the next point is 01100. The optimization function is evaluated at this point:

$$1\,Z1 + 2\,Z2 + 4\,Z3 + 3\,Z4 + 6\,Z5 - 16$$
$$= 1\,(0) + 2\,(0) + 4\,(1) + 3\,(1) + 6\,(0) - 16$$
$$= -9$$

This is the same value as the optimum to date, so the two solutions are maintained as the best solutions to date. It is known that this point has the best optimization function value of all the points in this column inside the feasible region, so that further study of the branch of the tree representing this process can be terminated. It therefore remains to discover the next possible new column header.

```
                01100
         −          1
                ─────
                01011
Compare         01100
                ─────
                01111
         +          1
                ─────
New number      10000
```

Using this process the points 01101, 01110, and 01111 are skipped from consideration. The next point investigated is 10000. This is found to be infeasible. The next feasible point is 10010. The optimization function evaluation of 10010 is -8. This is less than the two previously discovered optimal points, so it is rejected. The binary skipping procedure is again invoked:

$$
\begin{array}{r}
10010 \\
- \quad 1 \\
\hline
10001
\end{array}
$$

$$
\text{Compare} \quad
\begin{array}{r}
10010 \\
\hline
10011 \\
+ \quad 1 \\
\hline
\end{array}
$$

New number   10100

This might have been a new column header, but not in this example. As it checks out as feasible, it is evaluated. (Note, it is evaluated because we might have a new column header, and it must be a feasible point. We cannot assume that because the next sequential point is feasible, it is proceeding further into the feasible region along the same column.) Its value is -6 and it is rejected. A new point for investigation is sought,

$$
\begin{array}{r}
10100 \\
- \quad 1 \\
\hline
10011
\end{array}
$$

$$
\text{Compare} \quad
\begin{array}{r}
10100 \\
\hline
10111 \\
+ \quad 1 \\
\hline
\end{array}
$$

New number   11000

This is a new column header, and is infeasible. The process is repeated until point 11111 is reached.

At the completion of the process, two optimal solution points have been reached:

Z5	Z4	Z3	Z2	Z1	X1	X2
0	1	1	0	0	3	2
0	0	1	1	1	0	3

Both have an optimization function value of $9.00.

While this method might appear long-winded and tedious, it does have three tremendous advantages:

1. No computations other than additions and subtractions are involved.

2. It is easily implemented on a digital computer.

3. It may be used on nonlinear formulations.

The 0,1 formulation is useful, as some computer routines will handle 0,1 variables but have more difficulty with decimal-ranged variables.

COMPUTER SOLUTION

LINDO has an option that allows variables restricted to values 0 or 1 to be identified by the INTEGER command. This takes either of two alternative forms: INTEGER (variable-name) or INTEGER n.

The first alternative, the recommended form, identifies the variable by name, the other identifies the first n variables as 0,1 integers. Although this second form is more powerful, it requires the user to be aware of the exact order in which the variables are encountered in the input to LINDO.

The integer restriction on the variables is removed by the command INTEGER 0. The solution method used is a branch and bound approach. Since the variables are defined as 0,1, problems should be specified in this form. For the example used in this chapter, prepared by a microcomputer or other editor and placed on file, we would use

```
MAX  Y1 + 2 Y2 + 4 Y3 + 3 Y4 + 6 Y5
SUBJECT TO
2) Y1 + 2 Y2 + 4 Y3 + Y4 + 2 Y5 <= 6
3) Y1 + 2 Y2 + 4 Y3 + 2 Y4 + 4 Y5 <= 7
4) Y1 + 2 Y2 + 4 Y3 + 4 Y4 + 8 Y5 <= 12
END
```

We decide to use the second of the two optional integer definitions here:

```
: INTEGER 5
: GO

LP OPTIMUM FOUND AT STEP 3
OBJECTIVE FUNCTION VALUE
1)  9.5
     VARIABLE              VALUE           REDUCED COST
        Y1                  .0                 .0
        Y2                  .0                 .0
        Y3                  .5                 .0
        Y4                  .5                 .0
        Y5                 1.0                 .0


        ROW          SLACK OR SURPLUS     DUAL PRICES
        2)               1.5                  .0
        3)                .0                  .5
        4)                .0                  .5
NO. ITERATIONS= 3
BRANCHES= 0 DETERM.= 8.00E 0
SET Y4 TO 1 AT 1 BND= 9.5 TWIN= 9.0
SET Y3 TO 0 AT 2 BND= 9.5 TWIN= 9.5
SET Y5 TP 1 AT 3 BND= 9.0 TWIN= 9.0
NEW INTEGER SOLUTION AT BRANCH E PIVOT 9

OBJECTIVE FUNCTION VALUE
1)  9.0
     VARIABLE              VALUE           REDUCED COST
        Y1                  .0                 .0
        Y2                  .0                 .0
        Y3                  .0                 .0
        Y4                 1.0                1.0
        Y5                 1.0                2.0


        ROW          SLACK OR SURPLUS     DUAL PRICES
        2)               3.0                  .0
        3)               1.0                  .0
        4)                .0                 1.0
NO. ITERATIONS= 9
BRANCHES= 3 DETERM.= 2.00E 0
BOUND ON OPTIMUM: 9.500000
DELETE Y5 AT LEVEL 3
DELETE Y3 AT LEVEL 2
DELETE Y4 AT LEVEL 1
ENUMERATION COMPLETE. BRANCHES= 3 PIVOTS= 9
LAST INTEGER SOLUTION IS THE BEST FOUND
```

CHAPTER 9 PROBLEMS

The solutions to Problems 1 through 5 will be found at the end of the chapter.

1. Key Pit is a private security organization that is constantly seeking new security methods for its clients. A spinoff from aerospace development are the newly available drone patrolling devices, which fire upon unauthorized intruders. These are activated and home in on any person not carrying a current ID card equipped with the current IFF code.

Drone Operations Group (DOG) can supply two types of drone to Key Pit. These are JIM (Jolly Intruder Macerator) and FRED (Friendly Robot Elimination Device).

As a result of the popularity of the JIM drone with other organizations, DOG has only 11 JIM drones available to supply to Key Pit. However, Key Pit can have as many FRED drones as it needs.

Each JIM drone costs $30,000 and each FRED drone costs $20,000. A JIM drone can be operated and maintained by one person, but each FRED drone requires a crew of three. The state strictly controls the holding of drone missile launchers by individuals and private organizations, and Key Pit has obtained a state drone missile launcher license to hold and operate 15 of the devices.

Key Pit has a budget of $360,000 for drone operations, and DOG will train up to 27 Key Pit operatives.

If each JIM drone is assessed by Key Pit to be 14 percent more effective than the FRED drone, how should the controller of Key Pit allocate funds for drone purchase?

DOG has indicated that a new FRED II drone will shortly become available, at the same price, but will be two-and-a-half times as effective as the JIM drone. If this is true, how would it affect Key Pit's plans?

2. The El Toro Chips company has decided to produce portable nuclear power plants. One, intended for yachters, is water cooled, and the other, designed for campers, is air cooled.

The yachter's plant uses 30 pounds of cadmium and 3 radons of uranium. The camper's power plant incorporates 6 pounds of cadmium and 2 radons of uranium.

The building cost of the marine unit is $4,000 and the building cost of the land-based unit is $8,000. Either unit will sell for $12,000.

El Toro Chips has raised $72,000 for the project and an inventory of 130 pounds of cadmium and 24 radons of uranium.

For the initial trial, the company is confident that it can sell its initial production (at least 10 of each have been requested). How many of each unit should be produced with current inventory, and what profit is expected?

3. The solution to the party drinks problem of Chapter 1 (Problem 3) was not an exact integer. What is the true integer solution?

4. Use LINDO to discover the integer solution to the following problem:

Maximize   5 A + 3 B + 2 C
subject to   2.5 A   + 3.4 C < 6
                2 A + 6 B + 4 C < 10

A,B,C are 0,1 variables.

5. What is the best integer solution to the following formulation:

Maximize   0.6 A + 2.4 B + 4.5 C
subject to   12 A + 2 B + 6 C < 120
                15 A + 5 B + 3 C < 150
                              B   < 26
                       2 B +   C > 30

HOMEWORK PROBLEMS

6. Find the integer solution to

Maximize   8 X1 + 3 X2
subject to   X1 + 6 X2 ⩾ 60
                4 X1 + 2 X2 ⩽ 80
              − 18 X1 + 5 X2 ⩾ 108

7. What is the best integer solution to

Maximize   4 A + 5 B + 2 C
subject to   3 A +   B + 5 C < 6
                2 A + 2 B + 4 C < 10

A,B,C are all 0,1 variables.

8. What is the best integer solution to Problem 14, Chapter 1 (General Custard's advances)?

9. What is the best integer solution to the problem

Maximize $3 X1 + 2 X2 + 6 X3$
subject to $X1 + X2 + 4 X3 \leqslant 6$
                        $X1 + 2 X2 + X3 \leqslant 7$
                        $X1 \quad 4 X2 + 2 X3 \leqslant 12$

10. The Marines have been called on to transport 750 troops to secure a strategic section of country against foreign insurgents. The nearest base has two types of helicopter available for the task, the Sea Elephant and the Air Python. Each Sea Elephant can transport 200 men and equipment with a crew of 5; each Air Python can carry half that load with a crew of 2. The Sea Elephant requires 600 gallons of fuel and the Air Python needs 350 gallons. The base currently has 2,500 gallons of fuel in its bunkers. Unfortunately, due to an outbreak of Alaskan influenza, only 18 crewmembers are available to fly the helicopters. When the Marines arrive at their destination, the Sea Elephant is considered to be twice as useful as the Air Python in support of their activities. How should the airlift be effected?

CHAPTER 9 SOLUTIONS

1. The LP formulation is:

Maximize 1.14 JIM $+ 1.00$ FRED
subject to $3 \quad$ JIM $+ 2 \quad$ FRED $< 36 \quad$ Cash
                        $1 \quad$ JIM $+ 3 \quad$ FRED $< 27 \quad$ Operatives
                        $1 \quad$ JIM $\qquad\qquad < 11 \quad$ DOG stock
                        $1 \quad$ JIM $+ 1 \quad$ FRED $< 15 \quad$ License

The LP solution is

$\qquad$ 7.71428 $\quad$ JIM
$\qquad$ 6.42857 $\quad$ FRED

For the integer solution via partial enumeration, let

$\qquad$ FRED $= Z1 + 2 Z2 + 4 Z3 + 8 Z4$
$\qquad$ JIM $\; = Z5 + 2 Z6 + 4 Z7 + 8 Z8$

Then initial objective 1 is

Minimize

$Z1 + 2\ Z2 + 4\ Z3 + 8\ Z4 + 1.14\ Z5 + 2.28\ Z6 + 4.56\ Z7 + 9.12\ Z8$

and objective 2 is

Minimize

$2.5\ Z1 + 5\ Z2 + 10\ Z3 + 20\ Z4 + Z5 + 2\ Z6 + 4\ Z7 + 8\ Z8$

subject to

$$2\ Z1 + 4\ Z2 + 8\ Z3 + 16\ Z4 + 3\ Z5 + 6\ Z6 + 12\ Z7 + 24\ Z8 > 39$$
$$Z5 + 2\ Z6 + 4\ Z7 + 8\ Z8 > 4$$
$$Z1 + 2\ Z2 + 4\ Z3 + 8\ Z4 + Z5 + 2\ Z6 + 4\ Z7 + 8\ Z8 > 15$$
$$3\ Z1 + 6\ Z2 + 12\ Z3 + 2\ Z4 + Z5 + 2\ Z6 + 4\ Z7 + 8\ Z8 > 33$$

Feasible Solution $Z1 - Z8$	FRED	JIM	Obj 1	Obj 2
01101111	9	0	9	22.5
01111100	8	3	11.42	23
10001001	7	6	13.84	23.5*
10010111	6	8	15.12*	23
10100111	5	8	14.12	20.5
10110110	4	9	14.26	19
11000101	3	10	14.40	17.5
11010101	2	10	13.40	15
11100100	1	11	13.54	13.5
11110100	0	11	12.54	11

This solution indicates that 6 FRED drones and 8 JIM drones should be acquired under the initial conditions. However, should the new FRED II drone come up to the manufacturer's expectations, then Key Pit should revise its policy to 7 FRED II drones and 6 JIM drones.

It would therefore be sensible for Key Pit to order 6 JIM drones with an option on a further 2 JIM drones pending the outcome of the FRED II trials.

The solution using the cutting-plane technique is illustrated using the condensed tableaux of the simplex technique:

	X1	X2		
X3	3	2	36	
X4	1	3	27	
X5	1	0	11	*
X6	1	1	15	
X0	− 1.14	− 1	0	

*   D = 1

	X5	X2		
X3	− 3	2	3	*
X4	− 1	3	16	
X1	1	0	11	
X6	− 1	1	4	
X0	1.14	− 1	12.54	

*   D = 1

	X5	X3		
X2	− 3	1	3	
X4	7	− 3	23	*
X1	2	0	22	
X6	1	− 1	5	
X0	− 0.72	1	28.08	

*   D = 2

	X4	X3	
X2	3	− 1	45
X5	2	− 3	23
X1	− 2	3	54
X6	− 1	− 2	6
X0	0.72	2.42	106.56

D = 7

Rearranging the final tableau, eliminating D by dividing through and assembling the nonbasic variables to the right of the equality signs, gives

$$
\begin{array}{rrrrr}
 & & & X4 & X3 \\
X0 = & 15.22 & - & 0.103 & - & 0.346 \\
\hline
X2 = & 6.43 & + & 0.43 & + & 0.14 \\
X5 = & 3.29 & - & 0.29 & + & 0.43 \\
X1 = & 7.71 & + & 0.29 & - & 0.43 \\
X6 = & 0.86 & + & 0.14 & + & 0.29 \\
\end{array}
$$

Noting that the shadow price of X4 (0.103) is lower than that of X3, we see that it is less costly to move within the X4 than the X3 constraint. Hence, when reducing the solution to integers, we will come off the X4 constraint and send X4 positive, reducing X2 to the integer value 6 by means of the artificial variable X7.

This operation may be seen by breaking the equation for X2 into two parts:

$$
\begin{array}{rrrr}
X2 = & 6 & + & X3 - X7 \\
X7 = & -0.43 & + 0.86\ X3 & + 0.43\ X4 \\
\end{array}
$$

and

$$
X4 = \quad 1 \quad - 2 \quad X3 \quad + 2.33\ X7
$$

In this way X4 has become a basic variable and X7 a constraint variable, so, pivoting about X4, the tableau becomes:

	X7	X3	
X2	-1	1	6
X5	-0.66	1	3
X1	0.66	-1	8
X6	0.33	0	1
X4	2.33	-2	1
X0	0.24	0.14	15.12

$$D = 1$$

This solution is optimal because it states that all the basic variables are being produced at integer levels, the solution being

$$X1 = 8 \quad : \quad X2 = 6$$
$$\text{(JIM)} \qquad \text{(FRED)}$$

In the event that FRED II comes up to the manufacturer's expectation, then optimization function 2 will be appropriate:

	X1	X2		
X3	3	2	36	
X4	1	3	27	*
X5	1	0	11	
X6	1	1	15	
X0	− 1	− 2.5	0	

*   D = 1

	X1	X4		
X3	7	− 2	54	*
X2	1	1	27	
X5	3	0	33	
X6	2	− 1	18	
X0	− 0.5	2.5	0	

*   D = 3

	X3	X4	
X1	3	− 2	54
X2	− 1	3	45
X5	− 3	2	23
X6	− 2	− 1	6
X0	0.5	16.5	166.5

D = 7

The solution is:

X1 = 7.71
X2 = 6.43

This final tableau, rearranged as previously, becomes

		X3	X4
X0 =	23.8	− 0.07	− 2.36
X1 =	7.71	− 0.43	+ 0.29
X2 =	6.43	+ 0.14	− 0.43
X5 =	3.29	+ 0.43	− 0.29
X6 =	0.86	+ 0.29	+ 0.14

Here, as the shadow price of X3 is smaller than that of X4, we choose to make X3 positive and reduce X1 to the integer value 7 using the artificial variable X7, which becomes a constraint variable, while X3 enters the basis:

X1 =   7   +   X4 − X7
X7 = − 0.71 + 0.71 X4   + 0.43 X3

X3 =   1.67 − 1.67 X4   + 2.33 X7

This solution, however, requires X3, a basic variable, to be a noninteger, and therefore the total solution is a noninteger. We must therefore move within either of the holding contraints (X4 and X7) once more, the choice depending on the shadow prices. Substitution for X3 in the equation for X0 gives

X0 = 23.8 − 0.17 X7 − 2.2 X4

Thus X7 has the lower shadow price and we choose to increase X7, implying that we should increase X3 and reduce X1:

X1 =   6   +   X4 −   X7

X7 = − 1.71 + 0.71 X4 + 0.43 X3
X3 =   4   − 1.67 X4 + 2.33 X7

Pivoting about X3, we obtain the tableau:

	X7	X4	
X1	− 1	1	6
X2	0.33	− 0.66	7
X5	1	− 1	5
X6	0.66	− 0.33	2
X3	2.33	− 1.67	4
X0	0.17	2.24	23.5

$$D = 1$$

This solution is optimal because it states that all the basic variables are being produced at integer levels, the solution being

$$X1 = 6 \quad : \quad X2 = 7$$
$$\text{(JIM)} \qquad \text{(FRED)}$$

2. Noninteger solution:

$$X1 = 4.286 \qquad \text{Value \$145,714}$$
$$X2 = 5.143$$

Best integer solutions are

$$X1 = 6 \text{ or } X1 = 4 \qquad \text{Value \$140,000}$$
$$X2 = 4 \quad\quad X2 = 5$$

3.

Maximize $\quad\quad 7.6\,F + \ 9.6\,L$

subject to $\quad\quad 34\ \ F + 19\ \ L < 646$

$\quad\quad\quad\quad\quad 13\ \ F + 33\ \ L < 429$

$\quad\quad\quad\quad\quad\quad\quad\quad L < \ 10$

Noninteger solution:

$$F = 15.04$$
$$L = 7.07$$

The best integer solution is in fact the roundoff solution:

$$F = 15$$
$$L = 7$$

4. The LINDO session proceeds as follows:

```
look all
MAX
5 A + 3 B + 2 C
SUBJECT TO
2)  2.5 A + 3.4 C <= 6
3)  2 A + 6 B + 4 C <= 10
END

: integer a
: integer b
: integer c
: go
LP OPTIMUM FOUND AT STEP 3
              OBJECTIVE FUNCTION VALUE
1)  9.00000000
         VARIABLE              VALUE          REDUCED COST
             A              1.000000            -4.000000
             B              1.000000             .000000
             C               .500000             .000000
           ROW         SLACK OR SURPLUS      DUAL PRICES
            2)              1.800000             .000000
            3)               .000000             .500000


NO. ITERATIONS= 3
BRANCHES= 0 DETERM.= 4.000E 0
SET C TO 1 AT 1 BND= 9.0000000 TWIN= 9.0000000
SET B TO 0 AT 2 BND= 7.0000000 TWIN= 7.0000000
NEW INTEGER SOLUTION AT BRANCH 2 PIVOT 5


              OBJECTIVE FUNCTION VALUE
1)  7.00000000
```

```
     VARIABLE                VALUE        REDUCED COST
        A                 1.000000         -5.000000
        B                  .000000         -3.000000
        C                 1.000000         -2.000000
      ROW          SLACK OR SURPLUS        DUAL PRICES
       2)                  .100000          .000000
       3)                 4.000000          .000000


NO. ITERATIONS= 5
BRANCHES= 2 DETERM.= 1.000E 0
BOUND ON OPTIMUM: 9.000000
DELETE B AT LEVEL 2
  FLI C TO 0 WITH BOUND 9.000000
NEW INTEGER SOLUTION AT BRANCH 2 PIVOT 6

                OBJECTIVE FUNCTION VALUE
1)  8.00000000
     VARIABLE                VALUE        REDUCED COST
        A                 1.000000         -5.000000
        B                 1.000000         -3.000000
        C                  .000000         -2.000000
      ROW          SLACK OR SURPLUS        DUAL PRICES
       2)                 3.500000          .000000
       3)                 2.000000          .000000


NO. ITERATIONS= 6
BRANCHES= 2 DETERM.= 1.000E 0
BOUND ON OPTIMUM: 8.000000
DELETE C AT LEVEL 1
ENUMERATION COMPLETE. BRANCHES= 2 PIVOTS= 6
LAST INTEGER SOLUTION IS THE BEST FOUND
```

5. The best integer solution to the formulation is

Maximize $.6\,A + 2.4\,B + 4.5\,C$

subject to

$$
\begin{aligned}
12\,A + 2\,B + 6\,C &< 120 \\
15\,A + 5\,B + 3\,C &< 150 \\
B &< 26 \\
2\,B + C &> 30 \\
A &= 0 \\
B &= 21 \\
C &= 13
\end{aligned}
$$

10

Nonlinear Techniques

Thus far we have considered mathematical relations that are linear. What happens when these functions are no longer linear? This is best illustrated by an example.

First we shall consider the case where the curves of the graphs that describe the relations are all smooth and have no kinks or sudden breaks. As they are always smooth, they will always have a slope and therefore are mathematically differentiable. Furthermore, in the best cases, the slope of these graphs will be zero, because they cannot be further improved.

An important concept in nonlinear formulations is that of convexity. Convexity has already been considered earlier in the LP formulation in the context of a set of points that define the boundaries of the feasible region. In more mathematical terms, a convex curve, or set of points, is never overestimated by a straight line drawn between any two allowable points. (See Figures 10.1 and 10.2 for illustration.) If the convex function is always underestimated by linear interpolation, then it is known as a *strictly convex function*. If it is never overestimated, then it is known as a *convex function*. The difference is that a convex function allows straight lines to be included in the set, whereas the strictly convex function does not.

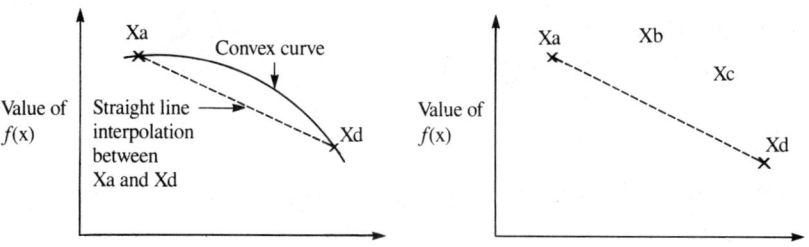

Figure 10.1 and Figure 10.2

Without constraints, the optimum value of any convex function will be found at the point where the function has no slope. Thus in n dimensions, the function will have no slope in any of the n directions indicated by each of the n variables. This is equivalent to saying that each of the differentials with respect to each of the variables will be zero. Mathematically this is represented as

$$\frac{dF}{dX1} = 0, \qquad \frac{dF}{dX2} = 0, \qquad \frac{dF}{dX3} = 0, \ldots \qquad \frac{dF}{dXn} = 0$$

for a function F dependent upon X1 . . . Xn, written as $F(X1 \ldots Xn)$. (Note that d represents partial differential; the conventional mathematical notation is ∂.)

For example, if we wish to optimize the profit P of a pair of products A and B, expressed by the relationship

Profit (\$) =   2 × (sales of A) × (sales of B)
+ 4 × (sales of B)
− 2 × (sales of A) × (sales of A)
− 1 × (sales of B) × (sales of B)

This would be expressed mathematically as

$$P = 2\,XA\,XB + 4\,XB - 2\,XA^2 - XB^2$$

where XA and XB are the sales of products A and B. Differentiating P with respect to XA and XB, respectively, we have

$$\frac{dP}{dXA} = 2\,XB - 4\,XA$$

and

$$\frac{dP}{dXB} = 2\,XA + 4 - 2\,XB$$

At the optimum, these two expressions for the profit gradient will be zero. Thus

$$2\,XB - 4\,XA = 0 \text{ and } 2\,XA + 4 - 2\,XB = 0$$

Solving these simultaneous equations gives

$$XA = 2,\ XB = 4$$

Thus selling two units of A and four units of B will result in maximum profit of

$$P = 2\,(\,2 \times 4\,) + (\,4 \times 4\,) - 2\,(\,2 \times 2\,) - (\,4 \times 4\,) = \$8$$

LAGRANGE MULTIPLIERS

The technique of differentiating a smooth function in order to discover the optimal value works unless the function is constrained within some boundaries. When constraints are added, the procedure is not quite so straightforward. Lagrange multipliers allow us to handle this complication. There are two forms of the technique: the more simple is useful when dealing with equality constraints; the more complex form is available for inequality constraints such as those encountered in the linear programming formulations.

Lagrange Multipliers with Equality-Constrained Problems

Suppose that we want to discover values of Xi ($i = 1, \ldots$ n) that optimize some function of these variables $f(Xi)$ as we previously did in the unconstrained case, but this time subject to a set of equality constraints, $Gj(Xi) = 0$, where $j = 1, \ldots$ m. Each of these $Gj(Xi)$ is another function. This may appear somewhat daunting to the practical decision maker who is not primarily a mathematician. An example will clarify the procedure.

Let us return to the difficulties of the edible fats manufacturer. Instead of a linear optimization function, which was used to optimize profit, let us now use an optimization function like the one used on the sales demonstration above:

$$\text{Profit (\$)} = \quad 2 \times \text{(sales of A)} \times \text{(sales of B)}$$
$$+ \ 4 \times \text{(sales of B)}$$
$$- \ 2 \times \text{(sales of A)} \times \text{(sales of A)}$$
$$- \ 1 \times \text{(sales of B)} \times \text{(sales of B)}$$

where

A = Production tons of cooking oil

B = Production tons of margarine

$$f(X1, X2) = 2\,X1\,X2 + 4\,X2 - 2\,X1^2 - X2^2$$

For illustration, and to modify the problem further, let us say that there are no constraints on the material supply, the sole active constraint being manpower capability. There would be therefore only one constraint $Gi(X1, X2)$, which is

$$X1 + 4\,X2 - 12 \leqslant 0$$

Consideration of the problem indicates that, in order to maximize profit, the total manufacturing capability must be utilized. (This would not be true for all such problems.) For completeness then, we define the constraint $Gi(X1, X2)$ as a binding equality constraint, as we wish to utilize all manpower at the solution.

$$X1 + 4\,X2 - 12 = 0$$

Returning to generalities, the obvious step is to solve the constraint equation for one of its variables; for example, we could use

$$X1 = 12 - 4\,X2$$

Inserting this expression into the objective function, we obtain

$$f(X2) = 2(12 - 4\,X2)\,X2 + 4\,X2 - 2(12 - 4\,X2)^2 - X2^2$$
$$= -\,41\,X2^2 + 220\,X2 - 288$$

We have now obtained an unconstrained problem, which may be solved using the ordinary differentiation technique.

$$\frac{df(X2)}{X2} = -82\,X2 + 220 = 0$$

$$X2 = 2.6829$$

Substituting for X2 into the Gi(X1,X2) constraint,

$$X1 = 12 - 4\,X2$$
$$X1 = 1.2684$$

With a more complex problem the substitution method is not so convenient. In this case, a new function may be written that combines the optimization and constraint functions:

$$F(X_i, L_j) = f(X_i) + \sum_{j=1}^{m} L_j\,G_j\,(X_i)$$

where

$f(Xi)$ is the optimization function in n variables Xi

Gj(Xi) are the m equality constraints in n variables Xi

Lj   are scalar multipliers that assist in the solution. These are called Lagrange multipliers.

Therefore, if we can discover values of L1 . . . Lm so that the point X1 . . . Xn, where F(X1 . . . Xn) is optimized, and all the constraints Gj(X1 . . . Xn) = 0 are satisfied, the problem is solved. This might seem terribly complicated, but is quite simple really. We now have an unconstrained function, F(Xi,Lj), rather than a constrained one, and we can use the previously employed unconstrained techniques. Returning to the example,

$$f(X1,X2) = 2\,X1\,X2 + 4\,X2 - 2\,X1^2 - X2^2$$
$$G(X1,X2) = X1 + 4\,X2 - 12 = 0$$

We construct the Lagrangian function

$$F(X1,X2,L1) = 2\,X1\,X2 + 4\,X2 - 2\,X1^2 - X2^2$$
$$+ L1(X1 + 4\,X2 - 12)$$

At an optimum

$$\frac{dF}{dX1} = 0, \qquad \frac{dF}{dX2} = 0, \qquad \frac{dF}{dL1} = 0$$

$$\frac{dF}{dX1} = 2\,X2 - 4\,X1 + L1 = 0$$

$$\frac{dF}{dX2} = 2\,X1 + 4 - 2\,X2 + 4\,L1 = 0$$

$$\frac{dF}{dL1} = X1 + 4\,X2 - 12 = 0$$

These are solved as simultaneous equations as follows:

$$
\begin{array}{rl}
-\quad 8\,X2 + 16\,X1 - 4\,L1 & = 0 \\
4 - \quad 2\,X2 + \quad 2\,X1 + 4\,L1 & = 0 \\
\hline
4 - 10\,X2 + 18\,X1 \qquad\quad & = 0 \\
-\ 12 + \ 4\,X2 + \quad X1 \qquad\quad & = 0 \\
\hline
-104 \qquad\qquad + 82\,X1 \qquad\quad & = 0
\end{array}
$$

Thus

$$X1 = \frac{104}{82} = 1.268$$

$$X2 = \frac{220}{82} = 2.683$$

Hence the optimum point is at $X1 = 1.268$ tons, $X2 = 2.683$ tons, with a profit of \$7.12.

This method works with constraints that are not linear; for instance, if the capacity constraint were to be bent outward from the origin in the form of a quarter ellipse, the constraint equation would be in the form

$$\frac{X1^2}{7^2} + \frac{X2^2}{3.5^2} - 1 = 0$$

The numerical steps follow exactly as before.

Lagrange Multipliers with Inequality Constraints

Lagrange multipliers can be generalized to encompass inequality constraints such as $Gj(Xi) \leqslant Bj$. Thinking in terms of a single variable, we know that the optimum of some function $f(X)$, subject to a constraint $X \leqslant B$, will be attained either when $df(X)/dX = 0$ (that is, inside the constraint), or when $df(X)/dX \neq 0$ and $X = B$ (that is, on the constraint boundary). This may be expressed in the form of the Lagrangian equation as

$$\frac{dF}{dX} + L = 0 \; ; L \geqslant 0, \text{ and } L(X - B) = 0$$

In general, with inequality constraints $Gj(Xi) \leqslant Bj$, then either $Lj = 0$ or $Lj \geqslant 0$ with $Gj(Xi) = Bj$ for any constraint j; that is, $Lj(Gj(Xi)-Bj) = 0$. The Lagrange multiplier equations may thus be written as

$$\frac{dF}{dX_i} + \sum_{j=1}^{m} L_j \frac{dG_j}{dX_i} = 0 \qquad (i = 1 \ldots n) \qquad (j = 1 \ldots m)$$

$$Gj(X_i) \leqslant B_j$$

$$L_j \geqslant 0$$

$$\sum_{j=1}^{m} L_j (G_j(X_i) - B_j) = 0$$

This fearsome-looking set of relationships is known as the Kuhn-Tucker conditions. They are sometimes supplemented with some other condition relationships that qualify the constraints, making sure that the formulation is a sensible one and not inconsistent.

While we have succeeded in formulating the problem, it is quite another matter to find a solution to these relationships in practice. In general, it is not possible. However, in some special cases, which fortunately are of interest to the practical decision maker, solution methods are available. The most used special case is where the optimization function takes on the form we have used in this chapter, a quadratic form, with only squared variable terms, linear terms, and constants, and without higher powers or ratio terms. When this quadratic form of optimization function is applied to a set of linear constraints, the formulation is known as a quadratic programming problem. These are described in Chapter 11.

11

Quadratic Programming

Quadratic programming is the name given to optimization formulations that have linear constraints but with second-order or quadratic optimization functions. Two predominant solution methods are in practice, Wolfe's method and Beale's method.

WOLFE'S METHOD

This approach uses the concept of Lagrange multipliers and the Kuhn-Tucker theorems in the following way. Taking as our example the problem of Chapter 10, but using an inequality constraint, we have

Maximize	$4\,X2 + 2\,X1\,X2 - 2\,X1^2 - X2^2$	
subject to	$X1 + 4\,X2$	$\leqslant 12$
	$X1$	$\geqslant 0$
	$X2$	$\geqslant 0$

From the Kuhn-Tucker theorems we can form the Lagrangian function

$$F = (4\,X2 + 2\,X1\,X2 - 2\,X1^2 - X2^2)$$
$$- L\,(X1 + 4\,X2 - 12) - M1\,(-\,X1) - M2\,(-\,X2)$$

where L, M1, and M2 are Lagrange multipliers associated with the constraints. Thus

$$\frac{dF}{dX1} = -\,4\,X1 + 2\,X2 - L + M1 = 0$$

$$\frac{dF}{dX2} = 2\,X1 + 4 - 2\,X2 - 4\,L + M2 = 0$$

The problem is therefore to discover nonnegative values of U, X1, X2, L, M1, and M2 that satisfy

$$
\begin{aligned}
4\,X1 - 2\,X2 + \quad L - M1 \qquad\qquad &= 0 \\
-\,2\,X1 + 2\,X2 + 4\,L \qquad - M2 \quad\quad &= 4 \\
X1 + 4\,X2 \qquad\qquad\qquad + U &= 12
\end{aligned}
$$

with

$$U \cdot L = 0$$
$$X1 \cdot M1 = 0$$
$$X2 \cdot M2 = 0$$

This is formulated as a linear programming problem in Wolfe's method by introducing artificial variables Z1 and Z2, and starting the solution at $Z1 = 0$, $Z2 = 4$, and U (the slack variable associated with the inequality constraint) $= 12$. The method proceeds by reducing the sums of the infeasibilities in this solution by maximizing the function $-(Z1 + Z2)$ with the added restrictions that the $U \cdot L$, $X1 \cdot M1$, and $X2 \cdot M2$ pairs cannot be in the basis together. This is conveniently represented by the extended simplex form.

Maximize $\qquad\qquad\qquad\qquad\qquad\qquad\qquad\qquad -\,Z1 - Z2$

subject to
$$
\begin{aligned}
4\,X1 - 2\,X2 \quad + \quad L - M1 \quad + Z1 \qquad\quad &= 0 \\
-2\,X1 + 2\,X2 \quad + 4\,L \qquad - M2 \quad + Z2 &= 4 \\
1\,X1 + 4\,X2 + U \qquad\qquad\qquad\qquad\quad &= 12
\end{aligned}
$$

Ci	Cj Basis	B	0 X1	0 X2	0 U	0 L	0 M1	0 M2	-1 Z1	-1 Z2
-1	Z1	0	4	-2	0	1	-1	0	1	0
-1	Z2	4	-2	2	0	4	0	-1	0	1
0	U	12	1	4	1	0	0	0	0	0
Zj-Cj			-2	0	0	-5	1	1	0	0
			*			**				

Cannot enter
basis because
of U

Ci	Basis	B	X1	X2	U	L	M1	M2	Z1	Z2
0	X1	0	1	-0.5	0	0.25	-0.25	0	0.25	0
-1	Z2	4	0	1	0	4.5	-0.5	-1	0.5	1
0	U	12	0	4.5	1	-0.25	0.25	0	-0.25	0
Zj-Cj			0	-1	0	-4.5	0.5	1	0.5	0
				*		**				

Cannot enter
basis because
of U

Ci	Basis	B	X1	X2	U	L	M1	M2	Z1	Z2
0	X1	1.33	1	0	0.11	0.22	-0.22	0	0.22	0
-1	Z2	1.33	0	0	-0.22	4.55	-0.55	-1	0.55	1
0	X2	2.66	0	1	0.22	-0.05	0.05	0	-0.05	0
Zj-Cj			0	0	0.22	-4.55	-0.55	1	0.44	0
						**				

May now
enter basis

Ci	Basis	B	X1	X2	U	L	M1	M2	Z1	Z2
0	X1	1.27	1	0	0.12	0	-0.19	0.04	0.19	-0.14
0	L	0.29	0	0	-0.05	1	-0.12	-0.21	-0.12	0.21
0	X2	2.68	0	1	0.21	0	0.04	-0.01	-0.06	0.01

The infeasibilities have been removed, so a feasible solution has been found.
The value of the solution is $X1 = 1.268$, $X2 = 2.683$.

BEALE'S METHOD

Let us extend the constraints to the full constraint set we used in the linear
programming example of Chapter 1, the margarine manufacturer. However,
this time, we will use the quadratic optimization function used for the example
in Wolfe's method.

Maximize $4 X2 + 2 X1 X2 - 2 X1^2 - X2^2$

subject to $X1 +\ \ \ X2 \leqslant\ 6$

                             $X1 + 2 X2 \leqslant\ 7$

                             $X1 + 4 X2 \leqslant 12$

The optimization function is not particularly convenient to handle in the form depicted. However, we can rewrite the objective function so that squared terms do not appear. This is achieved by writing the coefficients in the form of a symmetric array, thus

Minimize $X0 = -\ (4 X2 + 2 X1 X2 - 2 X1^2 - X2^2\ )$

becomes

Minimize $X0 =$ $(0 + 0 X1 - 2 X2)$

                      $+\ \ \ (0 + 2 X1 - 1 X2) X1$

                      $+\ (-2 - 1 X1 + 1 X2) X2$

This looks as if it is a complex step, but it is quite simple, as you will easily grasp by trying out a few examples.

        The next step is to include the optimization function, written in this form with the constraint equations:

Equation Set 11.1   Minimize $X0 =$ $(\ \ \ 0 + 0 X1 - 2 X2)$

                               $+\ (\ \ \ 0 + 2 X1 - 1 X2) X1$

                                 $+\ (\ -2 - 1 X1 + 1 X2) X2$ (i)

                     $X3 =$ $6 -\ \ \ X1 -\ \ \ X2$ (ii)

                     $X4 =$ $7 -\ \ \ X1 - 2 X2$ (iii)

                     $X5 =$ $12 -\ \ \ X1 - 4 X2$ (iv)

depicting a solution, as in the LP case, at $X1 = 0$, $X2 = 0$ with an optimization function value of 0.

        This form of presentation also readily shows the value of the slopes in the X1 and X2 directions. (The slopes are twice the constant coefficients within the brackets multiplied by X1 and X2, respectively.) Therefore, at $X1 = 0$, $X2 = 0$, these are

$$\frac{dF}{dX1} = 0 \times 2\ ;\ \frac{dF}{dX2} = -2 \times 2 = -4$$

Therefore, to decrease X0 (increase profit), we must introduce more X2 into the solution.

Consideration of Figures 1.1 and 11.1, together with examination of these equations, shows that we wish to increase X2 until either a constraint is reached (at X2 = 3, Equation 11.1(iv)) or until the slope of the optimization function changes where $1/2(dF/dX2) = 0$. This occurs at X2 = 2, given by Equation 11.1(i).

The smaller of these two values indicates which effect occurs first with increasing values of X2. We therefore choose X2 = 2, which indicates that we should increase X2 until X2 = 2, equivalent to point M in Figure 11.1.

To update the equation set in order to represent this point M, we must introduce a further variable, U1, to force X2 = 2 and X1 = 0. This U1 is equal to the function multiplied by X2 in the X0 representation. U1 is known as a *free variable*.

$$U1 = -2 - X1 + X2$$
$$X2 = \quad 2 + X1 + U1$$

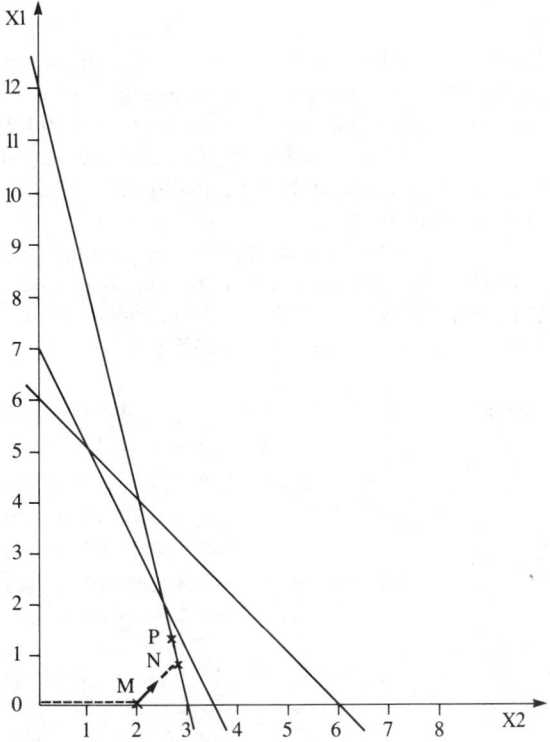

Figure 11.1

Updating the equations results in

Minimize $X0 = -2(2 + 1\,X1 + 1\,U1)$

$+(\ \ 0 + 2\,X1 - 1\,(2 + X1 + U1))\,X1$

$+(-2 - 1\,X1 + 1\,(2 + X1 + U1))(2 + X1 + U1)$

$X2 =\ \ \ 2\ \ \ +\ \ \ X1 + 1\,U1$

$X3 =\ \ \ 6\ \ \ -\ \ \ X1 - 1\,(2 + X1 + U1)$

$X4 =\ \ \ 7\ \ \ -\ \ \ X1 - 2\,(2 + X1 + U1)$

$X5 = 12\ \ \ -\ \ \ X1 - 4\,(2 + X1 + U1)$

or

Minimize $X0 =\ \ \ (-4 - 2\,X1\ \ \ \ \ \ \ \ \ )$

$+(-2 +\ \ \ X1\ \ \ \ \ \ \ \ \ )\,X1$

Equation Set 11.2 $+(\ \ \ \ +\ \ \ \ \ \ \ \ 1\,U1)\,U1$ (i)

$X2 =\ \ \ \ \ 2 + 1\,X1 + 1\,U1$ (ii)

$X3 =\ \ \ \ \ 4 - 2\,X1 - 1\,U1$ (iii)

$X4 =\ \ \ \ \ 3 - 3\,X1 - 2\,U1$ (iv)

$X5 =\ \ \ \ \ 4 - 5\,X1 - 4\,U1$ (v)

From this point $X1 = 0$, $X2 = 2$, the solution may be further improved by going along the direction of equal tangents to the optimization function until a constraint is reached. This is point N in Figure 11.1.

Equation 11.2(i) again indicates the direction, because the constant coefficient in the term multiplied by X2 is -2; that is, $1/2(dF/dX2) = -2$ at this point.

We can improve the solution by pivoting between X1 and X5. We do not introduce a new free variable as we did in the last step because the constraint $X1 + 4\,X2 \leqslant 12$ is encountered before the slope changes sign again. Pivoting between X1 and X5 gives

Minimize $X0 =\ \ \ (-6.56 + 0.24\,X5 + 0.96\,U1)$

$+(\ \ \ 0.24 + 0.04\,X5 + 0.16\,U1)\,X5$

Equation Set 11.3 $+(\ \ \ 0.96 + 0.16\,X5 + 1.64\,U1)\,U1$ (i)

$X2 =\ \ \ \ \ 2.8 - 0.20\,X5 + 0.20\,U1$ (ii)

$X3 =\ \ \ \ \ 2.40 + 0.40\,X5 + 0.60\,U1$ (iii)

$X4 =\ \ \ \ \ 0.60 + 0.60\,X5 - 0.40\,U1$ (iv)

$X1 =\ \ \ \ \ 0.80 - 0.20\,X5 - 0.80\,U1$ (v)

At this stage, we have increased X2 beyond the value for which we introduced the free variable U1 in order to constrain X2 to the value of 2. We must therefore reconsider the role of the variable U1. We know that U1 may take positive or negative values, so we examine the current equation set to see whether we can profitably reduce U1. We see that we can indeed reduce U1, making it negative, with a resulting improvement in the constant coefficient (6.56) of Equation 11.3(i) (as $dX0/dU1$ is positive, reducing U1 improves X0). We can therefore introduce a further variable U2:

$$U2 = \quad 0.96 + 0.16\,X5 + 1.64\,U1$$
$$U1 = -0.58 - 0.10\,X5 + 0.61\,U2$$

which gives

Minimize
$$
\begin{aligned}
X0 = \;& (\,6 \quad + 0.146\,X5 \qquad\qquad) \\
& +(\,0.146 + 0.024\,X5 - \qquad)\,X5 \\
& +(\qquad\qquad\qquad + 0.610\,U2)\,U2 \qquad\quad \text{(i)} \quad \textbf{Equation Set 11.4}
\end{aligned}
$$
$$X2 = \quad 2.683 - 0.220\,X5 + 0.122\,U2 \qquad \text{(ii)}$$
$$X3 = \quad 2.049 + 0.344\,X5 + 0.366\,U2 \qquad \text{(iii)}$$
$$X4 = \quad 0.834 + 0.639\,X5 - 0.244\,U2 \qquad \text{(iv)}$$
$$X1 = \quad 1.268 - 0.122\,X5 - 0.488\,U2 \qquad \text{(v)}$$

This is the same optimum point found in the equality-constrained Lagrange multiplier case and the single inequality-constrained example illustrating Wolfe's method of quadratic programming.

COMPUTER SOLUTION

The current version of LINDO contains an experimental quadratic programming option. It allows the use of optimization functions that contain the products of two variables. These may be expressed in the form

```
X*Y, X**2, 3*(X**2)
```

and so forth. However, we are reaching the point at which a practical decision maker, who is a novice mathematician, would be well advised to seek expert assistance with solution formulations.

LINDO requires the problem to be structured in the form of an LP problem, in the same manner as Wolfe's method. The Lagrangian function must be defined and the differentials worked out prior to formulating the input file. This requires the following for the problem just solved by Wolfe's method:

Maximize $2 X1 X2 + 4 X2 - 2 X1^2 - X2^2$

subject to $1 X1 + 1 X2 \leqslant 6$

$1 X1 + 2 X2 \leqslant 7$

$1 X1 + 4 X2 \leqslant 12$

$X1, X2 \geqslant 0$

The Lagrangian function is produced

$$F = \quad (2 X1 X2 + 4 X2 - 2 X1^2 - X2^2)$$
$$+ L1 (1 X1 + 1 X2 + 1 U1 - \quad 6)$$
$$+ L2 (1 X1 + 2 X2 + 1 U2 - \quad 7)$$
$$+ L3 (1 X1 + 4 X2 + 1 U3 - 12)$$

The first differentials with respect to the variables X1 and X2 are

$$\frac{dF}{dX1} = -4 X1 + 2 X2 + L1 + L2 + L3$$

$$\frac{dF}{dX2} = 2 X1 - 2 X2 + 4 + L1 + 2 L2 + 4 L3$$

L1, L2, L3 are the names of the Lagrange multipliers, or dual variables, associated with each constraint; we may, if we choose, call these Oil, Plant, Labor.

As in the procedure described under Wolfe's method, the linear program does not now have a specific objective function; one must be provided for the computer program. For this procedure, the optimization function takes the form of listing the variables and specifying row correspondence. The LINDO input file then becomes

```
MIN X1 + X2 + L1 + L2 + L3
ST
  4 X1 - 2 X2 + L1 + L2 + L3 > 0
 -2 X1 + 2 X2 - L1 - 2 L2 - 4 L3 > 4
X1 + X2 < 6
X1 + 2 X2 < 7
X1 + 4 X2 < 12
END
```

In order to run this as a quadratic programming program, the LINDO command QCP is used. The response is

```
ROW NO OF FIRST REAL CONSTRAINT :
```

Here the fourth row contains the first real constraint

```
> 4
```

Start the procedure with

```
: GO
```

The system will respond with

```
QP OPTIMUM FOUND AT STEP 4
       OBJECTIVE FUNCTION VALUE
          1) -7.12195122

VARIABLE       VALUE        REDUCED COST
X1           1.268293          .000000
X2           2.682927          .000000
L1            .000000         2.048780
L2            .000000          .365854
L3            .292683          .000000
```
(continued)

```
ROW          SLACK OR SURPLUS    DUAL PRICES
2)                   .000000      -1.268293
3)                   .000000      -2.682927
4)                  2.048780       .000000
5)                   .365854       .000000
6)                   .000000       .292683
NO. ITERATIONS  4
```

An important note: Care *must* be taken to ensure that non-negative variable values are used in this formulation; this includes the Lagrange multipliers. In the general case, it does not matter if these are formulated so that negative values result, but using LINDO, these must be nonnegative.

Note also that the optimal values of the variables X1 and X2 have been obtained and the value of the optimization function derived by the LINDO program has also been obtained. The value reported by LINDO is not the calculated value of the strange optimization function used in the quadratic optimization formulation, but the value of the true optimization function, thus:

$$(2 \times 1.268 \times 2.683) + (4 \times 2.683) - (2 \times 1.268^2)$$
$$- (2.683^2) = 7.12195122$$

CHAPTER 11 PROBLEMS

The solution to Problems 1 through 5 will be found at the end of the chapter.

1. The controller of a company is considering investing up to $500,000 of the company's assets in two firms, A and B. Firm A requires $900,000 and firm B requires $300,000.

By considering the covariances of the expected profits between the company and those of the two firms, the controller constructs the function

$$C1(X) = X1^2 - 6\,X1 + 4\,X2^2 - 16\,X2 + 25$$

This function represents the variance of the investment where X1 and X2 are the amounts (in $100,000) invested in each firm A and B. To minimize the variance subject to the constraints imposed, what amounts would you advise the controller to invest? How would your analysis be altered if there were only $400,000 to invest?

2. A corporation is considering investing up to $150,000 in expanding its operations via takeover into an area dominated by two companies, one private and one public. The private company has 11 partners, each of whom requires $10,000 for his or her interests. Up to 200,000 shares in the public company may be purchased at $1 per share, in blocks of 10,000. However, the time taken to negotiate the acquisition of 10,000 shares is three times that required to deal with a single partner, and the corporation has time to negotiate only 90,000 share transfers if they were all in the public company. Each block of 10,000 shares gives two-thirds as much control over the area of operations as one private company partner's holding. To be effective, the corporation must hold at least the control equivalent of 180,000 shares. The profitability resulting from the corporation holding is assessed by consultants to be

$$X1^2 + X2^2 - 10\,X1\,X2 - 4\,X2 - 10\,X1 + 1500$$

where X1 is the number of partners' holdings and X2 is the number of blocks of 10,000 shares held. What purchasing policy would you advise?

3. What is the best way to invest in three possible opportunities, A, B, or C, to obtain a growth of at least 15 percent, while minimizing the variance in the return and not exceeding the budget constraint?

There are the restrictions that A or B must not constitute more than 70 percent of the portfolio and C must not exceed 50 percent of the portfolio. The expected returns on the opportunities are

A	35%
B	25%
C	10%

If the fraction of the portfolio devoted to each opportunity is XA, XB, and XC, the variance of the portfolio has been calculated to be:

$$4\,XA^2 + 3\,XB^2 + XC^2 + 2\,XA\,XB - XA\,XC - 0.7\,XB\,XC$$

4.

Maximize $-3\,X1^2 - 2\,X2^2 + 2\,X1\,X2 + X1$
subject to $10\,X1 + 4\,X2 \leqslant 40$
        $6\,X1 + 5\,X2 \leqslant 30$
        $5\,X1 + 8\,X2 \leqslant 40$

5.

Minimize $\qquad -4\,X1 - 6\,X2 + 2\,X1^2 + 2\,X2^2 + 2\,X1\,X2$

subject to $\qquad 2\,X1 + 4\,X2 \leqslant 4$

$\qquad\qquad 2\,X1 + X2 \leqslant 10$

HOMEWORK PROBLEMS

6. The Seer's Company sells forecasting machines. One machine is obtained from a subcontractor at a fixed profit of $56 each; the other is made by Seer's itself, and the profit on the model increases as the number of sales of this unit increases by 50¢ per unit from an initial base unit profit of $42. During the next time period, the company forecasts a demand of 2,000 units. Customer preparation time for each subcontracted unit is one hour and for the own-manufactured unit is 48 minutes. There are 1,000 hours of preparation time budgeted for the next period.

How much profit do you forecast that Seer's will make from these sales in the next period?

7.

Minimize $\qquad 2\,X1^2 + 2\,X2^2 - 2\,X1\,X2 - 6\,X1$

subject to $\qquadX1 + X2 \leqslant 2$

8.

Minimize $\qquad 5\,X1 + 6\,X2 - 10\,X1^2 - 15\,X2^2 + 2\,X1\,X2$

subject to $\qquad 6\,X1 + 2\,X2 \leqslant 120$

$\qquad\qquad 3\,X1 + 5\,X2 \leqslant 150$

$\qquad\qquad 25\,X1 + 30\,X2 \leqslant 7{,}500$

9.

Minimize $\qquad 9\,X^2 + 4\,Y^2 + Z^2 + 3\,X\,Y - X\,Z - 2\,Y\,Z$

subject to $\qquad 12\,X + 11\,Y + 10\,Z \leqslant 11.8$

$\qquad\qquad X + Y + Z = 1$

$\qquad\qquad X \qquad\qquad\quad \leqslant 0.6$

$\qquad\qquad\qquad Y \qquad\quad \leqslant 0.7$

$\qquad\qquad\qquad\qquad Z \geqslant 0.1$

10.

Minimize $9 X^2 + 4 Y^2 + Z^2 + 3 X Y - X Z - 2 Y Z$
subject to $12 X + 11 Y + 10 Z \leq 11.8$
$$X + Y + Z = 1$$
$$X \leq 0.6$$
$$Y \leq 0.7$$
$$Z \geq 0.1$$

CHAPTER 11 SOLUTIONS

1. The objective function is

Minimize $X1^2 - 6 X1 + 4 X2^2 - 16 X2 + 25$

which can be expressed as

Maximize $X1^2 + 6 X1 - 4 X2^2 + 16 X2 - 25$

The constraints are

$$X1 + X2 \leq 5 \qquad \text{Total investment}$$
$$X1 \leq 9 \qquad \text{Investment in A}$$
$$X2 \leq 3 \qquad \text{Investment in B}$$

Using the Kuhn-Tucker theorems,

$$
\begin{aligned}
F = \quad & (- X1^2 + 6 X1 - 4 X2^2 - 16 X2 - 25) \\
& - L1 (\quad X1 \qquad X2 - 5) \\
& - L2 (\qquad\qquad X2 - 3) \\
& - M1 (- X1) \\
& - M2 (\qquad - X2)
\end{aligned}
$$

where L1,L2,M1 and M2 are Lagrange multipliers. Differentiating,

$$\frac{dF}{dX1} = - 2 X1 + 6 - L1 + M1 = 0$$

$$\frac{dF}{dX2} = - 8 X2 + 16 - L1 - L2 + M2 = 0$$

The problem is now to discover those values of U1, X1, X2, L1, L2, M1, and M2 so that:

$$
\begin{array}{rcccccc}
2\,X1 & + L1 & & - M1 & & = & 6 \\
& 8\,X2 + L1 & + L2 & & - M2 & = & 16 \\
U1 & + X1 + & X2 & & & = & 5 \\
U2 & + & X2 & & & = & 3
\end{array}
$$

where

U1 is the slack variable associated with constraint 1 and

U2 is the slack variable associated with constraint 2

Introducing artificial variables Z1 and Z2 to form the maximization equation $-(Z1 + Z2)$, this can now be solved via extended simplex tableaux:

C_j			0	0	0	0	0	0	0	0	-1	-1	
													Q
C_i	Basis	B	U1	U2	X1	X2	L1	L2	M1	M2	Z1	Z2	
-1	Z1	6	0	0	2	0	1	0	-1	0	1	0	—
-1	Z2	16	0	0	0	8	1	1	0	-1	0	1	2 *
0	U1	5	1	0	1	1	0	0	0	0	0	0	5
0	U2	3	0	1	0	1	0	0	0	0	0	0	3
	Z_j		0	0	-2	-8	-2	-1	1	1	-1	-1	
	Z_j-C_j	-22	0	0	-2	-8	-2	-1	1	1	0	0	
						*							

C_j			0	0	0	0	0	0	0	0	-1	-1	
													Q
C_i	Basis	B	U1	U2	X1	X2	L1	L2	M1	M2	Z1	Z2	
-1	Z1	6	0	0	2	0	1	0	-1	0	1	0	3 *
0	X2	2	0	0	0	1	0.13	0.13	0	-0.13	0	0.13	—
0	U1	3	1	0	1	0	-0.13	-0.13	0	0.13	0	-0.13	3
0	U2	1	0	1	0	0	-0.13	-0.13	0	0.13	0	-0.13	—
	Z_j		0	0	-2	0	-1	0	1	0	-1	0	
	Z_j-C_j	-6	0	0	-2	0	-1	0	1	0	0	1	
						*							

Cj			0	0	0	0	0	0	0	0	-1	-1
Ci	Basis	B	U1	U2	X1	X2	L1	L2	M1	M2	Z1	Z2
0	X1	3	0	0	1	0	0.5	0	-0.5	0	0.5	0
0	X2	2	0	0	0	1	0.13	0.13	0	-0.13	0	0.13
0	U1	0	1	0	0	0	-0.63	-0.13	0.5	0.13	-0.5	-0.13
0	U2	1	0	1	0	0	-0.13	-0.13	0	0.13	0	-0.13
Zj			0	0	0	0	0	0	0	0	0	0
Zj-Cj		0	0	0	0	0	0	0	0	0	1	1

Q

This solution indicates that the controller should invest $300,000 in firm A and $200,000 in firm B.

If the controller has only $400,000 to invest, then constraint 1 becomes

$$X1 + X2 \leqslant 4$$

The Kuhn-Tucker theorems now result in

$$
\begin{aligned}
F = \quad & (- X1^2 + 6\,X1 - 4\,X2^2 - 16\,X2 - 25) \\
& - L1\,(\quad X1 + \quad X2 - 4\,) \\
& - L2\,(\qquad\qquad X2 - 3\,) \\
& - M1\,(- X1\,) \\
& - M2\,(\qquad\quad - \quad X2\,)
\end{aligned}
$$

The solution of this via the same method yields a solution of

$$X1 = 2.2$$
$$X2 = 1.8$$

or $220,000 and $180,000, respectively.

2.

Maximize	$X1^2 + X2^2 - 10\,X1\,X2 - 4\,X2 - 10\,X1 + 1500$		
subject to	$10{,}000\quad X1 + 10{,}000\,X2 \leq 150{,}000$		Budget
	$X1 \qquad\qquad\qquad \leq \quad 11$		Private company
	$\qquad\qquad X2 \leq \quad 20$		Public company
	$0.33\,X1 + \qquad X2 \leq \quad 9$		Time
	$1.5\;\;X1 + \qquad X2 \geq \quad 18$		Control

$$
\begin{aligned}
\text{Object} \quad &= \quad 1{,}500 - 5\,X1 - 2\,X2 \\
&\quad (-\;\;5 + 1\,X1 - 5\,X2\,)\,X1 \\
&\quad (-\;\;2 - 5\,X1 + 1\,X2\,)\,X2
\end{aligned}
$$

$$
\begin{aligned}
\text{subject to} \quad X3 &= \quad 15 - 1\,X1 - 1\,X2 \\
X4 &= \quad 11 - 1\,X1 \\
X5 &= \quad 27 - 1\,X1 - 3\,X2 \\
X6 &= \quad -36 + 3\,X1 + 2\,X2
\end{aligned}
$$

This solution formulation is infeasible, due to the negative quantity of variable $X6$ in the basis. The infeasibility may be reduced by pivoting $X6$ from the basis and pivoting $X1$ into the basis:

$$X1 = 12 - 0.66\,X2 + 0.33\,X6$$

Hence

$$
\begin{aligned}
\text{Object} \quad &= \quad 1{,}524 \quad\;\; - 65.67\,X2 - 2.33\,X6 \\
&\quad (-\,66.67 + \;\;8.11\,X2 - 1.89\,X6\,)\,X2 \\
&\quad (-\;\;2.33 - \;\;1.89\,X2 + 0.11\,X6\,)\,X6
\end{aligned}
$$

$$
\begin{aligned}
\text{subject to} \quad X3 &= \quad\;\; 3 \;\; - \;\; 0.33\,X2 - 0.33\,X6 \\
X4 &= -\;\;1 \;\; + \;\; 0.67\,X2 - 0.33\,X6 \\
X5 &= \quad 15 \;\; - \;\; 2.33\,X2 - 0.33\,X6 \\
X1 &= \quad 12 \;\; - \;\; 0.67\,X2 + 0.33\,X6
\end{aligned}
$$

This formulation remains infeasible. To reduce the infeasibility in this formulation, pivot between $X4$ and $X2$:

$$X2 = 1.5 + 1.5\,X4 + 0.5\,X6$$

which results in

Object   =   1342.25 − 81.75 X4 − 27.75 X6
                     (−81.75 + 18.25 X4 +   3.25 X6) X4
                     (−27.75 +   3.25 X4 + 0.25 X6) X6

subject to   X3 =   2.5   −   0.5 X4 −   0.5 X6
                X2 =   1.5   +   1.5 X4 +   0.5 X6
                X5 =   11.5 −   3.5 X4 −   1.5 X6
                X1 =   11   −   1   X4

        The final solution is therefore X1 $=$ 11 and X2 $=$ 1.5. The profit is \$1,342.25.

        3. The quadratic programming problem is

Minimize $4\,XA^2 + 3\,XB^2 + XC^2 + 2\,XA\,XB - XA\,XC - 0.7\,XB\,XC$
subject to $1.35\,XA + 1.25\,XB + 1.1\,XC > 1.15$
               XA +   XB +   XC = 1
               XA   < 0.7
                     XB   < 0.7
                           XC < 0.7

Placing this into LINDO format,

```
look all
MIN
XA + XB + XC + R + U + X1 + X2 + X3
SUBJECT TO
2)  8 XA + 2 XB - XC - 1.35 R + U + X1 >= 0
3)  2 XA + 6 XB - 0.7 XC - 1.25 R + U + X2 >= 0
4)  - XA - 0.7 XB + 2 XC - 1.1 R + U + X3 >= 0
5)  1.35 XA + 1.25 XB + 1.1 XC >= 1.15
6)  XA + XB + XC = 1
7)  XA <= 0.7
8)  XB <= 0.7
9)  XC <= 0.5
END
```

(continued)

```
: qcp
ROW OF FIRST CONSTRAINT=
> 5
: go
QP OPTIMUM FOUND AT STEP 8

                    OBJECTIVE FUNCTION VALUE
1) .593875000
        VARIABLE            VALUE          REDUCED COST
            XA              .215000          .000000
            XB              .285000          .000000
            XC              .500000          .000000
            R               .000000          .046500
            U             -1.790000          .000000
            X1              .000000          .485000
            X2              .000000          .415000
            X3             1.204500          .000000

          ROW         SLACK OR SURPLUS     DUAL PRICES
           2)              .000000         -.215000
           3)              .000000         -.285000
           4)              .000000         -.500000
           5)              .046500          .000000
           6)              .000000        -1.790000
           7)              .485000          .000000
           8)              .415000          .000000
           9)              .000000         1.204500
    NO. ITERATIONS= 8
```

4. The solution is at X1 = 0.2, X2 = 0.1.

5. The solution is at X1 = 0.333, X2 = 0.833. The objective function value is 4.1667.

12

General Constrained Optimization Case

Many techniques, loosely described as search techniques, have been designed to deal with the general constrained case where the optimization function is not easy to handle and the constraints do not allow for easy manipulation. Some approaches are a form of simulation by trial. Others, more sophisticated, use a form of "hill-climb" routine to discover the optimal solution. The method chosen to illustrate this approach is a very useful practical method, to be used when the previous methods cannot be applied. It is known as the simplex technique.

SIMPLEX TECHNIQUE

This technique must not be confused with the simplex technique of linear programming. This method is a multidimensional hill-climb technique where the function is evaluated at a symmetrical set of $n + 1$ points in an n-dimensional space. This set of points is known as a *simplex of points*. (The simplex of points used in the LP problem is the one that bounds the feasible region. Here, the simplex is not the boundary of the feasible region, but a set of points lying somewhere within the feasible region.) The worst point is "reflected" through the remainder and re-evaluated. The process repeats until the simplex cycles. The size of the simplex is adjusted until an answer is as near as desired in practice. This is, therefore, an approximation technique rather than an exact calculation as in the other methods.

The method is best described by an example. Consider the quadratic programming problem solved in Chapter 11:

Maximize	$2 X1 X2 + 4 X2 - 2 X1^2 - X2^2$
subject to	$X1 \quad + \quad X2 \leqslant 6$
	$X1 \quad + 2 X2 \leqslant 7$
	$X1 \quad + 4 X2 \leqslant 12$

This is a two-variable problem, and hence a two-dimensional relationship. For a problem in n dimensions, we need $n + 1$ points, or 3. These three points form a regular simplex in two dimensions, which for the two-dimensional case is represented by the vertices of an equilateral triangle (Figure 12.1). We arbitrarily assign to the side of this simplex the length of 2 units. (In a three-dimensional case, the regular simplex would be a regular tetrahedron, and in more dimensions would be represented by regular hypertetrahedra).

For each evaluation point of the simplex therefore, we have for

$$f(X1,X2) = 2 X1 X2 + 4 X2 - 2 X1^2 - X2^2$$

the following values:

Point	X1	X2	$f(X1,X2)$
1	0	0	0
2	0	2	4
3	1.732	1	0.464

Figure 12.1

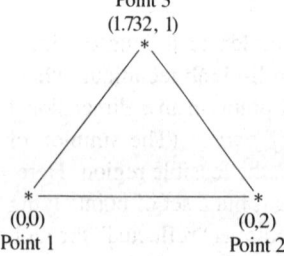

Point 3
(1.732, 1)

(0,0)
Point 1

(0,2)
Point 2

This set is then ranked in order of optimization function value:

Point	X1	X2	$f(X1,X2)$
2	0	2	4
3	1.732	1	0.464
1	0	0	0

      The worst of these is point 1, at X1 = 0, X2 = 0. This point is reflected across the boundary formed by the remaining points. (In higher dimensions, the inferior point is reflected through the centroid of the remaining points.) The result is shown in Figure 12.2.

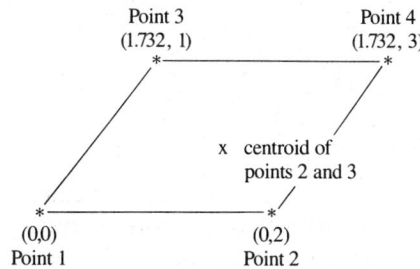

Point 3 (1.732, 1) Point 4 (1.732, 3) **Figure 12.2**

x centroid of points 2 and 3

(0,0) Point 1   (0,2) Point 2

THE UNCONSTRAINED CASE

In the unconstrained case, the method proceeds by replacing point 1 by point 4 in the table:

Point	X1	X2	$f(X1,X2)$
2	0	2	4
3	1.732	1	0.464
4	1.732	3	7.392

Reordering and repeating results in

Point	X1	X2	$f(X1,X2)$
4	1.732	3	7.392
2	0	2	4
3	1.732	1	0.464

Therefore we reflect point 3 across the other two (Figure 12.3). Evaluating point 5 and substituting it for point 3 gives

Point	X1	X2	$f(X1,X2)$
4	1.732	3	7.392
2	0	2	4
5	0	4	0

Figure 12.3

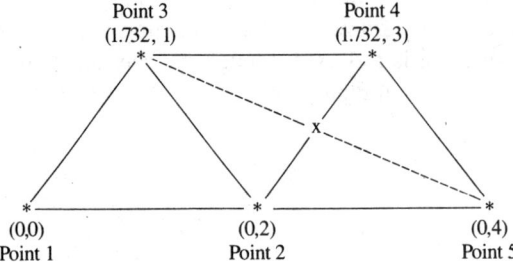

Point 3 (1.732, 1)   Point 4 (1.732, 3)

(0,0) Point 1   (0,2) Point 2   (0,4) Point 5

Point 5 has the minimum value of the optimization function. Any further reflection results in cycling between points 5 and 3. This is not the sole criterion for the simplex proceeding no further in an unconstrained case. In two dimensions, the triangle could cycle around one point in five steps, describing a hexagon (Figure 12.4). Thus if a point has remained in the table for five or more iterations, then the simplex has reached a local optimum.

Figure 12.4

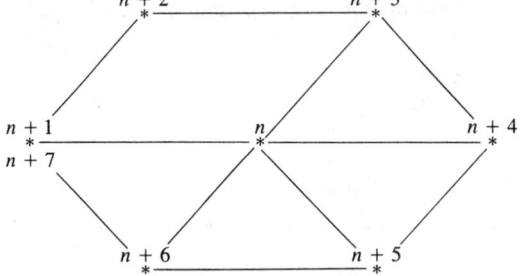

$n + 2$ $n + 3$

$n + 1$ n $n + 4$
$n + 7$

$n + 6$ $n + 5$

The method of achieving this is to reduce the simplex size. Generate a new simplex, based on the best point solution found to date. A usual step is to reduce the simplex size by half. The best point to date is point 4; thus, basing a new smaller simplex on this results in Figure 12.5:

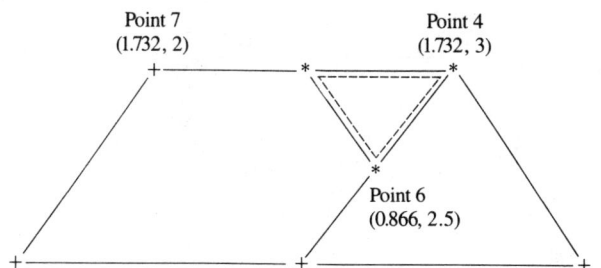

Figure 12.5

Point	X1	X2	$f(X1,X2)$
4	1.732	3	7.392
6	0.866	2.5	6.580
7	1.732	2	4.928

Proceeding as before, we obtain

Point	X1	X2	$f(X1,X2)$
4	1.732	3	7.392
6	0.866	2.5	6.580
8	0.866	3.5	6.312

The system is repeated until either the variation in the $f(X1,X2)$ function tends to zero or the simplex size tends to zero. Here the procedure terminates as X1 tends to 2 and X2 tends to 4, with the optimum value of the optimization function tending to a value of 8.

CONSTRAINED OPTIMIZATION

Let us consider the constrained problem solved in the quadratic programming case of Chapter 11. When this problem was unconstrained the first iteration produced a new point, point 4, shown in Figure 12.2. For the constrained case, we should first check to see if the new point falls within the feasible region. In this case, it does not: the constraint $X1 + 4X2 \leq 12$ is violated.

Here the initial simplex size chosen was too great. It is therefore reduced by dividing the simplex side by 2 and keeping the best point. The new, reduced simplex is based on point 2 (Figure 12.6):

Figure 12.6

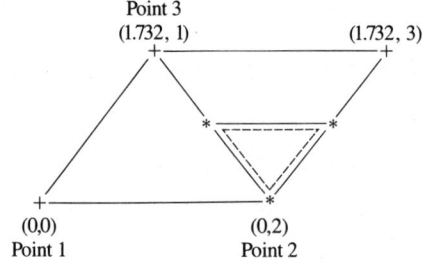

Point 3
(1.732, 1)   (1.732, 3)

(0,0)   (0,2)
Point 1   Point 2

Point	X1	X2	$f(X1,X2)$
2	0	2	4
4	0.866	1.5	4.848
5	0.866	2.5	6.580

The procedure continues, reflecting and reducing, as follows:

Point	X1	X2	$f(X1,X2)$
5	0.866	2.5	6.580
4	0.866	1.5	4.848
6	1.732	2	4.928

Point	X1	X2	$f(X1,X2)$
5	0.866	2.5	6.582
7	1.299	2.75	7.207
8	1.299	2.25	6.408

Notice that the best value to date is obtained with X1 = 1.299, X2 = 2.75. As will be observed, the solution is homing in on the values X1 = 1.268, X2 = 2.683.

13

Dynamic
Programming

The decision-tree concept has already been introduced in Chapter 9 in relation to integer programming. Decision trees are useful when analyzing a set of decisions that must be made sequentially. This set of decisions is normally referred to as a *policy*. Dynamic programming is a method of determining such a policy and is clearly allied to other optimization methods used under such circumstances, namely Pontryagin's Method and the calculus of variations. Here we shall limit the discussion to dynamic programming as devised by Bellman.

Quite simply, the basis of this technique is the observation that, from any state reached as a result of decisions taken from some initial state, the optimal policy of decisions needed to reach the end state is independent of the method by which the state in question was reached. Obvious? Not quite, since this excludes decision processess that employ feedback loops, because in such cases, the optimal policy from a state would not be independent of how that state was achieved.

Dynamic programming is a very powerful method, and is the first method we have considered that is capable of analyzing a sequential set of decisions. As a single decision may also be represented by a set of smaller decisions, both single and sequential aspects are covered.

Figure 13.1

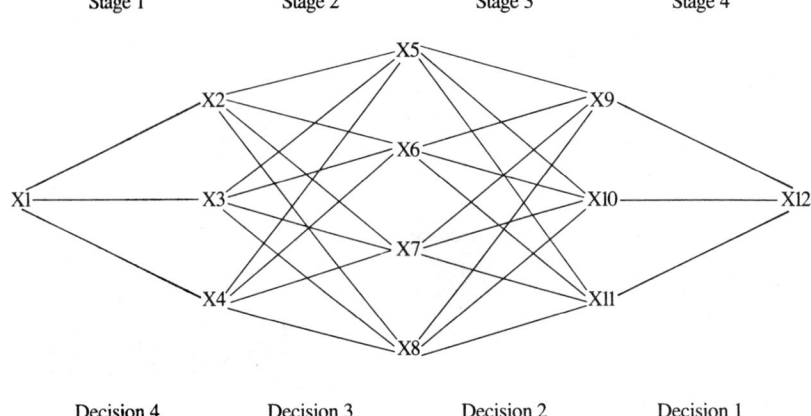

In order to illustrate the technique, consider the following problem. A traveler in a particular city wishes to travel to another city that is not directly connected to it, so that the traveler must make connections in other cities along the way. As there are alternate routes, incorporating different intermediate cities, the problem can become quite complex in the combinations available. The situation is illustrated by Figure 13.1. The traveler goes from city X1 to city X12, choosing a route via the intervening cities. The costs of these intervening connecting routes are as follows:

Table 13.1

		From	To	Cost	
		X1	X2	5	
	Stage 1	X1	X3	6	Decision 4
		X1	X4	3	
		X2	X5	4	
		X2	X6	2	
		X2	X7	5	
		X2	X8	3	
		X3	X5	6	
	Stage 2	X3	X6	4	Decision 3
		X3	X7	7	
		X3	X8	7	
		X4	X5	9	
		X4	X6	8	
		X4	X7	6	
		X4	X8	7	

	X5	X9	6	
	X5	X10	4	
	X5	X11	7	
	X6	X9	5	
	X6	X10	6	
	X6	X11	5	
Stage 3				Decision 2
	X7	X9	9	
	X7	X10	8	
	X7	X11	8	
	X8	X9	6	
	X8	X10	3	
	X8	X11	2	
	X9	X12	6	
Stage 4	X10	X12	5	Decision 1
	X11	X12	4	

Notice that the above table is compiled so that the traveler will, in sequence, travel from city X1, going via the stage 1 choice, then the stage 2 choice, followed by the stage 3 choice, and finally the stage 4 choice. The decisions, however, are listed in the reverse direction because the method normally proceeds by posing the question, "If this point has been reached, what is the optimal path to the end?"

The decision process followed is thus

From	Proceeding Via	Cost

Decision 1

From	Proceeding Via	Cost
X9	—	6
X10	—	5
X11	—	4

Table 13.2

Table 13.3 Decision 2

X5	X9	6 + 6 = 12
X5	X10	4 + 5 = 9 *
X5	X11	7 + 4 = 11
X6	X9	5 + 6 = 11
X6	X10	6 + 5 = 11
X6	X11	5 + 4 = 9 *
X7	X9	9 + 6 = 15
X7	X10	8 + 5 = 13
X7	X11	8 + 4 = 12 *
X8	X9	6 + 6 = 12
X8	X10	3 + 5 = 8
X8	X11	2 + 4 = 6 *

Thus for decision 2 the best alternatives are

X5	X10	9
X6	X11	9
X7	X11	12
X8	X11	6

Table 13.4 Decision 3

X2	X5	4 + 9 = 13
X2	X6	2 + 9 = 11
X2	X7	5 + 12 = 17
X2	X8	3 + 6 = 9 *
X3	X5	6 + 9 = 15
X3	X6	4 + 9 = 13 *
X3	X7	7 + 12 = 19
X3	X8	7 + 6 = 13 *
X4	X5	9 + 9 = 18
X4	X6	8 + 9 = 17
X4	X7	6 + 12 = 18
X4	X8	7 + 6 = 13 *

Thus for decision 3 the best alternatives are

X2	X8	9
X3	X6	13
	X8	13
X4	X8	13

Decision 4   Table 13.5

X1	X2	5 + 9 = 14	*
X1	X3	6 + 13 = 19	
X1	X4	3 + 13 = 16	

Thus for decision 4 the best alternative is

X1	X2	14

The overall solution for the minimum cost route is therefore X1—X2—X8—X11—X12, with a minimum cost of 14.

The process used to solve this problem is to replace the original problem with a series of smaller, sequential problems. The same procedure can be applied to a linear programming formulation, but that is not an efficient or even sensible way to solve such problems. For illustration, consider the three-dimensional problem of Chapter 1:

Maximize $4 X1 + 5 X2 + 3 X3$
subject to
$$X1 + X2 + 4 X3 \leqslant 6$$
$$X1 + 2 X2 + X3 \leqslant 7$$
$$X1 + 4 X2 + 2 X3 \leqslant 12$$
$$X1 \geqslant 0$$
$$X2 \geqslant 0$$

This problem can be turned into a sequential problem by deciding, for example:

choose X1, then choose X2, then choose X3

or

choose X2, then choose X1, then choose X3

and so forth. The simplex decision points defining the feasible region can be defined as

X1	6	5	2	2	0	0	0
X2	0	1	2.5	2.29	2.57	3	0
X3	0	0	0	0.43	0.86	0	1.5

Let us choose the sequence of choice X1, X2, X3. In ascending values of X1, we have, in tabular form:

Choose X1 from		0			2	5	6
Choose X2 from	0	2.57	3	2.29	2.5	1	0
Choose X3 from	1.5	0.86	0	0.43	0	0	0

The function to be maximized is $F(X1,X2,X3) = 4\,X1 + 5\,X2 + 3\,X3$. Having chosen X1 = 0, we must then choose either

$$X2 = 0 \text{ with } X3 = 1.5$$

or

$$X2 = 2.57 \text{ with } X3 = 0.86$$

or

$$X2 = 3 \text{ with } X3 = 0$$

Each of these decisions results in a point at which the optimization function may be evaluated. Incorporating these values into the table results in Table 13.6. The boxed areas designate the individual solutions for each set of variable choices.

Choose X1 from		0		2		5	6	**Table 13.6**
		(0)		(8)		(20)	(24)	
Choose X2 from	0	2.57	3	2.29	2.5	1	0	
	(0)	(12.86)	(15)	(11.43)	(12.5)	(5)	(0)	
Choose X3 from	1.5	0.86	0	0.43	0	0	0	
	(4.5)	(2.57)	(0)	(1.29)	(0)	(0)	(0)	
Contribution of policy X2,X3	(4.5)	(15.43) *	(15)	(12.71) *	(12.5)	(5) *	(0) *	
Contribution of policy X1,X2,X3		(15.43)		(20.71)		(25) *	(24)	

According to the principle of dynamic programming, having chosen X1, we must now choose that combination of X2 and X3 that maximizes the contribution. The optimal result is discovered if the sequential decisions are made in a different order, for example:

Choose X3		0				0.43	0.86	1.5	**Table 13.7**
		(0)				(1.29)	(2.57)	(4.5)	
Choose X1	6	5	2	0	2	0	0		
	(24)	(20)	(8)	(0)	(8)	(0)	(0)		
Choose X2	0	1	2.5	3	2.29	2.57	0		
	(0)	(5)	(12.5)	(15)	(11.43)	(12.86)	(0)		
(X1,X2)	(24)	(25) *	(20.5)	(15)	(19.43) *	(12.86) *	(0) *		
(X1,X2,X3)		(25) *				(20.71)	(15.43)	(4.5)	

This example, although illustrative, is unfair in that it does not really demonstrate the power of the technique. A better example is given by the following problem.

A farmer has X1 pounds of potatoes at the beginning of a period of N years. He may sell an amount Y1 of this X1 pounds for a price $P(Y1)$ (obviously $0 \leqslant Y1 \leqslant X1$). He may plant the remainder, $(X1 - Y1)$ pounds, which yield $A(X1 - Y1)$ in the second year. This yield factor will be assumed to be constant and greater than 1 over the N-year period.

The farmer wishes to maximize income over the N-year period, and in order to achieve this must determine the set of $Y1, \ldots YN$ that maximizes the total T where

$$T = \sum_{j=1}^{N} P\,(Y_j)$$

Let $Fi(Xj)$ be the total income obtainable when starting with Xj pounds of potatoes and using an optimal policy of planting and sales over the remaining $N - j$ years to the end of the N-year period. Consider the final period. All the potatoes harvested in year N will be sold. Hence

$$F_0\,(X_N) = P(X_N) = P(Y_N)$$

In the penultimate year, $N - 1$, he will have X_{N-1} pounds to plant and his total income for the last two years will be

$$P(Y_{N-1}) + F_0\,(A(X_{N-1} - Y_{N-1}))$$

because

$$A\,(X_{N-1} - Y_{N-1}) = X_N$$

Hence

$$F_1(X) = \underset{0 \leqslant Y \leqslant X}{\text{Max}} \quad [P(Y) + F_0\,(\,A\,(\,X - Y)]$$

And, in general

$$F_i(X) = \text{Max} \quad [P(Y) + F_{i-1}\,(A(X - Y)]$$

$0 \leqslant Y \leqslant X$

This fearsome-looking expression is, in fact, quite simple because it merely tells us to do what we did previously, which is to keep optimizing the best to date.

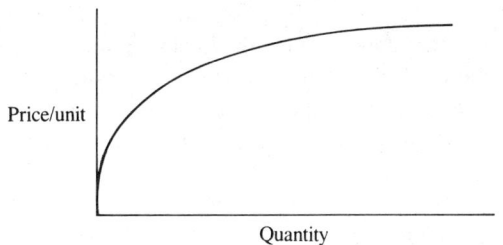

Figure 13.2

Some consideration of the structure of the problem will show that it is very dependent upon the shape of the price function. If the price function is linear (in other words, the 100th pound of potatoes will fetch the same price return as the tenth), then he should keep replanting the total crop and sell the lot at the end of the period. Similarly, if the function is concave upward (that is, the 100th will fetch more than the tenth), then he should again replant his yield until the end of the period. However, these are very unusual circumstances in practice; we would normally expect the price-quantity curve to have some form such as that in Figure 13.2. For illustration of such a form, we will use a price function where Y pounds will fetch a price P given by

$$P_Y = \$ (\log_{10} Y)$$

Furthermore, for illustration and simplicity we will assume A = 2.

Consider the problem of starting with 10 pounds of potatoes and maximizing income over three years. Table 13.8 shows the return expected from second-year policies of planting and selling potatoes in the second and third years.   **Table 13.8**

Final-Year Table of Returns on Second-Year Policies

X3	4	5	6	7	8	9	10	11	12	13	14	15	16
F0 (X3)	.60	.70	.78	.84	.90	.95	1.0	1.04	1.08	1.11	1.15	1.18	1.2

Y2	F1(Y2)												
2	.30				1.20[6]								
3	.48			**1.26[6]**			1.48[8]				1.62[10]		
4	.60	1.20[6]			**1.51[8]**				1.68[10]				
5	.70			1.48[8]				**1.70[10]**				1.84[12]		
6	.78					1.68[10]				**1.86[12]**				1.98[14]
7	.84			1.62[10]				1.84[12]				**1.99[14]**		
8	.90					1.81[12]				1.98[14]				**2.11[16]**
9	.95							1.95[14]				2.10[16]		
10	1													2.20[18]

In this table, the numbers in brackets are the amounts of potatoes considered in the final year decision. For example, if the farmer enters the final year with 12 pounds of potatoes, then he can choose to sell, say, 8 pounds at the beginning of the final year. This is represented by the value Y2 = 8. The return he receives from the sale of this 8 pounds is given by $F1(Y2)$, which is $\log_{10} 8 = 0.9$.

This 8 pounds sold from his 12 pounds will leave him with 4 pounds to plant, which will yield 8 pounds at the end of the period. As this is the period end, it will all be sold; therefore X3 = 8 and its return from the optimal policy of selling all the produce of the final year will be $F0(X3) = \log_{10} 8 = 0.9$. The return from this policy is therefore $0.9 + 0.9 = 1.81$ and this is placed in the Y2 = 8, X3 = 8 position of the table, together with the [12], indicating that this is the result of entering the final year with 12 pounds of potatoes.

All the alternative final-year policies are evaluated for different amounts of potatoes entering the final year. These are entered into the table and the best return for each amount discovered (these are marked in boldface). These values represent the best policy to adopt and the optimal return when entering the final year with this amount of potatoes. These entries are therefore transferred to the table representing the previous year's decision, Table 13.9.

Table 13.9

First-Year Table

X2		6	8	10	12	14	16	18
$F1(X2)$		1.255	1.505	1.699	1.857	1.989	2.107	2.209
Y1	$F2(Y1)$							
1	0							2.209[10]
2	0.301						2.408[10]	
3	0.477					**2.466[10]**		
4	0.602				2.459[10]			
5	0.699			2.398[10]				
6	0.778		2.283[10]					
7	0.843	2.098[10]						

In this table, only amounts of 10 pounds are considered, because that is the amount available at the beginning of this time period. The best alternative is to sell 3 pounds now, producing an immediate revenue of $0.477, and enter the final period with 14 pounds, which from Table 13.8 we know has an optimal return of $1.989 by selling 7 pounds and planting 7 pounds at the end of the second year, resulting in 14 pounds for sale at the end of the third year.

This result is:

Sell	3	lb now	for	$0.477
	7	lb next year	for	$0.843
and	14	lb at the end	for	$1.146
		Total return		$2.466

Incidentally, the tables also show the optimal two-year policy for 10 pounds of potatoes:

Sell	5	lb now	for	$0.699
and	10	lb at the end	for	$1.000
		Total return		$1.699

The tables may be extended to cover more years as follows: The first-year table is extended to cover more entries with corresponding extensions to the earlier tables, which result in the following return tables.

Return Tables

Table 13.10

(multiplied by 100; rounding errors included)

		4	5	6	7	8	9	10	11	12	13	14	15	16	17	18
		60	70	78	85	90	95	100	104	108	111	115	118	120	123	126
2	30	90	100	106	115	128	126	130	134	138	142	144	148	150	153	156
3	48	108	118	126	132	138	143	148	152	156	159	162	165	168	171	173
4	60	120	130	138	145	150	156	160	164	168	172	175	178	181	183	186
5	70	130	140	148	154	160	165	170	174	178	181	184	188	190	193	195
6	78	138	148	156	162	168	173	178	182	186	189	192	195	198	201	203
7	85	145	154	162	169	175	180	185	189	192	196	199	202	205	208	210
8	90	151	160	168	175	181	186	190	194	198	202	205	208	211	213	216
9	95	156	165	173	180	186	191	195	200	203	207	210	213	215	219	221
10	100	160	170	178	185	190	195	200	204	208	211	215	218	220	223	226
11	104	164	174	182	189	194	200	204	208	212	216	219	222	225	227	230
12	108	168	178	186	192	198	203	208	212	216	219	223	226	228	231	233
13	111	172	181	189	196	202	207	211	216	219	223	226	229	232	234	237
14	115	175	185	192	199	205	210	215	219	223	226	229	232	235	238	240

Table 13.11

		126	151	170	186	199	211	221
2	30	157	181	200	216	229	240	251
3	48	173	198	218	233	247	258	269
4	60	186	211	230	250	260	271	281
5	70	195	220	240	256	269	281	291
6	78	203	228	248	264	277	289	299
7	85	210	235	254	270	283	295	305
8	90	216	241	260	276	289	301	311
9	95	221	246	265	281	294	306	316
10	100	226	251	270	286	299	311	321

Table 13.12

		218	247	271	291	308	324
2	30	248	277	301	321	339	354
3	48	265	294	319	339	356	371
4	60	278	307	331	351	369	384
5	70	288	317	341	361	378	394
6	78	295	324	349	369	386	402
7	85	302	331	355	375	393	408
8	90	308	337	361	381	399	414
9	95	313	343	366	386	404	419
10	100	318	345	371	391	408	424

Table 13.13

		301	339	371	398	421	442
2	30	331	369	402	428	452	472
3	48	349	386	419	445	469	489
4	60	361	399	432	458	482	502
5	70	371	408	441	468	491	511
6	78	379	416	449	475	499	519
7	85	386	423	456	482	506	526
8	90	391	429	462	488	512	532
9	95	396	434	467	494	517	537
10	100	401	439	471	498	521	542

The tables can also be used if the problem is extended to cover more years. The final year is always the final year, and the penultimate year remains the penultimate year. Extra entries are made in the tables to cover other options. The result of extending the problem to four years is:

Sell	3 lb now	for $0.477
	5 lb next year	for $0.699
	9 lb in the third year	for $0.954
and	18 lb in the final year	for $2.209
	Total return	$3.385

The tables produced for this analysis could conveniently be generated with a spreadsheet program on a microcomputer. Such a program is available in LOTUS $1-2-3$ for the IBM PC, or in the integrated package format of LOTUS Symphony. Ashton-Tate's Framework will also provide a very nice environment for producing these tabular calculations. Virtually every microcomputer has available some "calc" (such as Visicalc) enabling these tables to be produced easily and quickly.

DYNAMIC PROGRAMMING WITH STOCHASTIC VARIABLES

Sequential decisions are not always made using deterministic variables. In many cases, the variables are probabilistic in nature, but dynamic programming can be used to solve such sequential decision problems too.

Consider the case of a person who has decided to sell her car. She has looked up the Blue Book price and has discovered that the car has a low listing of $1,500 and a high of $2,100. She has advertised the car in the local newspaper, and four people have phoned to make definite appointments. She calculates the probabilities of selling the car at a particular price to be

Price Range	Probability of Sale
> $2,100	0
$1,900–$2,100	0.2
$1,700–$1,900	0.35
$1,500–$1,700	0.45

She decides that she will definitely sell the car to one of the four people.

Consider the decisions to accept or reject a particular offer. Since she will definitely sell the car to one of the four, we consider first the fourth or last offer. If she has reached the stage of rejecting three previous offers, then her policy will be to accept the last, final offer. Thus, for the fourth bid,

$$F_4(X_4) = X_4$$

The expected value of a bid in isolation is calculated as the mean of the distribution of bids expected:

Range	Midpoint	Probability	
	Xi	Pi	PiXi
$2,100	$2,000	0.2	400
1,900	1,800	0.35	630
1,700	1,600	0.45	720
1,500			

Sum $1,750

The expected value of the final bid is therefore $1,750. For the third bid,

$$F_3(X_3) = \text{Max} \left(X_3 ;[(2000)(0.2) + (1800)(0.35) + (1600)(0.45)]\right)$$
$$= \text{Max} \left(X_3 ; 1750\right)$$

The value of the third bid is the better of the third offer, (X3), or of the fourth bid, which has an expected value of $1,750. The decision process for the third bid is therefore, "Accept third bid if greater than $1,750." For the second bid,

$$F_2(X_2) = \text{Max} \left(X_2 ;[(2000)(0.2) + (1800)(0.35) + (1750)(0.45)]\right)$$
$$= \text{Max} \left(X2 ; 1817.50\right)$$

The decision process for the second bid is, then, "Accept second bid if greater than $1,817.50." For the first bid, therefore,

$$F_1(X_1) = \text{Max} \left(X_1 ;[(2000)(0.2) + (1817.50)(0.35) + (1750)(0.45)]\right)$$
$$= \text{Max} \left(X1 ; 1823.63\right)$$

or, "Accept first bid if greater than $1,823.63." The optimal policy for the seller to adopt is then

1. Accept first bid if it is better than $1,823.62.

2. Accept next bid if it is better than $1,817.50.

3. Accept third bid if it is better than $1,750.

4. Accept last bid.

The approach used in the examples thus far has been to start the decision process at the end and work backward. It may have occurred to the reader that the system will also work for decisions made going forward. This is a true observation. Dynamic programming may be carried out in a forward pass approach, but the backward pass is normally adopted in practice because it has generally been found to be more efficient. Also, examination of the tables generated in the farmer's potato problem shows that the subproblems can be used to solve problems other than the original formulation, for example, for longer time periods or alternate start conditions.

The farmer problem indicates both the power of the technique and its weakness. For example, dynamic programming (DP) seems able to handle many practical problems that could not be handled by an LP type approach. However, many calculations are performed that are not used directly in the solution policy; imagine a multivariable problem such as is easily handled by LP being computed using multidimensional tables. Under these conditions the technique can rapidly become swamped under the weight of computation.

OPTIMAL POLICIES FOR AN INFINITE SEQUENCE OF DECISIONS

To solve the problems thus far, we have used sequences with both a start point and an end point. Thus, whether using forward or backward recursion, we had a place to start and a place to finish. What about ongoing repair and maintenance policies? Consider the following example.

A machine can be in one of two states: broken or working. If the machine is broken, there are two alternatives, repair at a cost of $50 or replace at a cost of $150. If the machine is working, there are two alternatives, perform preventive maintenance at a cost of $20 or ignore it at a cost of $0.

If a working machine operates throughout the week, the product produced has a value of $100. If the machine ends the week in a broken state, the product is scrapped at a cost of $10.

A new machine is guaranteed to work throughout the week. A repaired machine will work through the week with a probability of 0.8. A maintained machine will work through the week with a probability of 0.6. A nonmaintained machine will work through the week with a probability of 0.3.

A policy of repair/replace and maintenance decisions is needed for an infinite sequence of weekly operations.

Let B1 = a broken machine at week 1
B2 = a broken machine at week 2
W1 = a working machine at week 1
W2 = a working machine at week 2
X = an element of chance

Decision-tree elements may then be produced for the states produced from a broken machine at week 1, and also a working machine at week 1 ending at week 2, as shown in Figure 13.3.

Figure 13.3

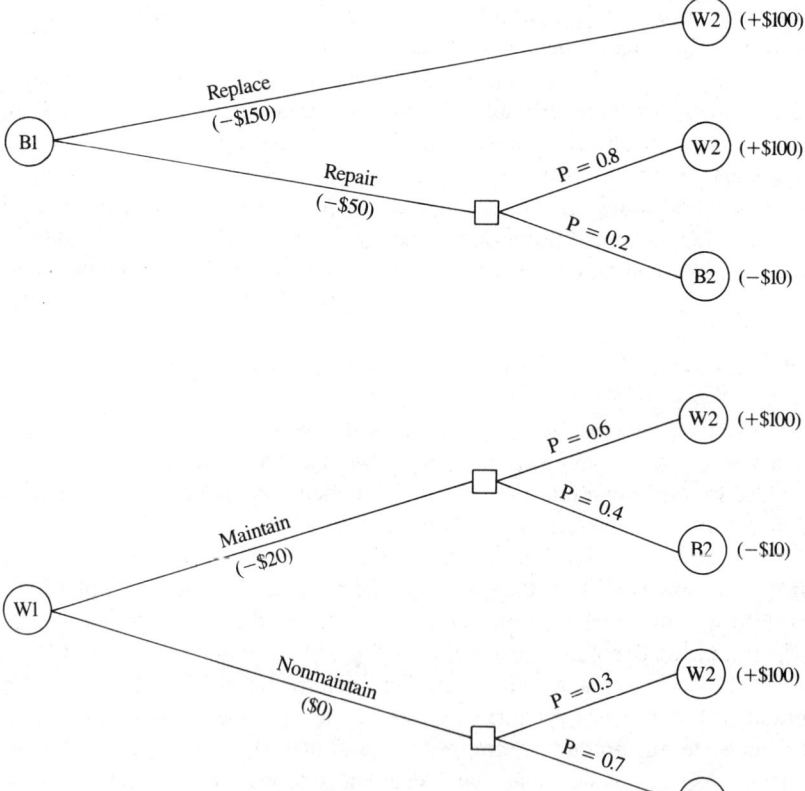

The infinite sequence of decisions is represented by each of these elemental decision trees being joined together so that a W tree is tacked onto the end of a W branch, and a B tree is tacked onto a B branch. The tree will then extend to infinity in both directions.

Since we do not know the values of either a broken or working machine, let

$$VB2 = \text{value of a broken machine at week 2}$$
$$VW2 = \text{value of a working machine at week 2}$$

Thus we can calculate the values of broken and working machines at week 1 from

$$VB1 = \text{Max} \left([VW2 - 50] ; [(0.8)(VW2 + 100) + (0.2)(VB2 - 10) - 50] \right)$$
$$VW1 = \text{Max} \left([(0.6)(VW2 + 100) + (0.4)(VB2 - 10) - 20]; \right.$$
$$\left. [(0.3)(VW2 + 100) + (0.7)(VB2 - 10)] \right)$$

There are too many terms in these expressions to allow these to be solved directly. However, by using the optimal policy, whatever that is, the average profit or return from the system will be some value, which we can call G. In a stable situation, the value of a machine in a particular state at week 1 will be the value of the machine in the same state at week 2 plus the value generated by the system in the intervening period, that is, G. Thus

$$VB1 = VB2 + G$$

and

$$VW1 = VW1 + G$$

Hence

$$VB2 + G = \text{Max} \left((VW2 - 50) ; [(0.8)(VW2 + 100) + (0.2)(VB2 - 10) - 50] \right)$$
$$VW2 + G = \text{Max} \left([(0.6)(VW2 + 100) + (0.4)(VB2 - 10) - 20]; \right.$$
$$\left. [(0.3)(VW2 + 100) + (0.7)(VB2 - 10)] \right)$$

These relationships represent alternative pairs of simultaneous equations. Because there are three variables for the pairs of simultaneous equations, no unique solutions exist. However, we are not particularly interested in the absolute values of the machines in their working or broken states at week 2, but rather in the difference in value between them. We may thus allow $VB2 = 0$, and, for convenience, set $VW2 = Y$. Thus Y indicates the extra value of a working machine at the end of week 1 above the value of a broken one at that time. Thus

$$G = \text{Max } ((Y - 50); [(0.8)(Y + 100) + (0.2)(-10) - 50])$$
$$Y + G = \text{Max } ([(0.6)(Y + 100) + (0.4)(-10) - 20];$$
$$[(0.3)(Y + 100) + (0.7)(-10)])$$

or

$$G = \text{Max } [(Y - 50); (0.8\,Y + 28)]$$
$$Y + G = \text{Max } [(0.6\,Y + 36); (0.3\,Y + 23)]$$

This represents four alternative combinations:

(1) $G = Y - 50$
       $Y + G = 0.6\,Y + 36$

(2) $G = Y - 50$
       $Y + G = 0.3\,Y + 23$

(3) $G = 0.8\,Y + 28$
       $Y + G = 0.6\,Y + 36$

(4) $G = 0.8\,Y + 28$
       $Y + G = 0.3\,Y + 23$

The solutions to these sets are

(1) $Y = 61.43$ $G = 11.43$
(2) $Y = 42.94$ $G = -7.06$
(3)   **$Y = 6.67$ $G = 33.33$**
(4) $Y = -3.33$ $G = 25.33$

The solution that maximizes the average weekly profit or return at $33.33 is the policy given by option (3), which, incidentally, can be checked for consistency as being the maximum of the alternate options. The optimal policy is therefore

Repair a broken machine

Maintain a working machine

Return from system $=$ \$33.33 per week

CHAPTER 13 PROBLEMS

The solutions to Problems 1 through 5 are found at the end of the chapter.

1. A special timber-curing kiln costs \$100,000 and is guaranteed for one year. A kiln has a four-year life, at the end of which it may be sold for scrap at \$5,000. A batch of timber must be cured for a year, after which it may be sold for a profit of \$100,000. If a kiln is out of service at the end of any year, it may be repaired or replaced by a new kiln. It is impractical to purchase and reassemble a used kiln.

At the end of the second year, an unserviceable two-year-old kiln costs \$20,000 to refurbish, with the probability of successful operation in the third year of 0.4. At the end of the third year, an unserviceable three-year-old kiln costs \$30,000 to refurbish, with the probability of successful operation in the fourth year of 0.1.

Maintenance costs \$2,000 on a serviceable kiln, with the probabilities of successful operation in the successive years of 0.9 in year 2, 0.7 in year 3, and 0.3 in year 4. Without maintenance, these probabilities are 0.7, 0.5, and 0.1, respectively. If a kiln fails during the curing operation, the batch may be sold for scrap at \$2,000. What would be your advice as to the optimal kiln replacement/maintenance policy? What is the average return per year that your policy would give?

If it is possible to acquire and renovate used kilns over two years old (including scrap kilns over four years old) at a renovation cost of \$50,000, with a probability of successful operation in the following year of 0.5, how would your analysis be affected? What value does your analysis suggest you would place on used, working kilns, one, two, and three years old?

2. Tim Carter owns a truck that is capable of a maximum payload of 20 tons. He has a contract offer to move four types of merchandise between two warehouse centers. This merchandise is palletized and will easily fit into the truck. The details of the merchandise are as follows:

Merchandise	Weight/Pallet (tons)	Profit/Pallet Shipped
1	8	$1,050
2	6	650
3	3	275
4	1.6	175

How should Carter load his truck for maximum profit on one trip?

3. Col. Yuri Roundoff is cultural attaché to the Peoples Republic Embassy and is due to be returned to his homeland following a tour of duty. He has five weeks in which to sell his custom-stretched Cadumerc-Rolls limo, which cannot be shipped with him.

The Red Book guide to prices says that given the distribution of stretched limo sales, he can expect bids on his limo with the following probabilities:

Bid Price	Probability
$20,000	0.5
30,000	0.2
40,000	0.2
50,000	0.1

At the end of each week, Col. Roundoff intends to consider the best of each week's bids and decide whether to accept it or wait and see what next week brings. What policy of acceptance should he follow?

4. Two jugs can hold 5 pints and 7 pints of water, respectively. What is the minimum number of operations needed to leave 3 pints of water in the 5-pint jug and nothing in the 7-pint jug ?

5. What is the shortest route through the following graph?

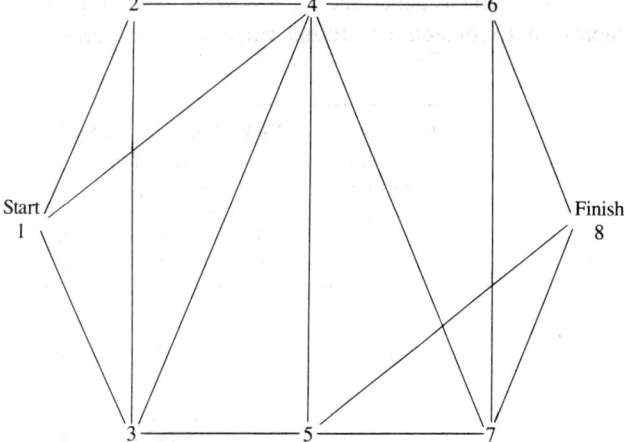

Figure 13.4

From	To							
	1	2	3	4	5	6	7	8
1	0	20	15	30	—	—	—	—
2	20	0	10	5	—	—	—	—
3	20	10	0	25	30	—	—	—
4	30	8	25	0	35	22	40	—
5	—	—	30	28	0	—	35	40
6	—	—	—	36	—	0	—	20
7	—	—	—	35	35	—	0	10
8	—	—	—	—	40	20	10	0

Note that the costs are not equidirectional. This would be typical of routes in an inner city where one-way systems are extensive, for example.

HOMEWORK PROBLEMS

6. The Columbia Enterprise Shuttle Bus Corporation has decided to expand operations at three major cities. The Board Chairman has decided to apportion money in blocks of $1 million to the newly appointed operations managers for these new sites. The decision support management systems group has generated the following evaluations of the proposed investments, in discounted net-present-value (NPV) criteria:

Investment Amount ($ million)	NPV of Returns Expected from Investment ($ million)		
	City 1	City 2	City 3
1	—	3	5
2	6	5	8
3	9	9	11
4	12	10	—

Formulate and solve as a dynamic programming problem.

7. What is the shortest route through the graph represented by the following table?

From	To							
	1	2	3	4	5	6	7	8
1	0	10	25	30	10	5	—	—
2	20	0	12	9	8	—	—	—
3	20	10	0	15	40	22	35	—
4	30	20	20	0	35	22	40	—
5	—	14	30	28	0	—	35	40
6	—	—	—	36	—	0	—	20
7	—	—	10	35	35	11	0	10
8	—	—	—	—	20	20	8	0

Note that the costs are not equidirectional. This would be typical of routes in an inner city where one-way systems are extensive.

8. A customs inspector suspects that one of a set of eight cases has been tampered with to smuggle a controlled substance into the country. As this case will not weigh the same as the others, what procedure would you use to determine the minimum number of weighings needed to find the culprit?

9. Five cannibals and five missionaries must cross a river, and there is only one canoe with a capacity of three persons. As the number of cannibals must never exceed the number of missionaries in any situation, how can the task be most efficiently accomplished? Generalize your approach to several cannibals and several missionaries with a boat of different capacity.

10. The rangers in Los Anglos National Forest have been notified that a hiker has not returned his wilderness permit at the expected time. The rangers have five search areas for this particular hiker, and it is thought that he will be in each of these areas with probabilities as listed:

Area	Probability
1	0.4
2	0.2
3	0.2
4	0.1
5	0.1

The time taken to search an area is

Area	Time (days)
1	2
2	2
3	1
4	1
5	1

If the hiker is known to have supplies to last for five days, how should the rangers plan their search, if they can only effectively search one area at a time?

CHAPTER 13 SOLUTIONS

1. First summarize the problem: timber-curing kilns cost $100,000, maximum life four years with scrap value of $5,000.

Broken kilns:		
Age (yrs)	Repair Cost ($)	Probability of Working Next Year
2	2,000	0.4
3	3,000	0.1

Kiln maintenance:			
Age (yrs)	Maintenance Cost	Probability of Working Next Year ($)	
		With	Without
		Maintenance	
1	2,000	0.9	0.7
2	2,000	0.7	0.5
3	2,000	0.3	0.1

Timber from failed kilns has scrap value of $2,000. New kilns have a one-year guarantee, which ensures sucessful operation for one year: The purchaser is compensated in full for failure in year 1 and is guaranteed starting year 2 with a working kiln.

Kilns may be scrapped and replaced at any time in their working life. Let

a brand-new working kiln be designated W0 with a value of h
a 1-year-old working kiln be designated W1 with a value of g
a 2-year-old working kiln be designated W2 with a value of a
a 2-year-old broken kiln be designated B2 with a value of b
a 3-year-old working kiln be designated W3 with a value of c
a 3-year-old broken kiln be designated B3 with a value of d
a 4-year-old working kiln be designated W4 with a value of e
a 4-year-old broken kiln be designated B4 with a value of f

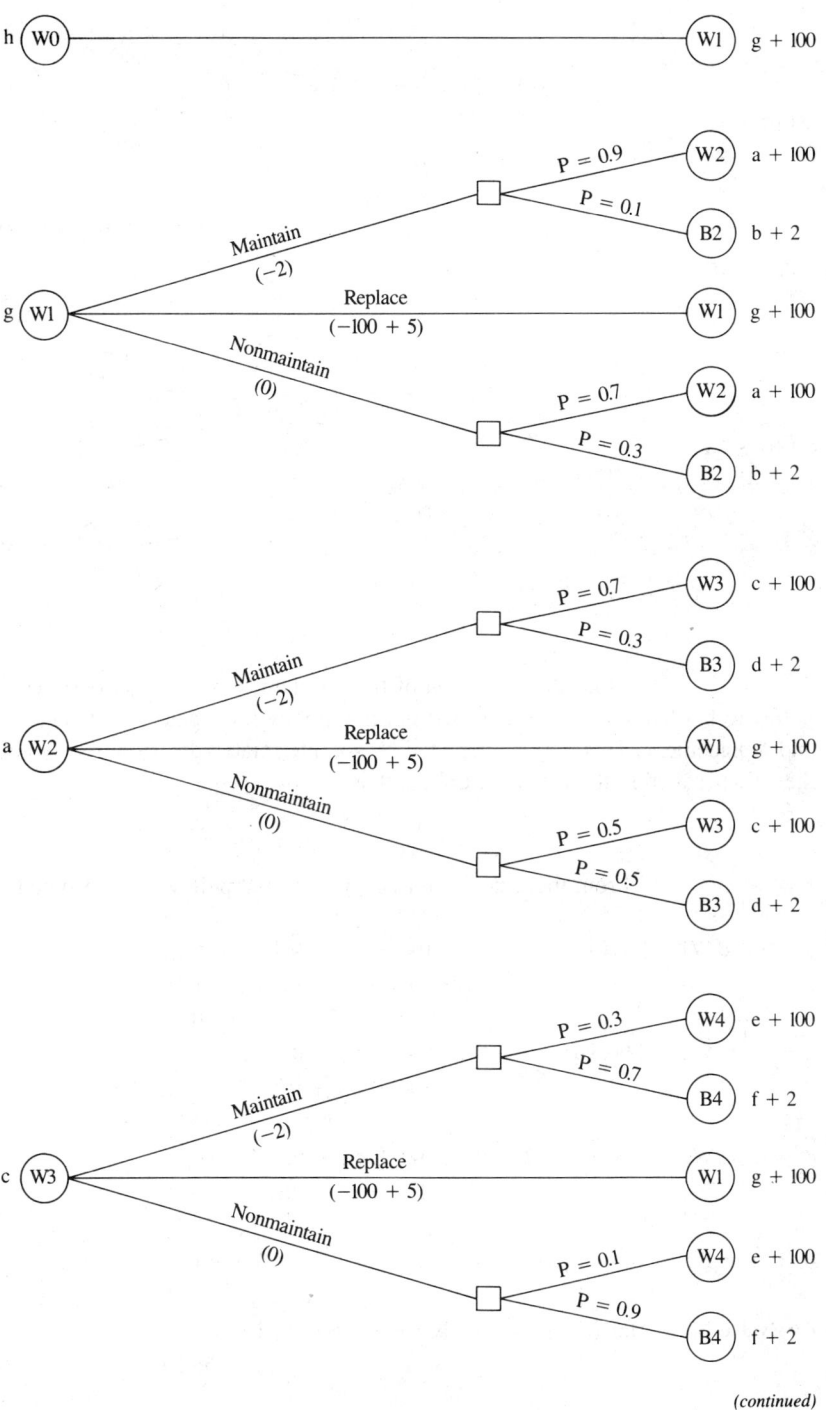

Figure 13.5

(continued)

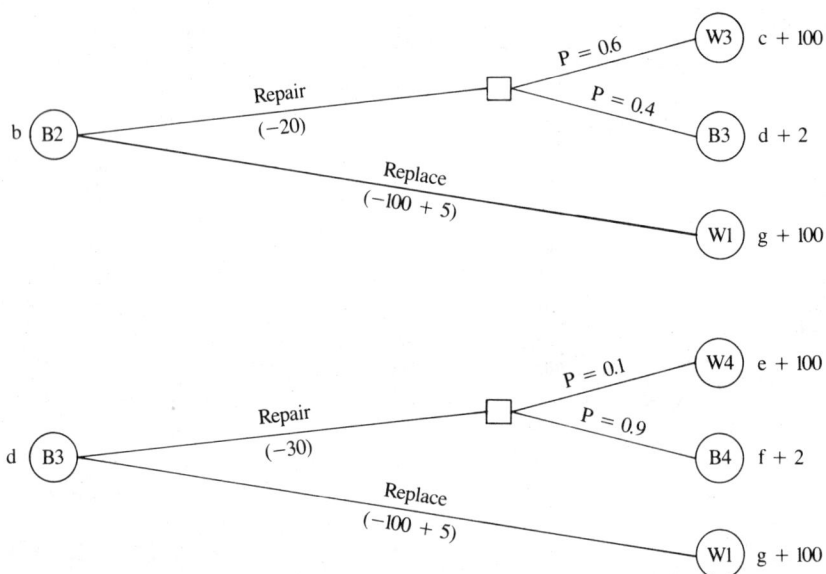

From the definition of the problem, $e = f = \$5,000$. The policy to be determined is that of the optimal policy for repair, maintenance, and replacement. This may be depicted by six interrelated decision trees with the values stated in thousands of dollars thus:

From these decision trees the optimal policy may be stated:

g = value W1 = Max ([− 2 + 0.9 (100 + a) + 0.1 (2 + b)];
         [0.7 (100 + a) + 0.3 (2 + b)];
         [g + 50])

a = value W2 = Max ([− 2 + 0.7 (100 + c) + 0.3 (2 + d)];
         [0.5 (100 + c) + 0.5 (2 + d)];
         [g + 50])

c = value W3 = Max ([− 2 + 0.3 (100 + c) + 0.7 (2 + f)];
         [0.1 (100 + c) + 0.9 (2 + f)];
         [g + 50])

b = value B2 = Max ([- 20 + 0.4 (100 + c) + 0.6 (2 + d)];
         [g + 50])

d = value B3 = Max ([- 30 + 0.1 (100 + c) + 0.9 (2 + f)];
         [g + 50])

Let us assume that the operations management of the kilns will smooth out the fluctuations in the revenue stream resulting from the operations and substitute an annual fixed amount Q, which is equivalent to the average kiln operation revenues. The goal of the management is therefore to maximize Q, the average annual return.

Solving these equations iteratively, we obtain the following optimal policy:

State of Kiln	Commencing Year of Operation		
	2	3	4
Working	Maintain	Maintain	Replace
Broken		Replace	Replace

The average annual profit achieved by the above policy =
$5,343.

The introduction of renovated kilns into the analysis adds the decision alternative to a new kiln:

$$p \; (W0) \quad \frac{\text{New kiln}}{-\;100\;+\;5} \qquad\qquad (W1) \quad g\;+\;100$$

$$\text{Scrap kiln}$$
$$(W) \quad +\;100$$
$$+ \quad 5$$

$$P = 0.5$$

$$q \; (Wr) \quad \frac{\text{Renovated kiln}}{-\;50}$$

$$P = 0.5$$

$$\text{Scrap kiln}$$
$$(B) \quad +\;2$$
$$+\;5$$

$$p = W1 + 5 - Q \qquad\qquad\qquad = \;+\;\$5{,}000$$
$$q = 0.5\,(\,100 + 5\,) + 0.5\,(\,2 + 5\,) - 50 - Q = \;-\;\$4{,}743$$

Therefore the availability of renovated kilns on the terms stated will not affect your analysis.

Value of kilns aged
1 year $53,430
2 years $20,170
3 years $ 5,000

2. Let the tonnage available on the truck for loading pallets of type i (i = 1 to 4) be W after some pallets have already been loaded on the truck. Loading Xi pallets of type i, the value of the Xi pallets is Xi Vi. Define $f_i(W)$ as the potential value with Xi pallets already loaded. Then

$$f_i(W) = \max_{0 < Xi < \frac{W}{Wi}} [\; Xi\; Vi + f_{i+1} (W - Xi\; Wi)],\; i > 2$$

Solution of this recursive relationship results in the solution:

Load two pallets each of types 1 and 4.
Total profit for the trip = $2,450.
Total weight on truck = 19.2 tons.

3. Col. Roundoff knows that if he has not sold his limo by week 5 he will have to accept whatever he is offered in the final week. In week 4 he computes

$$f_1 = \max \begin{cases} \text{Accept} & X \\ \text{Reject} & 0.5\,(20{,}000) + 0.2\,(30{,}000) \\ & + 0.2\,(40{,}000) + 0.1\,(50{,}000) \end{cases}$$

that is,

$$f_1 = \max\,[X;\; 29{,}000]$$

Thus if at the end of week 4 he has a bid better than $29,000, he should accept it. For week 3,

$$f_2 = \text{Max} \begin{cases} \text{Accept} & \text{X} \\ \\ \text{Reject} & 0.5\,(29,000) + 0.2\,(30,000) \\ & + 0.2\,(40,000) + 0.1\,(50,000) \end{cases}$$

that is,

$$f_2 = \text{Max } [X; \ 33,500]$$

Thus if at the end of week 3 he has a bid better than $33,500, he should accept it. For week 2,

$$f_3 = \text{Max} \begin{cases} \text{Accept} & \text{X} \\ \\ \text{Reject} & 0.5\,(33,500) + 0.2\,(33,500) \\ & + 0.2\,(40,000) + 0.1\,(50,000) \end{cases}$$

that is,

$$f_3 = \text{Max } [X; \ 36,450]$$

Thus at the end of week 2, if he has a bid better than $36,450 he should accept it. Similarly,

$$f_4 = \text{Max } [X; \ 38,515]$$

and

$$f_5 = \text{Max } [X; \ 40,051]$$

4.

Fill the 5-pint jug from the faucet.
Fill the 7-pint jug from the 5-pint jug.
Fill the 5-pint jug from the faucet.
Pour as much as possible from the 5-pint jug into the 7-pint jug.
Empty the 7-pint jug.

5. The optimum route is

1—2—4—6—8

Minimum cost = 67

14

Goal Programming and Stochastic Programming

In all of the problems considered so far, there has been only one objective to consider. In the problems of Chapter 1, profit (contribution to profit or minimization of cost) was the sole aim. Thus one objective function was developed that was optimized over a set of constraint relationships. Although maximizing profit is widely considered by classical economists and by some academics to be the theoretical sole purpose of a firm, many businesses frequently place higher priorities on noneconomic objectives. Firms often seek a "satisfactory" level of profit while also meeting other social or environmental considerations. Moreover, many public organizations have no profit motive, defense and environmental agencies or churches, for example. In other words, in the real world, people tend to be "satisficers" over several areas, rather than optimizers of a single variable such as profit. There is much more of a trade-off between specific constraints and objectives; a "satisficer" therefore attempts to achieve a satisfactory level of achievement for several objectives within certain boundaries. Consider the basic problem of the edible fats manufacturer:

Maximize	$4 X_1 + 5 X_2$	
subject to	$X_1 + X_2 \leq 6$	Sunflower oil
	$X_1 + 2 X_2 \leq 7$	Plant capacity
	$X_1 + 4 X_2 \leq 12$	Labor utilization
	$X_1 \geq 0$	
	$X_2 \geq 0$	

Rather than maximizing the profit from the operation, the manufacturer decides on a more sophisticated approach. It considers the total situation and formulates the following goals for the organization:

1. Generate at least $20 profit.

2. Avoid excessive use of sunflower oil, minimizing the purchase of new oil.

3. Reduce labor underutilization.

The manufacturer wants these goals to be achieved in the priority listed. If not all of the goals are feasible, then the earlier listed goals take precedence over the later listed goals.

In order to express these goals, we must consider a representation that allows for both underachievement and overachievement of the goal. In the straightforward linear programming approach, the constraints on the solution are expressed as being on one side or the other of the constraint boundary. For example, plant capacity

$$X1 + 2\,X2 \leqslant 7$$

is represented by the relationship

$$X1 + 2\,X2 + X4 = 7$$

where $X4$ is a positive slack variable indicating the underuse of the constraint commodity. Also, for the following constraint that arises in the dual formulation of this problem

$$Y1 + 2\,Y2 + 4\,Y3 \geqslant 5$$

the relationship is expressed with the use of a negative slack variable thus:

$$Y1 + 2\,Y2 + 4\,Y3 - Y5 = 5$$

indicating that the region for acceptable solutions lies above the constraint boundary

$$Y1 + 2\,Y2 + 4\,Y3 = 5$$

In the case of any goal, the goal may be exceeded or a solution may fall short of a goal. Thus, in expressing a goal, the mathematical relationships must take into account the possibilities of the goal being exceeded or falling short of it. This is achieved by the use of deviational variables. Consider the goals in order of their priority. First is that a satisfactory level of profit is $20. The profit level obtained is given by the expression

$$\text{Profit} = 4\,X1 + 5\,X2$$

If the goal is for the profit to be $20, then the goal may be unmet or exceeded. This is represented by the expression

$$4\,X1 + 5\,X2 + d_1^- - d_1^+ = 20$$

where there are two deviational variables: d_1^- is the underachievement of the $20 profit goal, and d_1^+ is the overachievement of the $20 profit goal. This expression is thus a constraint on the possible solutions that might be selected to satisfy the aims of management.

In order to determine whether it is desirable to overachieve or to underachieve a goal, an optimization function is constructed that reflects the merit of being on one side or other of the constraint. In this profit goal, we wish to minimize the underachievement of profit at the $20 level; thus it is desirable to minimize d_1^-, and the optimization function may be represented by

$$\text{Minimize } Z = P_1\, d_1^-$$

The second of the goals expressed as desirable by management is to avoid excessive use of sunflower oil. The original constraint on sunflower oil was

$$X1 + X2 \leqslant 6$$

For the goal of minimizing deviations from the current resource level of sunflower oil, the expression representing sunflower oil can be formed as

$$X1 + X2 + d_2^- - d_2^+ = 6$$

Here d_2^- and d_2^+ represent the deviations from the objectives of the second goal. In this case, it is desirable to be under the constraint boundary, so the optimization function will represent this desirability. In the same way that a negative slack variable represents the unused resource, a deviational variable with a negative coefficient represents the underutilization of the resource goal; in this relationship that variable is d_2^+. The optimization function now becomes

Minimize $Z = P_1\, d_1^- + P_2\, d_2^+$

The third goal is to reduce labor underutilization. The expression for this becomes

$$X1 + 4\, X2 + d_3^- - d_3^+ = 12$$

with the corresponding modification of the optimization function:

Minimize $Z = P_1\, d_1^- + P_2\, d_2^+ + P3\, d_3^-$

There is still a constraint on the solution—that of plant capacity. This is handled in the usual way for a simplex linear programming solution,

$$X1 + 2\, X2 \leqslant 7$$

and becomes, with the addition of a slack variable,

$$X1 + 2\, X2 + U = 7$$

The analysis has resulted in a linear program with an optimization function made up of four different, possibly conflicting, priority requirements. The problem may be expressed in the form

Minimize $Z = P_1\, d_1^- + P_2\, d_2^+ + P_3\, d_3^-$
subject to
$$4\, X1 + 5\, X2 + d_1^- - d_1^+ = 20$$
$$X1 + X2 + d_2^- - d_2^+ = 6$$
$$X1 + 4\, X2 + d_3^- - d_3^+ = 12$$
$$X1 + 2\, X2 + U = 7$$

This representation may be placed in the form of a simplex tableau, but as there is an objective function that is really three different objective functions, the simplex tableaux will be generated with three $Zj - Cj$ rows, each row expressing a different goal.

Tableau 14.1

Cj					$-P_1$		$-P_3$			$-P_2$	
Ci	Basis	B	X1	X2	d_1^-	d_2^-	d_3^-	U	d_1^+	d_2^+	d_3^+
$-P_1$	d_1^-	20	4	5	1				-1		
0	d_2^-	6	1	1		1				-1	
$-P_3$	d_3^-	12	1	4			1				-1 *
0	U	7	1	2				1			
	P_3	-12	-1	-4							1
Zj-Cj	P_2	0								1	
	P_1	-20	-4	-5					1		

**

The solution proceeds by taking the goals in order of priority. In this case, the first priority goal has not been achieved because the row applicable to the P1 goal has negative entries in its $Zj - Cj$, row, indicating a possible improvement. Therefore, ignoring the other goals for the present, the solution to the problem may be improved by pivoting into the basis the variable with the largest element in the $Zj - Cj$ row for the priority goal. Here it is the X2 column. The pivot row is discovered just as in the normal simplex tableau. This is then updated to

Tableau 14.2

Cj					$-P_1$		$-P_3$			$-P_2$	
Ci	Basis	B	X1	X2	d_1^-	d_2^-	d_3^-	U	d_1^+	d_2^+	d_3^+
$-P_1$	d_1^-	5	2.75	0	1		-1.25		-1		1.25*
0	d_2^-	3	0.75	0		1	$-.25$			-1	.25
0	X2	3	0.25	1			.25				$-.25$
0	U	1	0.5	0			$-.5$	1			.5
	P_3	0					1				
Zj-Cj	P_2	0								1	
	P_1	-5	-2.75				1.25		1		-1.25

**

This tableau indicates that by manufacturing 3 units of X2 with 0 units of X1, the primary objective of making at least \$20 profit has not been achieved. This is indicated by the $P1$ row of the multiple $Zj - Cj$ rows, where negative entries remain. Choosing the largest negative of these, the pivot column of X1 is indicated to enter the basis. Using the usual calculations to discover the appropriate pivot row, d_1^- is chosen to leave the basis, resulting in the next tableau:

Tableau 14.3

Cj						$-P_1$		$-P_3$		$-P_2$		
Ci	Basis	B	X1	X2	d_1^-	d_2^-	d_3^-	U	d_1^+	d_2^+	d_3^+
0	X1	1.82	1	0	.36		−.45		−.36		.45
0	d_2^-	1.64	0	0	−.27	1	−.08		.27	−1	−.08
0	X2	2.55	0	1	−.09		.36		.09		−.36
0	U	.09	0	0	−.18		−.28	1	.18		.28
	P3	0						1			
Zj-Cj	P2	0								1	
	P1	0			1						

Here, the three objectives have been satisfied. The solution with X1 = 1.82 and X2 = 2.55 has satisfied the goal of profit being at least \$20:

$$4 \times 1.82 + 5 \times 2.55 = \$ 20.03$$

The second goal, of avoiding excessive use of sunflower oil, has been realized, for

$$1.82 + 2.55 = 4.37 \text{ tons of sunflower oil,}$$

while the third goal, of using all labor,

$$1.82 + 4 \times 2.55 = 12.02 \text{ units of labor,}$$

has also been realized. (Note the rounding errors from the two-digit calculations.) The final constraint of plant capacity,

$$1.82 + 2 \times 2.55 = 6.92 \text{ units of plant capacity,}$$

has also been observed.

As it happens, all three goals could be met; in many cases, however, this will not be achievable. Consider the case where the $Z_j - C_j$ rows appear with the following coefficients

Tableau 14.4

		d_1^-	d_2^-		d_3^-
		.			
			.		
			.		
	P_3	-12	-1.5		-2
$Z_j - C_j$	P_2	-5	$-.8$	2	$-.5$
	P_1	0	1		

**

Here the first goal has been achieved: There are no negative entries in the $Z_j - C_j$ row for the $P1$ goal. However, the second- and third-ranking goals have not been met. In order to proceed toward goal $P2$, bringing into the basis the deviational variable d_1^- would provide the most effective means of satisfying $P2$, but it would be at the expense of the higher priority goal $P1$, which has already been achieved. However, there is a less efficient, but effective variable that can be brought into the basis without reducing the effectiveness of the primary goal: Variable d_3^- is therefore chosen to enter the basis.

Suppose the final $Z_j - C_j$ looks like this:

Tableau 14.5

		d_1^-	d_2^-		d_3^-
		.			
		.			
		.			
	P_3	-15	-2.5		-4
Z_j-C_j	P_2	0		3	1
	P_1	0	1		

The first- and second-priority goals have been achieved, but the third-priority goal has not. This is because of a conflict between these three goals. Had they been ordered differently in the $Z_j - C_j$ columns, then there would be a different solution, perhaps with this goal being achieved at the expense of one or both of the others.

STOCHASTIC PROGRAMMING

An approach closely allied to the goal programming approach of using multiple objective functions is to ask questions such as, "What do I do if I am not sure of the data?" For there is little advantage in spending time and expense in determining an accurate solution to an inaccurate model.

Random Coefficients in the Objective Function

Consider the problem where the coefficients in the objective function are not known specifically, but are subject to some randomness. By "random" it is not implied that utter chaos reigns, merely that they are subject to some probability distribution. This means that we do have some estimate of expected values of the random coefficients. We may, furthermore, have some estimate of the variability of these random coefficients. At this stage, a decision must be made concerning the method of analysis. If we are interested in the most likely or expected value of the objective function, it may be acceptable merely to insert the expected, or average, values of the random coefficients into the objective function and proceed as if it were a deterministic case.

An alternative case might be that we are less interested in the most likely values and much more interested in the variability of the values. Such a case arises when considering the risks involved in selecting an investment portfolio, for example. Markowitz has examined this case, subsequently considered by Sharpe and many others, but to illustrate the method we briefly consider the Markowitz approach.

Markowitz's approach to portfolio selection takes into account both the return and the risk of the selection. For this analysis risk is equated with variability. Markowitz makes no distinction between variability below the mean, which is normally considered bad, and variability above the mean, which is normally considered good. All variability is considered to be risk. Let

X_i = Fraction of total resources invested in security i

R_i = Return from security i (random variable)

M_i = Expected value (average) of Ri

S_{ii}^2 = Variance of Ri

S_{ij}^2 = Covariance of Ri with Rj

The expected value M of the portfolio containing n elements equals

$$X1\ M1\ +\ X2\ M2\ +\ \ldots\ +\ Xn\ Mn$$

The variance of the portfolio S^2 equals

$$X_1^2 S_{11}^2 + X_2^2 S_{22}^2 + \ldots + X_n^2 S_{nn}^2 + 2 X1\ X2\ S_{12}^2$$
$$+ 2\ X1\ X3\ S_{13}^2 + \ldots \text{etc.}$$

The problem is thus

Maximize $M - A\ S^2$ (where the weighting factor $0 < A < \text{infinity}$)

subject to $X1 + X2 + \ldots Xn = 1$

$Xi \geqslant 0$

This is a quadratic programming problem; for example, assume

Widget Co.	$M1 = 0.04$	
Jason Inc.	$M2 = 0.05$	
Bumble Bros.	$M3 = 0.06$	
	$S12 = 0$	$S11 = 0$
	$S13 = 0$	$S22 = 0.001$
	$S23 = 0.0001$	$S33 = 0.002$

That is,

Maximize $0.04\ X1 + 0.05\ X2 + 0.06\ X3$
$$- A\ (\ 0.001\ X2^2 + 0.002\ X3^2 + 2 \times 0.0001\ X2\ X3\ )$$

subject to $X1 + X2 + X3 = 1$

$X1 \qquad\qquad > 0$

$\qquad X2 \qquad > 0$

$\qquad\qquad X3 > 0$

for a stated value of A based upon the investor's risk aversion. Thus, a LINDO file can easily be generated, as in Chapter 11, by first generating the Lagrangian function and using the QCP command of LINDO to obtain a solution.

Care must be taken to ensure positive Lagrange multipliers, as previously stated in Chapter 11; furthermore, the selection of a value for the weighting factor A needs considerable care. The solution obtained from the computer can be quite sensitive to variations in this parameter.

A further point must be stressed. Quite precise statistical information is needed to produce the coefficients used in the quadratic program, and the data required for the production of these coefficients may not be available. Much consideration continues to be given to this point by academics.

Random Elements in the Constraints

Let us next consider cases where there are random elements in the constraints. A decision must be made as to the relative importance of the constraint boundaries and the random elements. For important constraints, mean values might be appropriate. With random elements having greater importance, perhaps a simulation technique might be more appropriate.

If the distributions of the elements are discrete, it might be possible to formulate one very large linear program that could be solved using a decomposition technique. Expert assistance would definitely be required for this approach. Such formulations come into the class of multistage models. As an example of such a formulation, consider the problem

Maximize	$C1\ X1 + 5\ X2$
subject to	$X1 + X2 \leq B1$
	$X1 + A1\ X2 \leq B2$
	$X1 + A2\ X2 \leq B3$

Here A1, A2, B1, B2, B3, and C1 are all subject to random variation. For purposes of illustration, assume a small, discrete probability distribution of three states, thus:

Table 14.1

State						Probability of Occurrence Pi
A11	A12	B11	B12	B13	C11	P1
2	3	5	6	10	3	0.3
A21	A22	B21	B22	B23	C21	P2
2	4	6	7	12	4	0.5
A31	A32	B31	B32	B33	C31	P3
3	6	6	7	10	5	0.2
				Total probability		1.0

The constraint values generated by these discrete distributions are then optimized using an objective function, which also contains a nondeterministic element. This is normally handled by the use of the expected or mean value of the coefficient. This example then becomes, in numerical terms,

$$\text{Maximize } E(C1) \quad X1 + 5\,X2$$
$$\text{(i.e. } 3.9\,X1 + 5\,X2\text{)}$$

subject to	$X1 + X2 \leqslant 5$	
	$X1 + 2\,X2 \leqslant 6$	when State 1 occurs
	$X1 + 3\,X2 \leqslant 10$	
	$X1 + X2 \leqslant 6$	
	$X1 + 2\,X2 \leqslant 7$	when State 2 occurs
	$X1 + 4\,X2 \leqslant 12$	
	$X1 + X2 \leqslant 6$	
	$X1 + 3\,X2 \leqslant 7$	when State 3 occurs
	$X1 + 6\,X2 \leqslant 10$	

Note: $E(C1)$ is the extracted value, or expectation, of the variable C1.

This might be solved as a multistage problem, using the dynamic programming approach of Chapter 13. To use that approach, first select the value of a variable, say X1, and then select the value of X2 when the probability state becomes known.

Even this tiny example is quite a complex linear programming formulation. A practical case would become much more complex and would normally require the use of some special insight or "twist" to effect a practical solution.

A more common method of dealing with uncertainty is to use the chance-constrained approach of Charnes and Cooper. This method is somewhat analogous to the problem of designing a statistical significance test. Restrictions are imposed upon the probabilities that a constraint will be violated. This appears odd at first, because in a constrained optimization problem the constraints are never violated—otherwise it is not truly constrained. If the constraints may be violated, then perhaps an unconstrained problem with a penalty function approach would be more appropriate and easier to implement. However, although an unconstrained approach is mathematically more attractive, the chance-constrained method has many followers in practice. This is due to the fact that it may be easier to supply the information needed for the chance-constrained approach than the precise data required for an unconstrained optimization function approach. The deterministic problem

Maximize $4 X1 + 5 X2$

subject to $X1 + X2 \leqslant 6$

$X1 + 2 X2 \leqslant 7$

$X1 + 4 X2 \leqslant 12$

may be turned into a chance-constrained problem when, for example, the constraint on sunflower oil may not be exceeded with a certain probability, say 0.9. In other words, we expect to violate this constraint 10 percent of the time. This chance-constrained constraint may be expressed as

$$P (X1 + X2 \leqslant 6) \geqslant 0.9$$

This is equivalent to a deterministic constraint written as

$$X1 + X2 \leqslant B$$

where B is the largest number so that the probability

$$P (6 \geqslant B) \geqslant 0.9$$

This value of B is called the *fractile* of the probability distribution of the supply of sunflower oil.

For illustration, let the probability distribution of the supply of sunflower oil (Y1) be represented by the discrete distribution

$$P (Y1 = 2) = 0.1$$
$$P (Y1 = 4) = 0.2$$
$$P (Y1 = 6) = 0.5$$
$$P (Y1 = 8) = 0.1$$
$$P (Y1 = 10) = 0.1$$

This is shown graphically in Figure 14.1. In order, therefore, to be 90 percent sure of not violating the constraint, we must choose that value of Y1, Y1B, which has 90 percent of the area of the graph lying to its right. For this case, the linear constraint that allows us not to violate the sunflower oil constraint 90 percent of the time (to be wrong only 10 percent of the time) is

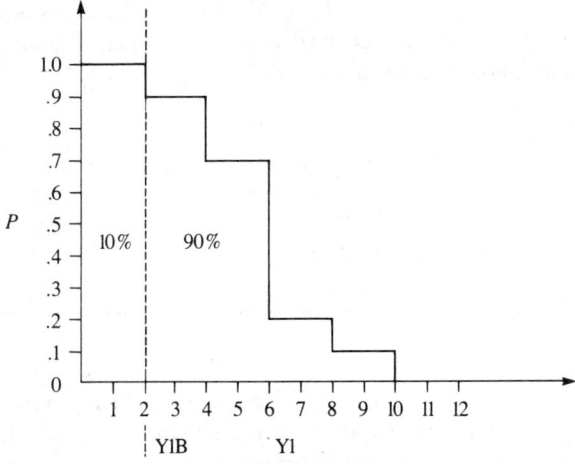

Figure 14.1

$$X1 + X2 \leqslant 2$$

Note the difference of using the mean value in this case,

$$X1 + X2 \leqslant 6$$

which would be violated 50 percent of the time.

CHAPTER 14 PROBLEMS

The solutions to Problems 1 through 5 are found at the end of the chapter.

1. Con-Pak produces two types of portable computers, the BT model, which will run IBM software, and the CT, which emulates a Cray. Con-Pak has two production lines: The BT line can make 4 computers per hour and the CT can make 3 per hour. Regular production capacity is based on a 40-hour week, and the contribution to profit from each computer is $1,000.

X. Con, managing director, wants to know how long to run each production line next week given a production goal of 360 machines. This is his first priority. Other considerations, in decreasing order of importance, are not to run the BT line for more than ten hours overtime, to avoid under-utilization of regular work on the lines, and to limit the total overtime worked on the lines. How long must each production line run next week?

2. The St. Vitus Dance Academy is planning next semester's activities. It has received a county education grant of $50,000 for dance education. Classes are scheduled in three areas:

Type of Class	Cost	Instructors	Enrollment
Classical	$7,000	5	150
Modern	3,000	9	280
Break	1,000	4	100

The principal of the academy has a pool of 40 instructors with County Footwork Skills accreditation and, in setting up the class rosters, takes into account the following, in order of importance:

a. Any money not spent on dance education must be returned to the county.

b. The county has stipulated that St. Vitus must have a minimum enrollment of 2,000 to maintain its accreditation.

c. The principal wishes to avoid having to recruit further dance instructors.

d. The county supervisor has recently issued a directive that the frequency of class offerings should be in the ratio of 10:6:2, but that this should be weighted according to enrollment demand. What is the academy's optimal class schedule?

3. Lord Elpass is considering opening his ancestral stately home to the public. He has decided that he will have two separate operations, one for the gardens and the other for the house itself. There is a staff of 6 retainers; each can either conduct tours of 20 people around the gardens on 2-hour tours costing $2 per person, or conduct tours of 10 people around the house on 1-hour tours costing $5 per person.

His Lordship wishes to open the grounds of the stately home from 10 A.M. to 6 P.M. and the house itself from 2 P.M. to 4 P.M., and by arrangement outside these hours. The local chamber of commerce has suggested that two busloads of tourists may be expected each morning, and two busloads each afternoon. Each bus has a capacity of 40 tourists.

His Lordship wishes to make as much money as possible, but wishes to minimize the number of people tramping through his home. He also wishes, with a lesser priority, to avoid asking his retainers to work overtime, and with least priority, to avoid disappointing tourists who have come to visit. What is his optimal tour schedule?

4. For the following LP formulation,

Maximize	$8 X1 + 6 X2$	
subject to	$5 X1 + 2 X2 \leqslant 20$	Constraint 1
	$6 X1 + 5 X2 \leqslant 30$	Constraint 2

Modify the LP to a goal-programming solution where the goals are, in order of preference,

a. The objective function should achieve at least 30.

b. Fully utilize the resource represented by constraint 1.

c. Avoid obtaining extra resources for constraint 2.

5. Ben Hooly manufactures mass storage devices for personal computers. It currently offers a 4-gigabyte box, an 8-gigabyte box, and a 16-gigabyte box.

Manufacturing time for the 4G box is 8 hours; the 8G box takes 12 hours; and the 16G box takes 20 hours to produce.

Standard monthly production capacity is 1,200 hours. The maximum monthly sales figures for the 4, 8, and 16G boxes are 50, 60, and 70, respectively.

Ben Hooly's best customer, IBN, has ordered 30 4G boxes and 40 8G boxes. The company's goals are to

a. Keep IBN happy.

b. Meet the sales estimates.

c. Not underutilize production capacity.

Find Ben Hooly's optimal manufacturing strategy.

HOMEWORK PROBLEMS

6. Archie Miedeez is considering his production schedule of deodorants. His base brand, Eureka, currently has a demand of 600 cases per week, and his up-market brand, Pon-Go, which he has just tested in Australia, has a demand of 400 cases per week.

The factory has a capacity of 1,300 worker-hours per week under normal schedule conditions. Each case of Eureka takes 1 worker-hour to produce with a profit contribution of $10. A case of Pon-Go consumes 2 worker-hours and produces a profit contribution of $15.

Archie's goals are, in order of importance,

a. Maintain a weekly profit of $11,000.

b. Keep his factory busy.

c. Satisfy consumer demand.

What is Archie's best production strategy?

7. Solve the following goal-programming formulation:

Minimize

$$P_1 d_1^+ + P_2 d_2^- + P_3 d_3^+ + P_4 d_1^-$$

subject to

$$X1 + 2 X2 + d_1^- - d_1^+ = 40$$
$$4 X1 + 5 X2 + d_2^- - d_2^+ = 160$$
$$4 X1 + 3 X2 + d_3^- - d_3^+ = 120$$

8. Eileen Dover (Chapter 1, Problem 7) has decided that she would like her goals for her "Cliff Hangar" launch to be, in order of preference,

a. Not to exceed her advertising budget.

b. Reach at least 4 million potential customers.

c. Have some exposure in all media.

How will this affect her campaign?

9. Milo Brandman, the president of the Lone Star Company (Chapter 1, Problem 19), has decided that he wants Lone Star to be active in all financial areas, and, after that, will be satisfied with a profit of $200,000 on Lone Star's activities. How will this affect its distribution of loans?

10. The Board of Blockwell (Chapter 2, Problem 5) has issued directives that its products will have at least a 10 percent domestic content. This can be interpreted that at least 10 percent of production is in-house.

As the primary goal of the department, the Board has issued a budget of $600,000 for production. The secondary goal is to satisfy the production demand, and the final goal is to meet domestic content requirements. How are Blockwell's production strategies affected?

CHAPTER 14 SOLUTIONS

1.

Minimize $\quad P_1d_1^- + P_2d_4^+ + P_3d_2^- + P_3d_3^- + P_4d_2^+ + P_4d_3^+$

subject to
$$4\,XBT + 3\,XCT + d_1^+ - d_1^+ = 360$$
$$XBT \qquad\qquad + d_2^- - d_2^+ = 40$$
$$XCT + d_3^- - d_3^+ = 40$$
$$d_2^+ + d_4^- - d_4^+ = 10$$

Simplex tableau:

Cj					$-P_1$	$-P_3$	$-P_3$			$-P_4$	$-P_4$	$-P_2$	
	Basis	B	XBT	XCT	d_1^-	d_2^-	d_3^-	d_4^-	d_1^+	d_2^+	d_3^+	d_4^+	
Ci													
$-P_1$	d_1^-	360	4	3	1				-1				
$-P_3$	d_2^-	40	1			1				-1			*
$-P_3$	d_3^-	40		1			1				-1		
0	d_4^-	10						1		1		-1	
Zj−Cj													
	P_4	0								1	1		
	P_3	-80	-1	-1						1	1		
	P_2	0										1	
	P_1	-360	-4	-3					1				
			*										

Cj					$-P_1$	$-P_3$	$-P_3$			$-P_4$	$-P_4$	$-P_2$	
	Basis	B	XBT	XCT	d_1^-	d_2^-	d_3^-	d_4^-	d_1^+	d_2^+	d_3^+	d_4^+	
Ci													
$-P_1$	d_1^-	200		3	1	-4			-1	4			
0	XBT	40	1			1				-1			
$-P_3$	d_3^-	40		1			1				-1		
0	d_4^-	10						1		1		-1	*
Zj−Cj													
	P_4	0								1	1		
	·P_3	-40		-1		-1					1		
	P_2	0										1	
	P_1	-200		-3		-4			1	-4			
											*		

Cj				-P₁	-P₃	-P₃			-P₄	-P₄	-P₂	
Ci / Basis	B	XBT	XCT	d_1^-	d_2^-	d_3^-	d_4^-	d_1^+	d_2^+	d_3^+	d_4^+	
$-P_1$ d_1^-	160		3	1	-4		-4	-1			4	*
0 XBT	50	1			1		-1				1	
$-P_3$ d_3^-	40		1			1			-1			
$-P_4$ d_2^+	10						1		1		-1	
Zj − Cj												
P_4	-10							-1		1	1	
P_3	-40		-1		1					1		
P_2	0										1	
P_1	-200		-3		4		4	1			-4	

*

Cj				-P₁	-P₃	-P₃			-P₄	-P₄	-P₂	
Ci / Basis	B	XBT	XCT	d_1^-	d_2^-	d_3^-	d_4^-	d_1^+	d_2^+	d_3^+	d_4^+	
$-P_2$ d_4^+	40		3/4	1/4	-1		-1	-1/2			1	
0 XBT	90	1	3/4	1/4				-1/2				
$-P_3$ d_3^-	40		1			1			-1			*
$-P_4$ d_2^+	50		3/4	1/4	-1			-1/2	1			
Zj − Cj												
P_4	-50		-3/4	-1/4	-1			1/2		1		
P_3	-40		-1		1					1		
P_2	-40		-3/4	-1/4	1		1	1/2				
P_1	0			1								

*

At this point, the primary goal has been achieved. Thus, the secondary goal can now be optimized:

Cj				$-P_1$	$-P_3$	$-P_3$			$-P_4$	$-P_4$	$-P_2$	
C_i \ Basis	B	XBT	XCT	d_1^-	d_2^-	d_3^-	d_4^-	d_1^+	d_2^+	d_3^+	d_4^+	
$-P_2$ d_4^+	10			1/4	−1	−3/4	−1	−1/2		3/4	1	*
0 XBT	60	1		1/4		−3/4		−1/2		3/4		
0 XCT	40		1			1				−1		
$-P_4$ d_2^+	20			1/4	−1	−3/4		−1/2	1	3/4		
Zj − Cj												
P_4	−10			−1/4	1	3/4		1/2		1/4		
P_3	0				1	1						
P_2	−10			−1/4	1	3/4	1	1/2		−3/4		
P_1	0			1								

**

Note that although the first and third goals have been achieved, the second goal, which ranks higher in the preference scale than the third goal, has not been satisfied.

Cj				$-P_1$	$-P_3$	$-P_3$			$-P_4$	$-P_4$	$-P_2$
C_i \ Basis	B	XBT	XCT	d_1^-	d_2^-	d_3^-	d_4^-	d_1^+	d_2^+	d_3^+	d_4^+
$-P_4$ d_3^+	13.3			3	−4/3	−1		−4/3	−2/3		1	4/3
0 XBT	50	1			1		1				−1
0 XCT	43.3		1	3	−4/3			−4/3	−2/3			4/3
$-P_4$ d_2^+	10							1	−1/4	1		−1
Zj − Cj												
P_4	−23.3			−3	−4/3	1	1/3	11/12				−1/3
P_3	0				1	1						
P_2	0											1
P_1	0			1								

Now the first three goals have been satisfied, but the fourth goal cannot be achieved without compromising the first three, higher-ranking goals.

2. The formulation for the St. Vitus Dance Academy is

Minimize

$$P_1d_1^- + P_2d_2^- + P_3d_3^+ + 1.5\,P_4d_4^- + 2.8\,P_4d_5^- + P_4d_6^-$$

subject to

$7{,}000\,X1 + 3{,}000\,X2 + 1{,}000\,X3 + d_1^-$		$= 50{,}000$	(Funds)
$150\,X1 + 280\,X2 + 100\,X3 + d_2^- - d_2^+ =$		$2{,}000$	(Enrollment)
$5\,X1 + 9\,X2 + 4\,X3 + d_3^- - d_3^+ =$		40	(Instructors)
$X1 + d_4^- - d_4^+ =$		$10\ R$	Supervisor
$X2 + d_5^- - d_5^+ =$		$6\ R$	Ratios
$X3 + d_6^- - d_6^+ =$		$2\ R$	

3. If Lord Elpass decides to have XGi garden tours and XHi house tours, then he can have two groups of garden tours in the morning and two groups of garden tours in the afternoon. He also has a choice of two groups of house tours instead of the first afternoon garden tour group.

House tour				1	2		
Garden tour	1		2		3	4	
	10	12	2		3	4	6
	am	noon			pm		

10 − 12	XG1	$+ d_1^- - d_1^+ = 6$
12 − 2	XG2	$+ d_2^- - d_2^+ = 6$
2 − 3	XG3 + XH1	$+ d_3^- - d_3^+ = 6$
3 − 4	XG3 + XH2	$+ d_4^- - d_4^+ = 6$
4 − 6	XG4	$+ d_5^- - d_5^+ = 6$
Outside hours	XG5 + XH3	$+ d_6^- - d_6^+ = 6$
Morning coaches	XG1 + XG2	$+ d_7^- - d_7^+ = 4$
Afternoon coaches	XG3 + XG4 + XH1 + XH2	$+ d_8^- - d_8^+ = 4$

Lord Elpass, however, must now be asked to be more specific in his goal formulation, for he has stated that he wishes to make as much money as possible and minimize his house tours. These are two mutually exclusive criteria. He may optimize one, given a satisfactory level on the other, or he may attempt to satisfy levels on these criteria in order, and then continue to attempt to satisfy the lower-level criteria.

4. The goal-programming formulation is

Minimize $P_1 d_1^- + P_2 d_2^- + P_3 d_3^+$

subject to $8\,X1 + 6\,X2 + d_1^- - d_1^+ = 30$

$5\,X1 + 2\,X2 + d_2^- - d_2^+ = 20$

$6\,X1 + 5\,X2 + d_3^- - d_3^+ = 30$

The solution is

$X1 = 0.714$

$X2 = 4.286$

5. Ben Hooly's formulation is to let the numbers of storage devices be

G4   Number of 4-gigabyte devices produced;

G8   Number of 8-gigabyte devices produced;

G16   Number of 16-gigabyte devices produced;

Assuming that the IBN order is included in the sales estimates, then

Minimize $P_1 d_1^- + P_1 d_2^- + P_2 d_3^- + P_2 d_4^- + P_2 d_5^- + P_3 d_6^-$

subject to

$$
\begin{aligned}
G4 \qquad\qquad\qquad\quad + d_1^- - d_1^+ &= 30 \\
G8 \qquad\qquad\quad + d_2^- - d_2^+ &= 40 \\
G4 \qquad\qquad\qquad\quad + d_3^- - d_3^+ &= 50 \\
G8 \qquad\qquad\quad + d_4^- - d_4^+ &= 60 \\
G16 + d_5^- - d_5^+ &= 30 \\
8\,G4 + 12\,G8 + 20\,G16 + d_6^- - d_6^+ &= 1200
\end{aligned}
$$

15

Interfacing LINDO Models with MPS; Conclusions

In the Introduction, we outlined the reasons for using the LINDO package to illustrate the use of computer packages for the solution of linear programming formulations. LINDO is particularly attractive, not only because it is easily accessible using remote links between minicomputers and mainframes where the program is installed, but also because its input file structure is appropriate to other packages that might be encountered by people solving linear programs.

The other class of LP applications packages is found in those large computer systems still existing that run in batch-mode number-crunching sessions. These packages were developed when much data processing input was designed for magnetic cards, and carried over to magnetic tape data transfer, which also used card images. When cards are used for data input, it is helpful to structure matrix input according to specific areas of cards, such as columns. Cards can then be prepared off line and, with batch processing, alterations made off line to single cards before a further batch is run. The input for these large number crunchers is therefore structured toward rows and columns of the linear programming matrix, rather than the mathematical expressions used by LINDO. Also, areas of the card can be set aside to sequentially number the cards, to provide for the likely event that cards are inadvertently shuffled out of order. A common industry standard for such applications programs is the MPS applications package from IBM (full details can be found in the IBM manuals).

Should a prospective linear programmer discover that it is desirable to use one of these, then the files that have been prepared for LINDO can be converted to MPS format. LINDO has two built-in MPS applicable commands:

RMPS: Retrieve an MPS format file.

SMPS: Save the current formulation in an MPS format file.

The MPS applications package offers several options, some of which are available in LINDO, some not. Again, prospective users of any applications package must familiarize themselves with the package via the manufacturer's documentation.

An MPS format file can contain up to five sections. All MPS format files contain at least three sections, entitled ROWS, COLUMNS, and RHS.

1. The ROWS section indicates the type of each row (the type of constraint). The code used is

L $\leq$
E $=$
G $\geq$
N   Not constrained (that is, objective function row)

2. The COLUMNS section lists each nonzero element of the matrix. This is preceded by the column name and the row in which it appears.

3. The RHS section details the nonzero right-hand-side elements. These are preceded by the name of the row in which the element appears. A name must also be given to the right-hand side.

To illustrate the conversion, take the edible fats manufacturer's problem (from Chapter 1) in LINDO format:

```
MAX 4 X1 + 5 X2
SUBJECT TO
2)  X1 +   X2 <= 6
3)  X1 + 2 X2 <= 7
4)  X1 + 4 X2 <= 12
END
```

The procedure to convert this formulation is the following set of commands and responses:

```
:SMPS
FILENAME:
>MPSEX [your filename]
:QUIT
```

The file MPSEX can then be saved in the host computer's workspace as a permanent file or can be telecommunicated back to your microcomputer and captured on disk using the same name or any other you like. Suppose you keep the same name for the file captured in your IBM PC. Subsequent listing of the characters stored in the file MPSEX on your machine will display

```
>A type mpsex
NAME          LINDO GENERATED MPS FILE
ROWS
    N ROW10001
    L ROW10002
    L ROW10003
    L ROW10004
COLUMNS
    X1          ROW10001          4.00000
    X1          ROW10002          1.00000
    X1          ROW10003          1.00000
    X1          ROW10004          1.00000
    X2          ROW10001          5.00000
    X2          ROW10002          1.00000
    X2          ROW10003          2.00000
    X2          ROW10004          4.00000
RHS
    RHS         ROW10002          6.00000
    RHS         ROW10003          7.00000
    RHS         ROW10004         12.00000
ENDATA
>A
```

This file can then be transmitted to a mainframe using the MPS applications package.

One further point will help sophisticated users of LINDO: The developers have provided a method of interfacing LINDO with user-written FORTRAN subroutines. A dummy subroutine named USER is included within LINDO. Since most linear programmers accessing the package will not have FORTRAN experience, this will not be discussed here. Those who can make use of this feature are referred to *Linear, Integer, and Quadratic Programming with LINDO* by Linus Schrage (Scientific Press, 1984). This document also contains other LINDO commands that have not been discussed specifically in this book.

When manipulating large files, and also when using higher transmission speeds, data transfer via communications links (direct up- and downloading) may not be reliable, because errors might be introduced in the transfer. An alternative method, illustrated by Crosstalk XVI (the package discussed in the text), is capable of error checking in file transfers. The XMODEM protocol, using the Christiansen method, is available; however, the host computer must also be equipped to handle it. Other packages are available, such as KERMIT and MITE, which are supported by many host computers. Contact with your host computer system staff is strongly advised at this stage.

CONCLUSIONS

Mathematical programming is perhaps the most available and certainly one of the most powerful decision support tools available to today's decision maker. This text has attempted to illustrate the relative ease of access to considerable mathematical computing power by the use of inexpensive personal computing hardware, pointing out both the power and the pitfalls awaiting the unwary in this field. It might be useful here to reiterate some important points and add a few new ones.

First, define your problem. This is by far the major difficulty confronting the would-be mathematical programmer. Having surmounted the initial hurdle of "getting started," the difficulty of problem recognition and definition can only be conquered by experience. Practice makes perfect; there is no substitute for hard work in this area. Attempt the problems presented in the text. Avoid the temptation to look at the solution too soon. Carefully study the examples in the text, then attempt the problems at the end of the chapter. Many texts try to be cookbooks of rules that allow students to grind through the process of generating simplex tableaux by hand. This system may help you pass certain university examinations, but despite the merits of disciplining the mind (like those ascribed to Latin declensions), the practical advantages for real decision making are limited. Since machines are available to take the drudgery out of arithmetic manipulations, the decision maker is best applying his or her talents to defining the problem and interpreting the solution.

For problems even lightly camouflaged in a situational context, some people have difficulty extracting the right formulation. The most common fault lies in confusing pieces of the primal formulation with parts of the dual formulation. Keep in mind what it is that you are doing. Which are the variables over which you have decision control? Which are your resources? Clear thinking and practice will overcome these initial difficulties. A further word of warning: Many students see solutions as "obvious" when presented with a problem and its solution, but without the solution, the same problem appears difficult. Again, avoid the temptation to look at the answer too soon.

The difficulty of model formulation cannot be overemphasized. A nonmathematician understands the problem to be solved but may be unable to express it in the correct mathematical terms. The mathematician, on the other hand, well understands the language but may have difficulty in understanding the problem. This interface has been the cause of more difficulties in practice than any other factor. However, more and more practical decision makers are gaining access to user-friendly software and powerful desktop and portable computers with convenient links to mainframes. With these tools, and a study of this text, the difficulties can be reduced. At least the reader of this text will know what questions to ask the experts when difficulties are encountered. (Of course, the quality of the "expert" advice may also be assessed.)

One not-so-obvious difficulty to avoid occurs with some time-dependent formulations. These arise when one works with a time horizon, for example in budgetary systems. Attention must be paid to carrying over functions from one planning period to the next; otherwise there is a danger of producing a poor starting position for the next planning period. Imagine if a plan to maximize income over a two-year period were to be adopted with no consideration for the following periods. The plan would have the enterprise behaving as though it were going out of business at the end of the planning period, running down stocks and inventories with no provision for the future. Such a policy will cause fluctuations and cycles that are avoidable. A rolling horizon does not, in itself, avoid the problem.

Having successfully defined the problem, odd effects may still arise. The difficulties encountered are dependent on the way in which the problem is being solved. Most commercial computer routines are adequately debugged, but some cheap (or free) microcomputer software is questionable. In manual solution methods, which the cheap microcomputer software usually emulates, several difficulties can arise:

1. *Linear simplex:* The most common fault is choosing the incorrect pivot row—either making the new variable infeasible (dividing by a negative number to obtain Q), or making one or more other variables infeasible by not choosing the smallest Q row. The other major difficulty lies in degeneracy, which occurs when one or more of the basic variables take the value zero. This implies that the optimization function is unchanged and the iterations cycle. (Actually, cycling can occur if two or more of the basic variables correspond to the minimum ratio computed for the choice of variable eliminated from the solution.) Degeneracy is a somewhat contrived state when found in examples constructed to illustrate it, but it can be encountered in reality. It can be overcome by the use of a perturbation technique—adding a small value into the equations so that the iteration increases by this small amount.

2. *Transportation:* Again the difficulty here is degeneracy. Purists might argue that employing a value of zero does not allow the solution to improve. They can therefore use the same technique but employ an infinitely small amount, commonly portrayed as an e or ϵ . The rest of us can use zero.

3. *Allocation:* The cause of difficulty here, in a manual technique, is the inability to draw the smallest number of lines through the zeros. Once that is mastered, there is nothing else to go wrong.

4. *Duality:* Mixtures of different types of constraints and variables allowed to take different signs tend to bother students. This problem dissolves if the relationships are written in the standard Tucker format. Also, check to see whether the primal system is consistent. A nonconsistent primal formulation will have an unbounded solution to the dual, and vice versa.

5. *Parametric programming:* On the objective function, there is no use seeking the values of optimization coefficients that will pivot to a particular solution point if the ratios of the objective tend to a limit that gives an optimization function gradient outside those of the constraints around the desired point. Most reputable computer routines allow parametric options to be taken on linear programming formulations; some of the less competent do not.

6. *Nonlinear techniques:* Being more complex, these techniques allow more mistakes, both in formulation and execution. The problems normally arise from nonconvexity of the constraints and nonconcavity of the optimization function. When using LINDO to solve quadratic programming formulations, it is essential that the Lagrangian is formulated so that positive Lagrange multipliers result. When solving by hand, negative Lagrange multipliers can be recognized and dealt with. (In LINDO, these must be positive.) Following the text will generally avoid problems.

Integer programming produces a great deal of extra computation over the linear case. LINDO only allows for 0,1 integer variables, thus causing increased effort in problem formulation. The question to be asked in both the integer and quadratic cases is, "Is it worth the extra effort?"

Dynamic programming is a very powerful technique. With more and more people becoming accustomed to spreadsheet manipulations, it is a convenient way of carrying out the computations, especially in the case of single-variable or small-size problems. However, as complexity increases, so do computation and storage demands—by several orders of magnitude. Careful problem definition must be made to avoid being swamped.

Stochastic techniques either require much computation or, as in chance-constrained programming, require soul-searching answers to the question "What am I doing?" In goal programming, many academics consider the deviational variables to be dubious because of the assumption of equality of deviational differences. However, most practitioners find the representation useful, and the closely allied method of multiobjective programming is available in LINDO using the methods described in the text.

Overall comments regarding a change to the nonlinear case boil down to the question "What is gained by taking into account these extra refinements over the straight linear programming formulation?" In some cases the linear assumption will be very wrong. However, in many cases, the errors in the assumptions made on the rather fuzzy data available far outweigh the extra precision provided by the refinements. In practice, you will be surprised by just how far the linearity assumption will take you, particularly if you are aware of your assumptions. In many cases, the substitution of a series of linear problems for a nonlinear problem by the use of piece-wise linearization (see Figure 15.1) can be much more efficient than a direct attack as a nonlinear problem.

Figure 15.1

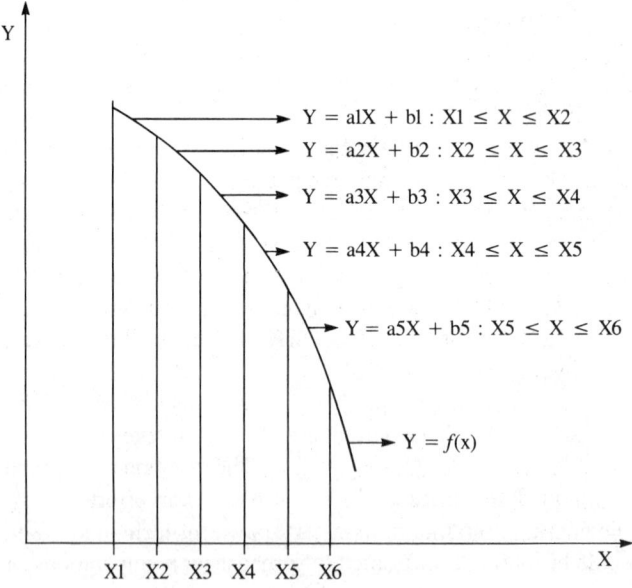

$Y = a1X + b1 : X1 \leq X \leq X2$

$Y = a2X + b2 : X2 \leq X \leq X3$

$Y = a3X + b3 : X3 \leq X \leq X4$

$Y = a4X + b4 : X4 \leq X \leq X5$

$Y = a5X + b5 : X5 \leq X \leq X6$

$Y = f(x)$

X1   X2   X3   X4   X5   X6

Because there is generally far less computation required in the linear case, this allows for more calculations at the same cost. "Prodding" the sensitivity of linear models via parametric programming by a manager who understands the decision situation and the strengths of the assumptions could, therefore, produce a better decision than the more complex analysis of a precise, but perhaps ill-defined, nonlinear model.

Alternative Approaches to Simplex Linear Programming

In recent years, alternatives to the simplex method of solving LP problems have been proposed. The two most notable are Khachian's "Russian algorithm" (described in Russian in the Russian journal *Doklady,* Jan. 1979) and the Bell Labs' proposal (developed by Karmarkar, described in IEEE's Spectrum, Dec. 1985, pp. 54–55). Khachian's method proposes a polynomial-time algorithm, taking the form of an iterative process where half-ellipsoids are sought within an ellipsoid that contains the smallest residual for the current iteration. A new ellipsoid is then circumscribed around the half-ellipsoid and a new half-ellipsoid calculated that contains the smallest residual for this next iteration. The process terminates when the residual becomes zero or after a preset number of iterations. This sounds extremely complex, and much computation of ellipsoids is involved. The process is essentially a successive approximation technique. The claimed advantage for this approach is that linear programming problems should be solvable by this approach in polynomial time rather than the exponential time scale of the simplex algorithm.

Tests have not demonstrated the claimed superiority of this approach. One experiment by George Dantzig (the original theorist in linear programming) indicated that a sample problem could be solved by the simplex algorithm in 30 minutes but required an estimated 50 million years for Khachian's method. This is because commercial simplex routines are considerably more efficient than crude exponential time estimates.

In the Bell Labs approach, the proposal is to distort the linear spaces into a single unconstrained nonlinear problem. The academic jury is still out on this claim at this time; but the examples that show off the method's strengths have clearly been structured in its favor. The method calls for inversions of matrices, and the examples have used matrices that are diagonal and easily inverted. Representative test problems are not easy to design, although discussions with colleagues have indicated a strong demand from academic researchers for such a set.

At this time, these contenders are not sufficiently convincing to remove the simplex method from its domination of the mathematical programming field. Further research might change this, but it is much more likely that these new developments will find practical application in nonlinear and dynamic programming rather than in linear programming.

In the last analysis, the powerful tools described in the text, which are available to you through your microcomputer, are wonderful parts of your decision-making armory. But, in the end, *it is your decision.*

Project

The following applications problem is suitable for use as a semester project, with a report submitted by the end of the semester. The report should be detailed, with recommendations supported by computer output where needed.

THE BAKE SHOPPE

The Bake Shoppe is a small family firm owning two bakeries. Their basic products are

1. Loaves of bread

2. Croissants

3. Cakes

Any of these products can be made at either bakery location. The locations do not need to follow the same production plan, and each has outlets to the firm's own speciality shop, to a franchised store, and to the local supermarket.

Currently the supply of loaves of bread to the franchised store is specified by a contract that must be met, and the supermarket does not handle cakes.

The resources of the operation, basically ovens and labor, are fixed in the short term. The labor force of skilled bakers and laborers all live in the area and can work at either bakery.

The owners believe that the business has expanded to a point at which it is becoming difficult to make the correct decisions both in the short term and for the long-term expansion of the business. In particular, for the short term they wish to optimize their contribution to profit on a weekly basis, and for the long-term, they would like advice on their available options, with a detailed cost analysis.

As an addendum to your report, the directors would like an explanation of your reasoning in reaching your conclusions and some discussion of the likely impact of such things as inflation and other intangibles.

Recipes

1 lb of bread:
    0.8 lb flour
    1.5 oz baking fat
           other costs, excluding labor = $0.05

1 lb of croissants
    0.75 lb flour
    1.0 oz baking fat
    1.3 oz butter
    1   egg
           other costs and materials, excluding labor = $0.14

1 lb of cakes
    0.7 lb flour
    2 oz   butter
    2   eggs
           other costs and materials, excluding labor = $0.26

Supplies

	Weekly available	Cost
Grade 1 flour	8,000 lb	0.16/lb
Grade 2 flour	10,000 lb	0.18/lb
Grade 3 flour	9,000 lb	0.20/lb
Baking fat	unlimited	0.09/oz
Butter	400 lb	0.04/oz
Eggs	unlimited	0.07 each

Sales (Weekly Demand Maxima)

	Own Store		Franchise		Supermarket	
	Demand	Price	Demand	Price	Demand	Price
	lb	$/lb	lb	$/lb	lb	$/lb
Loaves	5,000	0.65	6,000	0.65	12,000	0.80
Croissants	1,000	1.28	1,000	1.28	6,000	1.00
Cakes	500	2.40	600	2.60	?	?

In the Own Store, it has been noted that there are never any sales of croissants if bread has sold out, and that cake sales have never exceeded 5 percent of bread sales.

Croissants can be purchased for resale to the supermarket at a contribution of 5¢/lb.

Production Data

Each bakery operates a 6-day week. Bakery 1 has 30 ovens. Each oven can process 300 lb of loaves, 40 lb of croissants, or 35 lb of cake in one day (production can be pro rata split). Bakery 2 has 20 ovens. Each oven can process 250 lb of bread, 35 lb of croissants, or 28 lb of cake per day.

There are 8 bakers and 7 general workers. The dough is prepared by the bakers, each of whom can prepare either 1,000 lb of loaves, 200 lb of croissants, or 100 lb of cake per day, or pro rata amounts.

The products are processed through the ovens by the general workers, each of whom can process 2,000 lb of loaves, 500 lb of croissants, or 100 lb of cake per day, or pro rata amounts.

Bakers earn $100 per day and general workers earn $40 per day. In addition, general workers have agreed to work overtime at only 50 percent above normal rates.

Quality Control

Product	Vitamins (units/lb)	Impurities (units/lb)
Loaves	> 15	< 550
Croissants	> 11	< 460
Cakes	> 8	< 450

Flour specifications:

Grade 1	10	750
Grade 2	20	600
Grade 3	30	520

Transport Costs (cents/lb)

		Own Store	Franchise	Supermarket
Loaves				
	Bakery 1	1.0	1.0	0.9
	Bakery 2	0.8	0.9	2.8
Croissants				
	Bakery 1	4.0	4.0	3.2
	Bakery 2	2.8	3.0	2.6
Cakes				
	Bakery 1	6.0	6.0	6.5
	Bakery 2	5.5	5.0	6.1

In addition, there are transport costs for flour to Bakery 1 of 0.65¢ per pound and to Bakery 2 of 0.42¢ per pound in addition to the purchase price.

Appendix 1

LINDO Solution Output

SOLUTION TO PROBLEM 9, CHAPTER 4

```
look all
 MIN      6 AA +  9 AB +  6 AC +  4 AD +  6 AE +  9 AF +  2 AG +  3 AH +  AI
        + 2 AJ +  7 AK +  8 AL +  6 AM +  AN +  2 AO +  AP +  6 AQ +  8 AR
        + 2 AS +  3 AT +  3 BA +  8 BB +  9 BC +  9 BD +  BE +  2 BF +  6 BG
        + 4 BH +  7 BI +  8 BJ +  2 BK +  3 BL +  3 BM +  8 BN +  6 BO +  7 BP
        + 3 BQ +  6 BR +  8 BS +  3 BT +  4 CA +  5 CB +  3 CC +  7 CD +  2 CE
        + 5 CF +  6 CG +  2 CH +  8 CI +  9 CJ +  4 CK +  6 CL +  3 CM +  5 CN
        + 2 CO +  5 CP +  6 CQ +  3 CR +  2 CS +  4 CT +  2 DA +  5 DB +  8 DC
        + 2 DD +  4 DE +  2 DF +  4 DG +  3 DH +  5 DI +  7 DJ +  2 DK +  9 DL
        + 9 DM +  6 DN +  7 DO +  4 DP +  5 DQ +  3 DR +  5 DS +  2 DT +  2 EA
        + 3 EB +  5 EC +  3 ED +  4 EE +  5 EF +  6 EG +  3 EH +  4 EI +  2 EJ
        + 6 EK +  7 EL +  8 EM +  5 EN +  4 EO +  2 EP +  3 EQ +  5 ER +  2 ES
        + 7 ET +  3 FA +  5 FB +  FC +  3 FD +  5 FE +  7 FF +  3 FG +  8 FH
        + 6 FI +  9 FJ +  4 FK +  3 FL +  6 FM +  7 FN +  8 FO +  4 FP +  FQ
        + 2 FR +  FS +  3 FT +  3 GA +  5 GB +  3 GC +  2 GD +  5 GE +  4 GF
        + 5 GG +  3 GH +  8 GI +  7 GJ +  6 GK +  9 GL +  6 GM +  7 GN +  3 GO
        + 6 GP +  4 GQ +  6 GR +  2 GS +  3 GT +  7 HA +  6 HB +  8 HC +  7 HD
        + 9 HE +  6 HF +  7 HG +  7 HH +  8 HI +  6 HJ +  5 HK +  9 HL +  4 HM
        + 7 HN +  3 HO +  6 HP +  7 HQ +  9 HR +  6 HS +  3 HT +  5 IA +  6 IB
        + 3 IC +  4 ID +  5 IE +  3 IF +  7 IG +  8 IH +  9 II +  4 IJ +  5 IK
        + 3 IL +  2 IM +  6 IN +  7 IO +  5 IP +  8 IQ +  9 IR +  3 IS +  5 IT
        + 6 JA +  7 JB +  8 JC +  9 JD +  5 JE +  4 JF +  3 JG +  5 JH +  7 JI
        + 5 JJ +  8 JK +  9 JL +  4 JM +  3 JN +  5 JO +  7 JP +  8 JQ +  9 JR
        + 6 JS +  4 JT +  5 KA +  6 KB +  4 KC +  7 KD +  8 KE +  9 KF +  5 KG
```

(continued)

```
+ 4 KH + 8 KI + 6 KJ + 7 KK + 7 KL + 4 KM + 9 KN + 9 KO + 3 KP
+ 3 KQ + 3 KR + 10 KS + 5 KT + 6 LA + 7 LB + 9 LC + 12 LD
+ 3 LE + 4 LF + 8 LG + 7 LH + 9 LI + 6 LJ + 7 LK + 4 LL + 3 LM
+ 6 LN + 7 LO + 3 LP + 6 LQ + 4 LR + 2 LS + LT + 5 MA + 6 MB
+ 7 MC + 4 MD + 3 ME + 7 MF + 8 MG + 9 MH + 4 MI + 5 MJ + 3 MK
+ 5 ML + 2 MM + MN + 2 MO + 8 MP + 9 MQ + 5 MR + 4 MS + 5 MT
+ 4 NA + 5 NB + 6 NC + 3 ND + 7 NE + 3 NF + 4 NG + 7 NH + 8 NI
+ 9 NJ + 3 NK + 4 NL + 2 NM + 3 NN + 12 NO + 5 NP + 6 NQ
+ 7 NR + 8 NS + 3 NT + 6 OA + 7 OB + 4 OC + 3 OD + 5 OE + 6 OF
+ 4 OG + 7 OH + 8 OI + 5 OJ + 6 OK + 7 OL + 3 OM + 4 ON + 5 OO
+ 2 OP + 7 OQ + 8 OR + 4 OS + 5 OT + 2 PA + 3 PB + 4 PC + 2 PD
+ 5 PE + 7 PF + 8 PG + 4 PH + 3 PI + 2 PJ + 4 PK + 3 PL + 5 PM
+ 6 PN + 6 PO + 7 PP + 8 PQ + 4 PR + 5 PS + 3 PT + 5 QA + 6 QB
+ 7 QC + 4 QD + 5 QE + 3 QF + 4 QG + 7 QH + 2 QI + 8 QJ + 9 QK
+ 4 QL + 2 QM + 3 QN + QO + 4 QP + 6 QQ + 7 QR + 4 QS + 7 QT
+ 5 RA + 6 RB + 8 RC + 9 RD + 4 RE + 3 RF + 2 RG + 5 RH + 6 RI
+ 7 RJ + 8 RK + 5 RL + 6 RM + 3 RN + 2 RO + 3 RP + 2 RQ + RR
+ 4 RS + 5 RT + 4 SA + 5 SB + 6 SC + 7 SD + 4 SE + 5 SF + 6 SG
+ 3 SH + 4 SI + 8 SJ + 6 SK + 5 SL + 4 SM + 3 SN + SO + 3 SP
+ 4 SQ + 2 SR + 5 SS + 6 ST + 5 TA + 6 TB + 3 TC + 7 TD + 8 TE
+ 9 TF + 5 TG + 7 TH + 4 TI + 8 TJ + 9 TK + 3 TL + 2 TM + TN
+ 5 TO + 4 TP + 5 TQ + 2 TR + 4 TS + 6 TT + 5 UA + 3 UB + 6 UC
+ 7 UD + 8 UE + 9 UF + 4 UG + 3 UH + 2 UI + 5 UJ + 6 UK + 3 UL
+ 5 UM + 2 UN + 7 UO + 8 UP + 9 UQ + 3 UR + 2 US + UT + 5 VA
+ 6 VB + 7 VC + 4 VD + 3 VE + 2 VF + 5 VG + 6 VH + 7 VI + 5 VJ
+ 4 VK + 3 VL + 2 VM + 6 VN + 7 VO + 8 VP + 9 VQ + 4 VR + 5 VS
+ 3 VT + 4 WA + 5 WB + 7 WC + 8 WD + 5 WE + 4 WF + 6 WG + 3 WH
+ 5 WI + 4 WJ + 2 WK + WL + 4 WM + 3 WN + 2 WO + 6 WP + 7 WQ
+ 8 WR + 5 WS + 4 WT + 4 XA + 3 XB + 6 XC + 7 XD + 3 XE + 5 XF
+ 7 XG + 8 XH + 9 XI + 4 XJ + 3 XK + 2 XL + XM + 5 XN + 4 XO
+ 7 XP + 8 XQ + 6 XR + 5 XS + 3 XT
```

SUBJECT TO
```
  2) AA + AB + AC + AD + AE + AF + AG + AH + AI + AJ + AK + AL
     + AM + AN + AO + AP + AQ + AR + AS + AT = 10
  3) BA + BB + BC + BD + BE + BF + BG + BH + DI + BJ + BK + BL
     + BM + BN + BO + BP + BQ + BR + BS + BT = 5
  4) CA + CB + CC + CD + CE + CF + CG + CH + CI + CJ + CK + CL
     + CM + CN + CO + CP + CQ + CR + CS + CT = 12
  5) DA + DB + DC + DD + DE + DF + DG + DH + DI + DJ + DK + DL
     + DM + DN + DO + DP + DQ + DR + DS + DT = 8
  6) EA + EB + EC + ED + EE + EF + EG + EH + EI + EJ + EK + EL
     + EM + EN + EO + EP + EQ + ER + ES + ET = 10
  7) FA + FB + FC + FD + FE + FF + FG + FH + FI + FJ + FK + FL
     + FM + FN + FO + FP + FQ + FR + FS + FT = 30
  8) GA + GB + GC + GD + GE + GF + GG + GH + GI + GJ + GK + GL
     + GM + GN + GO + GP + GQ + GR + GS + GT = 13
```

```
 9) HA + HB + HC + HD + HE + HF + HG + HH + HI + HJ + HK + HL
    + HM + HN + HO + HP + HQ + HR + HS + HT = 25
10) IA + IB + IC + ID + IE + IF + IG + IH + II + IJ + IK + IL
    + IM + IN + IO + IP + IQ + IR + IS + IT = 14
11) JA + JB + JC + JD + JE + JF + JG + JH + JI + JJ + JK + JL
    + JM + JN + JO + JP + JQ + JR + JS + JT = 30
12) KA + KB + KC + KD + KE + KF + KG + KH + KI + KJ + KK + KL
    + KM + KN + KO + KP + KQ + KR + KS + KT = 25
13) LA + LB + LC + LD + LE + LF + LG + LH + LI + LJ + LK + LL
    + LM + LN + LO + LP + LQ + LR + LS + LT = 53
14) MA + MB + MC + MD + ME + MF + MG + MH + MI + MJ + MK + ML
    + MM + MN + MO + MP + MQ + MR + MS + MT = 9
15) NA + NB + NC + ND + NE + NF + NG + NH + NI + NJ + NK + NL
    + NM + NN + NO + NP + NQ + NR + NS + NT = 10
16) OA + OB + OC + OD + OE + OF + OG + OH + OI + OJ + OK + OL
    + OM + ON + OO + OP + OQ + OR + OS + OT = 34
17) PA + PB + PC + PD + PE + PF + PG + PH + PI + PJ + PK + PL
    + PM + PN + PO + PP + PQ + PR + PS + PT = 30
18) QA + QB + QC + QD + QE + QF + QG + QH + QI + QJ + QK + QL
    + QM + QN + QO + QP + QQ + QR + QS + QT = 10
19) RA + RB + RC + RD + RE + RF + RG + RH + RI + RJ + RK + RL
    + RM + RN + RO + RP + RQ + RR + RS + RT = 6
20) SA + SB + SC + SD + SE + SF + SG + SH + SI + SJ + SK + SL
    + SM + SN + SO + SP + SQ + SR + SS + ST = 12
21) TA + TB + TC + TD + TE + TF + TG + TH + TI + TJ + TK + TL
    + TM + TN + TO + TP + TQ + TR + TS + TT = 11
22) UA + UB + UC + UD + UE + UF + UG + UH + UI + UJ + UK + UL
    + UM + UN + UO + UP + UQ + UR + US + UT = 13
23) VA + VB + VC + VD + VE + VF + VG + VH + VI + VJ + VK + VL
    + VM + VN + VO + VP + VQ + VR + VS + VT = 20
24) WA + WB + WC + WD + WE + WF + WG + WH + WI + WJ + WK + WL
    + WM + WN + WO + WP + WQ + WR + WS + WT = 5
25) XA + XB + XC + XD + XE + XF + XG + XH + XI + XJ + XK + XL
    + XM + XN + XO + XP + XQ + XR + XS + XT = 10
26) AA + BA + CA + DA + EA + FA + GA + HA + IA + JA + KA + LA
    + MA + NA + OA + PA + QA + RA + SA + TA + UA + VA + WA
    + XA <= 12
27) AB + BB + CB + DB + EB + FB + GB + HB + IB + JB + KB + LB
    + MB + NB + OB + PB + QB + RB + SB + TB + UB + VB + WB
    + XB <= 24
28) AC + BC + CC + DC + EC + FC + GC + HC + IC + JC + KC + LC
    + MC + NC + OC + PC + QC + RC + SC + TC + UC + VC + WC
    + XC <= 20
29) AD + BD + CD + DD + ED + FD + GD + HD + ID + JD + KD + LD
    + MD + ND + OD + PD + QD + RD + SD + TD + UD + VD + WD
    + XD <= 35
```

(continued)

```
      30) AE + BE + CE + DE + EE + FE + GE + HE + IE + JE + KE
          + LE + ME + NE + OE + PE + QE + RE + SE + TE + UE + VE + WE
          + XE <= 25
  31) AF + BF + CF + DF + EF + FF + GF + HF + IF + JF + KF + LF
      + MF + NF + OF + PF + QF + RF + SF + TF + UF + VF + WF
      + XF <= 20
  32) AG + BG + CG + DG + EG + FG + GG + HG + IG + JG + KG + LG
      + MG + NG + OG + PG + QG + RG + SG + TG + UG + VG + WG
      + XG <= 24
  33) AH + BH + CH + DH + EH + FH + GH + HH + IH + JH + KH + LH
      + MH + NH + OH + PH + QH + RH + SH + TH + UH + VH + WH
      + XH <= 28
  34) AI + BI + CI + DI + EI + FI + GI + HI + II + JI + KI + LI
      + MI + NI + OI + PI + QI + RI + SI + TI + UI + VI + WI
      + XI <= 13
  35) AJ + BJ + CJ + DJ + EJ + FJ + GJ + HJ + IJ + JJ + KJ + LJ
      + MJ + NJ + OJ + PJ + QJ + RJ + SJ + TJ + UJ + VJ + WJ
      + XJ <= 50
  36) AK + BK + CK + DK + EK + FK + GK + HK + IK + JK + KK + LK
      + MK + NK + OK + PK + QK + RK + SK + TK + UK + VK + WK
      + XK <= 25
  37) AL + BL + CL + DL + EL + FL + GL + HL + IL + JL + KL + LL
      + ML + NL + OL + PL + QL + RL + SL + TL + UL + VL + WL
      + XL <= 60
  38) AM + BM + CM + DM + EM + FM + GM + HM + IM + JM + KM + LM
      + MM + NM + OM + PM + QM + RM + SM + TM + UM + VM + WM
      + XM <= 23
  39) AN + BN + CN + DN + EN + FN + GN + HN + IN + JN + KN + LN
      + MN + NN + ON + PN + QN + RN + SN + TN + UN + VN + WN
      + XN <= 30
  40) AO + BO + CO + DO + EO + FO + GO + HO + IO + JO + KO + LO
      + MO + NO + OO + PO + QO + RO + SO + TO + UO + VO + WO
      + XO <= 36
  41) AP + BP + CP + DP + EP + FP + GP + HP + IP + JP + KP + LP
      + MP + NP + OP + PP + QP + RP + SP + TP + UP + VP + WP
      + XP <= 32
  42) AQ + BQ + CQ + DQ + EQ + FQ + GQ + HQ + IQ + JQ + KQ + LQ
      + MQ + NQ + OQ + PQ + QQ + RQ + SQ + TQ + UQ + VQ + WQ
      + XQ <= 25
  43) AR + BR + CR + DR + ER + FR + GR + HR + IR + JR + KR + LR
      + MR + NR + OR + PR + QR + RR + SR + TR + UR + VR + WR
      + XR <= 15
  44) AS + BS + CS + DS + ES + FS + GS + HS + IS + JS + KS + LS
      + MS + NS + OS + PS + QS + RS + SS + TS + US + VS + WS
      + XS <= 13
  45) AT + BT + CT + DT + ET + FT + GT + HT + IT + JT + KT + LT
      + MT + NT + OT + PT + QT + RT + ST + TT + UT + VT + WT
      + XT <= 28
END
```

```
:  go
LP OPTIMUM FOUND AT STEP 64
OBJECTIVE FUNCTION VALUE
  1)  808.000000
            VARIABLE              VALUE            REDUCED COST
               AA               .000000              4.000000
               AB               .000000              7.000000
               AC               .000000              5.000000
               AD               .000000              2.000000
               AE               .000000              4.000000
               AF               .000000              7.000000
               AG               .000000              1.000000
               AH               .000000              1.000000
               AI              6.000000               .000000
               AJ               .000000               .000000
               AK               .000000              5.000000
               AL               .000000              6.000000
               AM               .000000              5.000000
               AN              4.000000               .000000
               AO               .000000              2.000000
               AP               .000000               .000000
               AQ               .000000              5.000000
               AR               .000000              7.000000
               AS               .000000              1.000000
               AT               .000000              3.000000
               BA               .000000              2.000000
               BB               .000000              7.000000
               BC               .000000              9.000000
               BD               .000000              8.000000
               BE              5.000000               .000000
               BF               .000000              1.000000
               BG               .000000              6.000000
               BH               .000000              3.000000
               BI               .000000              7.000000
               BJ               .000000              7.000000
               BK               .000000              1.000000
               BL               .000000              2.000000
               BM               .000000              3.000000
               BN               .000000              8.000000
               BO               .000000              7.000000
               BP               .000000              7.000000
               BQ               .000000              3.000000
               BR               .000000              6.000000
               BS               .000000              8.000000
               BT               .000000              4.000000
               CA               .000000              2.000000
               CB               .000000              3.000000
```

(continued)

CC	.000000	2.000000
CD	.000000	5.000000
CE	8.000000	.000000
CF	.000000	3.000000
CG	.000000	5.000000
CH	4.000000	.000000
CI	.000000	7.000000
CJ	.000000	7.000000
CK	.000000	2.000000
CL	.000000	4.000000
CM	.000000	2.000000
CN	.000000	4.000000
CO	.000000	2.000000
CP	.000000	4.000000
CQ	.000000	5.000000
CR	.000000	2.000000
CS	.000000	1.000000
CT	.000000	4.000000
DA	.000000	.000000
DB	.000000	3.000000
DC	.000000	7.000000
DD	8.000000	.000000
DE	.000000	2.000000
DF	.000000	.000000
DG	.000000	3.000000
DH	.000000	1.000000
DI	.000000	4.000000
DJ	.000000	5.000000
DK	.000000	.000000
DL	.000000	7.000000
DM	.000000	8.000000
DN	.000000	5.000000
DO	.000000	7.000000
DP	.000000	3.000000
DQ	.000000	4.000000
DR	.000000	2.000000
DS	.000000	4.000000
DT	.000000	2.000000
EA	.000000	.000000
EB	.000000	1.000000
EC	.000000	4.000000
ED	.000000	1.000000
EE	.000000	2.000000
EF	.000000	3.000000
EG	.000000	5.000000
EH	.000000	1.000000
EI	.000000	3.000000

EJ	10.000000	.000000
EK	.000000	4.000000
EL	.000000	5.000000
EM	.000000	7.000000
EN	.000000	4.000000
EO	.000000	4.000000
EP	.000000	1.000000
EQ	.000000	2.000000
ER	.000000	4.000000
ES	.000000	1.000000
ET	.000000	7.000000
FA	.000000	1.000000
FB	.000000	3.000000
FC	20.000000	.000000
FD	.000000	1.000000
FE	.000000	3.000000
FF	.000000	5.000000
FG	.000000	2.000000
FH	.000000	6.000000
FI	.000000	5.000000
FJ	.000000	7.000000
FK	.000000	2.000000
FL	.000000	1.000000
FM	.000000	5.000000
FN	.000000	6.000000
FO	.000000	8.000000
FP	.000000	3.000000
FQ	10.000000	.000000
FR	.000000	1.000000
FS	.000000	.000000
FT	.000000	3.000000
GA	.000000	1.000000
GB	.000000	3.000000
GC	.000000	2.000000
GD	13.000000	.000000
GE	.000000	3.000000
GF	.000000	2.000000
GG	.000000	4.000000
GH	.000000	1.000000
GI	.000000	7.000000
GJ	.000000	5.000000
GK	.000000	4.000000
GL	.000000	7.000000
GM	.000000	5.000000
GN	.000000	6.000000
GO	.000000	3.000000

(continued)

GP	.000000	5.000000
GQ	.000000	3.000000
GR	.000000	5.000000
GS	.000000	1.000000
GT	.000000	3.000000
HA	.000000	2.000000
HB	.000000	1.000000
HC	.000000	4.000000
HD	.000000	2.000000
HE	.000000	4.000000
HF	.000000	1.000000
HG	.000000	3.000000
HH	.000000	2.000000
HI	.000000	4.000000
HJ	.000000	1.000000
HK	.000000	.000000
HL	.000000	4.000000
HM	.000000	.000000
HN	.000000	3.000000
HO	25.000000	.000000
HP	.000000	2.000000
HQ	.000000	3.000000
HR	.000000	5.000000
HS	.000000	2.000000
HT	.000000	.000000
IA	.000000	2.000000
IB	.000000	3.000000
IC	.000000	1.000000
ID	.000000	1.000000
IE	.000000	2.000000
IF	.000000	.000000
IG	.000000	5.000000
IH	.000000	5.000000
II	.000000	7.000000
IJ	.000000	1.000000
IK	.000000	2.000000
IL	1.000000	.000000
IM	13.000000	.000000
IN	.000000	4.000000
IO	.000000	6.000000
IP	.000000	3.000000
IQ	.000000	6.000000
IR	.000000	7.000000
IS	.000000	1.000000
IT	.000000	4.000000
JA	.000000	2.000000
JB	.000000	3.000000

JC	.000000	5.000000
JD	.000000	5.000000
JE	.000000	1.000000
JF	.000000	.000000
JG	24.000000	.000000
JH	.000000	1.000000
JI	.000000	4.000000
JJ	.000000	1.000000
JK	.000000	4.000000
JL	.000000	5.000000
JM	.000000	1.000000
JN	6.000000	.000000
JO	.000000	3.000000
JP	.000000	4.000000
JQ	.000000	5.000000
JR	.000000	6.000000
JS	.000000	3.000000
JT	.000000	2.000000
KA	.000000	1.000000
KB	.000000	2.000000
KC	.000000	1.000000
KD	.000000	3.000000
KE	.000000	4.000000
KF	.000000	5.000000
KG	.000000	2.000000
KH	1.000000	.000000
KI	.000000	5.000000
KJ	.000000	2.000000
KK	.000000	3.000000
KL	.000000	3.000000
KM	.000000	1.000000
KN	.000000	6.000000
KO	.000000	7.000000
KP	.000000	.000000
KQ	15.000000	.000000
KR	9.000000	.000000
KS	.000000	7.000000
KT	.000000	3.000000
LA	.000000	3.000000
LB	.000000	4.000000
LC	.000000	7.000000
LD	.000000	9.000000
LE	12.000000	.000000
LF	.000000	1.000000
LG	.000000	6.000000
LH	.000000	4.000000

(continued)

L I	.000000	7.000000
L J	.000000	3.000000
L K	.000000	4.000000
L L	.000000	1.000000
L M	.000000	1.000000
L N	.000000	4.000000
L O	.000000	6.000000
L P	.000000	1.000000
L Q	.000000	4.000000
L R	.000000	2.000000
L S	13.000000	.000000
L T	28.000000	.000000
M A	.000000	3.000000
M B	.000000	4.000000
M C	.000000	6.000000
M D	.000000	2.000000
M E	.000000	1.000000
M F	.000000	5.000000
M G	.000000	7.000000
M H	.000000	7.000000
M I	.000000	3.000000
M J	.000000	3.000000
M K	.000000	1.000000
M L	.000000	3.000000
M M	.000000	1.000000
M N	9.000000	.000000
M O	.000000	2.000000
M P	.000000	7.000000
M Q	.000000	8.000000
M R	.000000	4.000000
M S	.000000	3.000000
M T	.000000	5.000000
N A	.000000	1.000000
N B	.000000	2.000000
N C	.000000	4.000000
N D	10.000000	.000000
N E	.000000	4.000000
N F	.000000	.000000
N G	.000000	2.000000
N H	.000000	4.000000
N I	.000000	6.000000
N J	.000000	6.000000
N K	.000000	.000000
N L	.000000	1.000000
N M	.000000	.000000
N N	.000000	1.000000
N O	.000000	11.000000

NP	.000000	3.000000
NQ	.000000	4.000000
NR	.000000	5.000000
NS	.000000	6.000000
NT	.000000	2.000000
OA	.000000	3.000000
OB	.000000	4.000000
OC	.000000	2.000000
OD	2.000000	.000000
OE	.000000	2.000000
OF	.000000	3.000000
OG	.000000	2.000000
OH	.000000	4.000000
OI	.000000	6.000000
OJ	.000000	2.000000
OK	.000000	3.000000
OL	.000000	4.000000
OM	.000000	1.000000
ON	.000000	2.000000
OO	.000000	4.000000
OP	32.000000	.000000
OQ	.000000	5.000000
OR	.000000	6.000000
OS	.000000	2.000000
OT	.000000	4.000000
PA	12.000000	.000000
PB	.000000	1.000000
PC	.000000	3.000000
PD	2.000000	.000000
PE	.000000	3.000000
PF	.000000	5.000000
PG	.000000	7.000000
PH	.000000	2.000000
PI	.000000	2.000000
PJ	16.000000	.000000
PK	.000000	2.000000
PL	.000000	1.000000
PM	.000000	4.000000
PN	.000000	5.000000
PO	.000000	6.000000
PP	.000000	6.000000
PQ	.000000	7.000000
PR	.000000	3.000000
PS	.000000	4.000000
PT	.000000	3.000000
QA	.000000	2.000000

(continued)

QB	.000000	3.000000
QC	.000000	5.000000
QD	.000000	1.000000
QE	.000000	2.000000
QF	.000000	.000000
QG	.000000	2.000000
QH	.000000	4.000000
QI	.000000	.000000
QJ	.000000	5.000000
QK	.000000	6.000000
QL	.000000	1.000000
QM	.000000	.000000
QN	.000000	1.000000
QO	10.000000	.000000
QP	.000000	2.000000
QQ	.000000	4.000000
QR	.000000	5.000000
QS	.000000	2.000000
QT	.000000	6.000000
RA	.000000	3.000000
RB	.000000	4.000000
RC	.000000	7.000000
RD	.000000	7.000000
RE	.000000	2.000000
RF	.000000	1.000000
RG	.000000	1.000000
RH	.000000	3.000000
RI	.000000	5.000000
RJ	.000000	5.000000
RK	.000000	6.000000
RL	.000000	3.000000
RM	.000000	5.000000
RN	.000000	2.000000
RO	.000000	2.000000
RP	.000000	2.000000
RQ	.000000	1.000000
RR	6.000000	.000000
RS	.000000	3.000000
RT	.000000	5.000000
SA	.000000	1.000000
SB	.000000	2.000000
SC	.000000	4.000000
SD	.000000	4.000000
SE	.000000	1.000000
SF	.000000	2.000000
SG	.000000	4.000000
SH	11.000000	.000000

SI	.000000	2.000000
SJ	.000000	5.000000
SK	.000000	3.000000
SL	.000000	2.000000
SM	.000000	2.000000
SN	.000000	1.000000
SO	1.000000	.000000
SP	.000000	1.000000
SQ	.000000	2.000000
SR	.000000	.000000
SS	.000000	3.000000
ST	.000000	5.000000
TA	.000000	3.000000
TB	.000000	4.000000
TC	.000000	2.000000
TD	.000000	5.000000
TE	.000000	6.000000
TF	.000000	7.000000
TG	.000000	4.000000
TH	.000000	5.000000
TI	.000000	3.000000
TJ	.000000	6.000000
TK	.000000	7.000000
TL	.000000	1.000000
TM	.000000	1.000000
TN	11.000000	.000000
TO	.000000	5.000000
TP	.000000	3.000000
TQ	.000000	4.000000
TR	.000000	1.000000
TS	.000000	3.000000
TT	.000000	6.000000
UA	.000000	2.000000
UB	.000000	.000000
UC	.000000	4.000000
UD	.000000	4.000000
UE	.000000	5.000000
UF	.000000	6.000000
UG	.000000	2.000000
UH	6.000000	.000000
UI	7.000000	.000000
UJ	.000000	2.000000
UK	.000000	3.000000
UL	.000000	.000000
UM	.000000	3.000000
UN	.000000	.000000

(continued)

UO	.000000	6.000000
UP	.000000	6.000000
UQ	.000000	7.000000
UR	.000000	1.000000
US	.000000	.000000
UT	.000000	.000000
VA	.000000	3.000000
VB	.000000	4.000000
VC	.000000	6.000000
VD	.000000	2.000000
VE	.000000	1.000000
VF	20.000000	.000000
VG	.000000	4.000000
VH	.000000	4.000000
VI	.000000	6.000000
VJ	.000000	3.000000
VK	.000000	2.000000
VL	.000000	1.000000
VM	.000000	1.000000
VN	.000000	5.000000
VO	.000000	7.000000
VP	.000000	7.000000
VQ	.000000	8.000000
VR	.000000	3.000000
VS	.000000	4.000000
VT	.000000	3.000000
WA	.000000	3.000000
WB	.000000	4.000000
WC	.000000	7.000000
WD	.000000	7.000000
WE	.000000	4.000000
WF	.000000	3.000000
WG	.000000	6.000000
WH	.000000	2.000000
WI	.000000	5.000000
WJ	.000000	3.000000
WK	.000000	1.000000
WL	5.000000	.000000
WM	.000000	4.000000
WN	.000000	3.000000
WO	.000000	3.000000
WP	.000000	6.000000
WQ	.000000	7.000000
WR	.000000	8.000000
WS	.000000	5.000000
WT	.000000	5.000000
XA	.000000	2.000000

XB	.000000	1.000000
XC	.000000	5.000000
XD	.000000	5.000000
XE	.000000	1.000000
XF	.000000	3.000000
XG	.000000	6.000000
XH	.000000	6.000000
XI	.000000	8.000000
XJ	.000000	2.000000
XK	.000000	1.000000
XL	.000000	.000000
XM	10.000000	.000000
XN	.000000	4.000000
XO	.000000	4.000000
XP	.000000	6.000000
XQ	.000000	7.000000
XR	.000000	5.000000
XS	.000000	4.000000
XT	.000000	3.000000

ROW	SLACK OR SURPLUS	DUAL PRICES
2)	.000000	-2.000000
3)	.000000	-1.000000
4)	.000000	-2.000000
5)	.000000	-2.000000
6)	.000000	-2.000000
7)	.000000	-2.000000
8)	.000000	-2.000000
9)	.000000	-5.000000
10)	.000000	-3.000000
11)	.000000	-4.000000
12)	.000000	-4.000000
13)	.000000	-3.000000
14)	.000000	-2.000000
15)	.000000	-3.000000
16)	.000000	-3.000000
17)	.000000	-2.000000
18)	.000000	-3.000000
19)	.000000	-2.000000
20)	.000000	-3.000000
21)	.000000	-2.000000
22)	.000000	-3.000000
23)	.000000	-2.000000
24)	.000000	-1.000000
25)	.000000	-2.000000
26)	.000000	.000000

(continued)

```
27)                 24.000000                    .000000
28)                   .000000                   1.000000
29)                   .000000                    .000000
30)                   .000000                    .000000
31)                   .000000                    .000000
32)                   .000000                   1.000000
33)                  6.000000                    .000000
34)                   .000000                   1.000000
35)                 24.000000                    .000000
36)                 25.000000                    .000000
37)                 54.000000                    .000000
38)                   .000000                   1.000000
39)                   .000000                   1.000000
40)                   .000000                   2.000000
41)                   .000000                   1.000000
42)                   .000000                   1.000000
43)                   .000000                   1.000000
44)                   .000000                   1.000000
45)                   .000000                   2.000000
NO. ITERATIONS= 64
```

SOLUTION TO PROBLEM 10, CHAPTER 4

```
: go
LP OPTIMUM FOUND AT STEP 158
OBJECTIVE FUNCTION VALUE
1) 3653.00000
        VARIABLE              VALUE            REDUCED COST
           AA               .000000            3.000000
           AB             10.000000             .000000
           AC               .000000            4.000000
           AD               .000000            9.000000
           AE               .000000            3.000000
           AF               .000000             .000000
           AG               .000000            8.000000
           AH               .000000            6.000000
           AI               .000000            9.000000
           AJ               .000000            8.000000
           AK               .000000            2.000000
           AL               .000000            1.000000
           AM               .000000            3.000000
           AN               .000000            8.000000
           AO               .000000            7.000000
           AP               .000000            8.000000
           AQ               .000000            4.000000
```

AR	.000000	2.000000
AS	.000000	8.000000
AT	.000000	6.000000
BA	.000000	5.000000
BB	.000000	.000000
BC	2.000000	.000000
BD	.000000	3.000000
BE	.000000	7.000000
BF	.000000	6.000000
BG	.000000	3.000000
BH	.000000	4.000000
BI	.000000	2.000000
BJ	.000000	1.000000
BK	.000000	6.000000
BL	.000000	5.000000
BM	.000000	5.000000
BN	3.000000	.000000
BO	.000000	2.000000
BP	.000000	1.000000
BQ	.000000	6.000000
BR	.000000	3.000000
BS	.000000	1.000000
BT	.000000	5.000000
CA	.000000	4.000000
CB	.000000	3.000000
CC	.000000	6.000000
CD	.000000	5.000000
CE	.000000	6.000000
CF	.000000	3.000000
CG	.000000	3.000000
CH	.000000	6.000000
CI	.000000	1.000000
CJ	12.000000	.000000
CK	.000000	4.000000
CL	.000000	2.000000
CM	.000000	5.000000
CN	.000000	3.000000
CO	.000000	6.000000
CP	.000000	3.000000
CQ	.000000	3.000000
CR	.000000	6.000000
CS	.000000	7.000000
CT	.000000	4.000000
DA	.000000	7.000000
DB	.000000	4.000000
DC	.000000	2.000000

(continued)

DD	.000000	11.000000
DE	.000000	5.000000
DF	.000000	7.000000
DG	.000000	6.000000
DH	.000000	6.000000
DI	.000000	5.000000
DJ	.000000	3.000000
DK	.000000	7.000000
DL	8.000000	.000000
DM	.000000	.000000
DN	.000000	3.000000
DO	.000000	2.000000
DP	.000000	5.000000
DQ	.000000	5.000000
DR	.000000	7.000000
DS	.000000	5.000000
DT	.000000	7.000000
EA	.000000	6.000000
EB	.000000	5.000000
EC	.000000	4.000000
ED	.000000	9.000000
EE	.000000	4.000000
EF	.000000	3.000000
EG	.000000	3.000000
EH	.000000	5.000000
EI	.000000	5.000000
EJ	.000000	7.000000
EK	.000000	2.000000
EL	.000000	1.000000
EM	10.000000	.000000
EN	.000000	3.000000
EO	.000000	4.000000
EP	.000000	6.000000
EQ	.000000	6.000000
ER	.000000	4.000000
ES	.000000	7.000000
ET	.000000	1.000000
FA	.000000	5.000000
FB	.000000	3.000000
FC	.000000	8.000000
FD	.000000	9.000000
FE	.000000	3.000000
FF	.000000	1.000000
FG	.000000	6.000000
FH	.000000	.000000
FI	.000000	3.000000
FJ	26.000000	.000000

FK	.000000	4.000000
FL	.000000	5.000000
FM	.000000	2.000000
FN	.000000	1.000000
FO	4.000000	.000000
FP	.000000	4.000000
FQ	.000000	8.000000
FR	.000000	7.000000
FS	.000000	8.000000
FT	.000000	5.000000
GA	.000000	6.000000
GB	.000000	4.000000
GC	.000000	7.000000
GD	.000000	11.000000
GE	.000000	4.000000
GF	.000000	5.000000
GG	.000000	5.000000
GH	.000000	6.000000
GI	.000000	2.000000
GJ	.000000	3.000000
GK	.000000	3.000000
GL	13.000000	.000000
GM	.000000	3.000000
GN	.000000	2.000000
GO	.000000	6.000000
GP	.000000	3.000000
GQ	.000000	6.000000
GR	.000000	4.000000
GS	.000000	8.000000
GT	.000000	6.000000
HA	.000000	2.000000
HB	.000000	3.000000
HC	.000000	2.000000
HD	.000000	6.000000
HE	25.000000	.000000
HF	.000000	3.000000
HG	.000000	3.000000
HH	.000000	2.000000
HI	.000000	2.000000
HJ	.000000	4.000000
HK	.000000	4.000000
HL	.000000	.000000
HM	.000000	5.000000
HN	.000000	2.000000
HO	.000000	6.000000
HP	.000000	3.000000

(continued)

HQ	.000000	3.000000
HR	.000000	1.000000
HS	.000000	4.000000
HT	.000000	6.000000
IA	.000000	3.000000
IB	.000000	2.000000
IC	.000000	6.000000
ID	.000000	8.000000
IE	.000000	3.000000
IF	.000000	5.000000
IG	.000000	2.000000
IH	.000000	.000000
II	13.000000	.000000
IJ	.000000	5.000000
IK	.000000	3.000000
IL	.000000	5.000000
IM	.000000	6.000000
IN	.000000	2.000000
IO	.000000	1.000000
IP	.000000	3.000000
IQ	.000000	1.000000
IR	1.000000	.000000
IS	.000000	6.000000
IT	.000000	3.000000
JA	.000000	3.000000
JB	.000000	2.000000
JC	.000000	2.000000
JD	.000000	4.000000
JE	.000000	4.000000
JF	.000000	5.000000
JG	.000000	7.000000
JH	.000000	4.000000
JI	.000000	3.000000
JJ	.000000	5.000000
JK	.000000	1.000000
JL	30.000000	.000000
JM	.000000	5.000000
JN	.000000	6.000000
JO	.000000	4.000000
JP	.000000	2.000000
JQ	.000000	2.000000
JR	.000000	1.000000
JS	.000000	4.000000
JT	.000000	5.000000
KA	.000000	4.000000
KB	.000000	3.000000
KC	.000000	6.000000

KD	.000000	6.000000
KE	.000000	1.000000
KF	5.000000	.000000
KG	.000000	5.000000
KH	.000000	5.000000
KI	.000000	2.000000
KJ	.000000	4.000000
KK	.000000	2.000000
KL	.000000	2.000000
KM	.000000	5.000000
KN	7.000000	.000000
KO	.000000	.000000
KP	.000000	6.000000
KQ	.000000	7.000000
KR	.000000	7.000000
KS	13.000000	.000000
KT	.000000	4.000000
LA	.000000	2.000000
LB	.000000	1.000000
LC	18.000000	.000000
LD	35.000000	.000000
LE	.000000	5.000000
LF	.000000	4.000000
LG	.000000	1.000000
LH	.000000	1.000000
LI	.000000	.000000
LJ	.000000	3.000000
LK	.000000	1.000000
LL	.000000	4.000000
LM	.000000	5.000000
LN	.000000	2.000000
LO	.000000	1.000000
LP	.000000	5.000000
LQ	.000000	3.000000
LR	.000000	5.000000
LS	.000000	7.000000
LT	.000000	7.000000
MA	.000000	4.000000
MB	.000000	3.000000
MC	.000000	3.000000
MD	.000000	9.000000
ME	.000000	6.000000
MF	.000000	2.000000
MG	.000000	2.000000
MH	9.000000	.000000
MI	.000000	6.000000

(continued)

MJ	.000000	5.000000
MK	.000000	6.000000
ML	.000000	4.000000
MM	.000000	7.000000
MN	.000000	8.000000
MO	.000000	7.000000
MP	.000000	1.000000
MQ	.000000	1.000000
MR	.000000	5.000000
MS	.000000	6.000000
MT	.000000	4.000000
NA	.000000	8.000000
NB	.000000	7.000000
NC	.000000	7.000000
ND	.000000	13.000000
NE	.000000	5.000000
NF	.000000	9.000000
NG	.000000	9.000000
NH	.000000	5.000000
NI	.000000	5.000000
NJ	.000000	4.000000
NK	.000000	9.000000
NL	.000000	8.000000
NM	.000000	10.000000
NN	.000000	9.000000
NO	10.000000	.000000
NP	.000000	7.000000
NQ	.000000	7.000000
NR	.000000	6.000000
NS	.000000	5.000000
NT	.000000	9.000000
OA	.000000	1.000000
OB	7.000000	.000000
OC	.000000	4.000000
OD	.000000	8.000000
OE	.000000	2.000000
OF	.000000	1.000000
OG	.000000	4.000000
OH	9.000000	.000000
OI	.000000	.000000
OJ	.000000	3.000000
OK	.000000	1.000000
OL	9.000000	.000000
OM	.000000	4.000000
ON	.000000	3.000000
OO	.000000	2.000000
OP	.000000	5.000000

OQ	.000000	1.000000
OR	9.000000	.000000
OS	.000000	4.000000
OT	.000000	2.000000
PA	.000000	5.000000
PB	.000000	4.000000
PC	.000000	4.000000
PD	.000000	9.000000
PE	.000000	2.000000
PF	.000000	.000000
PG	24.000000	.000000
PH	.000000	3.000000
PI	.000000	5.000000
PJ	.000000	6.000000
PK	.000000	3.000000
PL	.000000	4.000000
PM	.000000	2.000000
PN	.000000	1.000000
PO	.000000	1.000000
PP	.000000	.000000
PQ	6.000000	.000000
PR	.000000	4.000000
PS	.000000	3.000000
PT	.000000	4.000000
QA	.000000	4.000000
QB	.000000	3.000000
QC	.000000	3.000000
QD	.000000	9.000000
QE	.000000	4.000000
QF	.000000	6.000000
QG	.000000	6.000000
QH	.000000	2.000000
QI	.000000	8.000000
QJ	.000000	2.000000
QK	10.000000	.000000
QL	.000000	5.000000
QM	.000000	7.000000
QN	.000000	6.000000
QO	.000000	8.000000
QP	.000000	5.000000
QQ	.000000	4.000000
QR	.000000	3.000000
QS	.000000	6.000000
QT	.000000	2.000000
RA	.000000	3.000000
RB	.000000	2.000000

(continued)

RC	.000000	1.000000
RD	.000000	3.000000
RE	.000000	4.000000
RF	.000000	5.000000
RG	.000000	7.000000
RH	.000000	3.000000
R I	.000000	3.000000
R J	.000000	2.000000
RK	6.000000	.000000
RL	.000000	3.000000
RM	.000000	2.000000
RN	.000000	5.000000
RO	.000000	6.000000
RP	.000000	5.000000
RQ	.000000	7.000000
RR	.000000	8.000000
RS	.000000	5.000000
RT	.000000	3.000000
SA	.000000	3.000000
SB	.000000	2.000000
SC	.000000	2.000000
SD	.000000	4.000000
SE	.000000	3.000000
SF	.000000	2.000000
SG	.000000	2.000000
SH	.000000	4.000000
S I	.000000	4.000000
S J	12.000000	.000000
SK	.000000	1.000000
SL	.000000	2.000000
SM	.000000	3.000000
SN	.000000	4.000000
SO	.000000	6.000000
SP	.000000	4.000000
SQ	.000000	4.000000
SR	.000000	6.000000
SS	.000000	3.000000
ST	.000000	1.000000
TA	.000000	4.000000
TB	.000000	3.000000
TC	.000000	7.000000
TD	.000000	6.000000
TE	.000000	1.000000
TF	2.000000	.000000
TG	.000000	5.000000
TH	.000000	2.000000
T I	.000000	6.000000

TJ	.000000	2.000000
TK	9.000000	.000000
TL	.000000	6.000000
TM	.000000	7.000000
TN	.000000	8.000000
TO	.000000	4.000000
TP	.000000	5.000000
TQ	.000000	5.000000
TR	.000000	8.000000
TS	.000000	6.000000
TT	.000000	3.000000
UA	.000000	4.000000
UB	.000000	6.000000
UC	.000000	4.000000
UD	.000000	6.000000
UE	.000000	1.000000
UF	13.000000	.000000
UG	.000000	6.000000
UH	.000000	6.000000
UI	.000000	8.000000
UJ	.000000	5.000000
UK	.000000	3.000000
UL	.000000	6.000000
UM	.000000	4.000000
UN	.000000	7.000000
UO	.000000	2.000000
UP	.000000	1.000000
UQ	.000000	1.000000
UR	.000000	7.000000
US	.000000	8.000000
UT	.000000	8.000000
VA	.000000	3.000000
VB	.000000	2.000000
VC	.000000	2.000000
VD	.000000	8.000000
VE	.000000	5.000000
VF	.000000	6.000000
VG	.000000	4.000000
VH	.000000	2.000000
VI	.000000	2.000000
VJ	.000000	4.000000
VK	.000000	4.000000
VL	.000000	5.000000
VM	.000000	6.000000
VN	.000000	2.000000
VO	.000000	1.000000

(continued)

VP	1.000000	.000000
VQ	19.000000	.000000
VR	.000000	5.000000
VS	.000000	4.000000
VT	.000000	5.000000
WA	.000000	3.000000
WB	.000000	2.000000
WC	.000000	1.000000
WD	.000000	3.000000
WE	.000000	2.000000
WF	.000000	3.000000
WG	.000000	2.000000
WH	.000000	4.000000
WI	.000000	3.000000
WJ	.000000	4.000000
WK	.000000	5.000000
WL	.000000	6.000000
WM	.000000	3.000000
WN	.000000	4.000000
WO	.000000	5.000000
WP	.000000	1.000000
WQ	.000000	1.000000
WR	5.000000	.000000
WS	.000000	3.000000
WT	.000000	3.000000
XA	.000000	4.000000
XB	.000000	5.000000
XC	.000000	3.000000
XD	.000000	5.000000
XE	.000000	5.000000
XF	.000000	3.000000
XG	.000000	2.000000
XH	10.000000	.000000
XI	.000000	.000000
XJ	.000000	5.000000
XK	.000000	5.000000
XL	.000000	6.000000
XM	.000000	7.000000
XN	.000000	3.000000
XO	.000000	4.000000
XP	.000000	1.000000
XQ	.000000	1.000000
XR	.000000	3.000000
XS	.000000	4.000000
XT	.000000	5.000000

ROW	SLACK OR SURPLUS	DUAL PRICES
2)	.000000	9.000000
3)	.000000	8.000000
4)	.000000	8.000000
5)	.000000	9.000000
6)	.000000	8.000000
7)	.000000	8.000000
8)	.000000	9.000000
9)	.000000	9.000000
10)	.000000	8.000000
11)	.000000	9.000000
12)	.000000	9.000000
13)	.000000	8.000000
14)	.000000	9.000000
15)	.000000	12.000000
16)	.000000	7.000000
17)	.000000	7.000000
18)	.000000	9.000000
19)	.000000	8.000000
20)	.000000	7.000000
21)	.000000	9.000000
22)	.000000	9.000000
23)	.000000	8.000000
24)	.000000	7.000000
25)	.000000	8.000000
26)	12.000000	.000000
27)	7.000000	.000000
28)	.000000	1.000000
29)	.000000	4.000000
30)	.000000	.000000
31)	.000000	.000000
32)	.000000	1.000000
33)	.000000	.000000
34)	.000000	1.000000
35)	.000000	1.000000
36)	.000000	.000000
37)	.000000	.000000
38)	13.000000	.000000
39)	20.000000	.000000
40)	22.000000	.000000
41)	31.000000	.000000
42)	.000000	1.000000
43)	.000000	1.000000
44)	.000000	1.000000
45)	28.000000	.000000

NO. ITERATIONS= 158
DO RANGE (SENSITIVITY) ANALYSIS? > y

(continued)

RANGES IN WHICH THE BASIS IS UNCHANGED

VARIABLE	CURRENT COEF	OBJ COEFFICIENT RANGES ALLOWABLE INCREASE	ALLOWABLE DECREASE
AA	6.000000	3.000000	INFINITY
AB	9.000000	INFINITY	.000000
AC	6.000000	4.000000	INFINITY
AD	4.000000	9.000000	INFINITY
AE	6.000000	3.000000	INFINITY
AF	9.000000	.000000	INFINITY
AG	2.000000	8.000000	INFINITY
AH	3.000000	6.000000	INFINITY
AI	1.000000	9.000000	INFINITY
AJ	2.000000	8.000000	INFINITY
AK	7.000000	2.000000	INFINITY
AL	8.000000	1.000000	INFINITY
AM	6.000000	3.000000	INFINITY
AN	1.000000	8.000000	INFINITY
AO	2.000000	7.000000	INFINITY
AP	1.000000	8.000000	INFINITY
AQ	6.000000	4.000000	INFINITY
AR	8.000000	2.000000	INFINITY
AS	2.000000	8.000000	INFINITY
AT	3.000000	6.000000	INFINITY
BA	3.000000	5.000000	INFINITY
BB	8.000000	.000000	INFINITY
BC	9.000000	.000000	1.000000
BD	9.000000	3.000000	INFINITY
BE	1.000000	7.000000	INFINITY
BF	2.000000	6.000000	INFINITY
BG	6.000000	3.000000	INFINITY
BH	4.000000	4.000000	INFINITY
BI	7.000000	2.000000	INFINITY
BJ	8.000000	1.000000	INFINITY
BK	2.000000	6.000000	INFINITY
BL	3.000000	5.000000	INFINITY
BM	3.000000	5.000000	INFINITY
BN	8.000000	1.000000	.000000
BO	6.000000	2.000000	INFINITY
BP	7.000000	1.000000	INFINITY
BQ	3.000000	6.000000	INFINITY
BR	6.000000	3.000000	INFINITY
BS	8.000000	1.000000	INFINITY
BT	3.000000	5.000000	INFINITY
CA	4.000000	4.000000	INFINITY
CB	5.000000	3.000000	INFINITY
CC	3.000000	6.000000	INFINITY

CD	7.000000	5.000000	INFINITY
CE	2.000000	6.000000	INFINITY
CF	5.000000	3.000000	INFINITY
CG	6.000000	3.000000	INFINITY
CH	2.000000	6.000000	INFINITY
CI	8.000000	1.000000	INFINITY
CJ	9.000000	INFINITY	1.000000
CK	4.000000	4.000000	INFINITY
CL	6.000000	2.000000	INFINITY
CM	3.000000	5.000000	INFINITY
CN	5.000000	3.000000	INFINITY
CO	2.000000	6.000000	INFINITY
CP	5.000000	3.000000	INFINITY
CQ	6.000000	3.000000	INFINITY
CR	3.000000	6.000000	INFINITY
CS	2.000000	7.000000	INFINITY
CT	4.000000	4.000000	INFINITY
DA	2.000000	7.000000	INFINITY
DB	5.000000	4.000000	INFINITY
DC	8.000000	2.000000	INFINITY
DD	2.000000	11.000000	INFINITY
DE	4.000000	5.000000	INFINITY
DF	2.000000	7.000000	INFINITY
DG	4.000000	6.000000	INFINITY
DH	3.000000	6.000000	INFINITY
DI	5.000000	5.000000	INFINITY
DJ	7.000000	3.000000	INFINITY
DK	2.000000	7.000000	INFINITY
DL	9.000000	INFINITY	.000000
DM	9.000000	.000000	INFINITY
DN	6.000000	3.000000	INFINITY
DO	7.000000	2.000000	INFINITY
DP	4.000000	5.000000	INFINITY
DQ	5.000000	5.000000	INFINITY
DR	3.000000	7.000000	INFINITY
DS	5.000000	5.000000	INFINITY
DT	2.000000	7.000000	INFINITY
EA	2.000000	6.000000	INFINITY
EB	3.000000	5.000000	INFINITY
EC	5.000000	4.000000	INFINITY
ED	3.000000	9.000000	INFINITY
EE	4.000000	4.000000	INFINITY
EF	5.000000	3.000000	INFINITY
EG	6.000000	3.000000	INFINITY
EH	3.000000	5.000000	INFINITY
EI	4.000000	5.000000	INFINITY

(continued)

EJ	2.000000	7.000000	INFINITY
EK	6.000000	2.000000	INFINITY
EL	7.000000	1.000000	INFINITY
EM	8.000000	INFINITY	1.000000
EN	5.000000	3.000000	INFINITY
EO	4.000000	4.000000	INFINITY
EP	2.000000	6.000000	INFINITY
EQ	3.000000	6.000000	INFINITY
ER	5.000000	4.000000	INFINITY
ES	2.000000	7.000000	INFINITY
ET	7.000000	1.000000	INFINITY
FA	3.000000	5.000000	INFINITY
FB	5.000000	3.000000	INFINITY
FC	1.000000	8.000000	INFINITY
FD	3.000000	9.000000	INFINITY
FE	5.000000	3.000000	INFINITY
FF	7.000000	1.000000	INFINITY
FG	3.000000	6.000000	INFINITY
FH	8.000000	.000000	INFINITY
FI	6.000000	3.000000	INFINITY
FJ	9.000000	1.000000	1.000000
FK	4.000000	4.000000	INFINITY
FL	3.000000	5.000000	INFINITY
FM	6.000000	2.000000	INFINITY
FN	7.000000	1.000000	INFINITY
FO	8.000000	1.000000	.000000
FP	4.000000	4.000000	INFINITY
FQ	1.000000	8.000000	INFINITY
FR	2.000000	7.000000	INFINITY
FS	1.000000	8.000000	INFINITY
FT	3.000000	5.000000	INFINITY
GA	3.000000	6.000000	INFINITY
GB	5.000000	4.000000	INFINITY
GC	3.000000	7.000000	INFINITY
GD	2.000000	11.000000	INFINITY
GE	5.000000	4.000000	INFINITY
GF	4.000000	5.000000	INFINITY
GG	5.000000	5.000000	INFINITY
GH	3.000000	6.000000	INFINITY
GI	8.000000	2.000000	INFINITY
GJ	7.000000	3.000000	INFINITY
GK	6.000000	3.000000	INFINITY
GL	9.000000	INFINITY	2.000000
GM	6.000000	3.000000	INFINITY
GN	7.000000	2.000000	INFINITY
GO	3.000000	6.000000	INFINITY
GP	6.000000	3.000000	INFINITY

GQ	4.000000	6.000000	INFINITY
GR	6.000000	4.000000	INFINITY
GS	2.000000	8.000000	INFINITY
GT	3.000000	6.000000	INFINITY
HA	7.000000	2.000000	INFINITY
HB	6.000000	3.000000	INFINITY
HC	8.000000	2.000000	INFINITY
HD	7.000000	6.000000	INFINITY
HE	9.000000	INFINITY	.000000
HF	6.000000	3.000000	INFINITY
HG	7.000000	3.000000	INFINITY
HH	7.000000	2.000000	INFINITY
HI	8.000000	2.000000	INFINITY
HJ	6.000000	4.000000	INFINITY
HK	5.000000	4.000000	INFINITY
HL	9.000000	.000000	INFINITY
HM	4.000000	5.000000	INFINITY
HN	7.000000	2.000000	INFINITY
HO	3.000000	6.000000	INFINITY
HP	6.000000	3.000000	INFINITY
HQ	7.000000	3.000000	INFINITY
HR	9.000000	1.000000	INFINITY
HS	6.000000	4.000000	INFINITY
HT	3.000000	6.000000	INFINITY
IA	5.000000	3.000000	INFINITY
IB	6.000000	2.000000	INFINITY
IC	3.000000	6.000000	INFINITY
ID	4.000000	8.000000	INFINITY
IE	5.000000	3.000000	INFINITY
IF	3.000000	5.000000	INFINITY
IG	7.000000	2.000000	INFINITY
IH	8.000000	.000000	INFINITY
II	9.000000	INFINITY	.000000
IJ	4.000000	5.000000	INFINITY
IK	5.000000	3.000000	INFINITY
IL	3.000000	5.000000	INFINITY
IM	2.000000	6.000000	INFINITY
IN	6.000000	2.000000	INFINITY
IO	7.000000	1.000000	INFINITY
IP	5.000000	3.000000	INFINITY
IQ	8.000000	1.000000	INFINITY
IR	9.000000	.000000	.000000
IS	3.000000	6.000000	INFINITY
IT	5.000000	3.000000	INFINITY
JA	6.000000	3.000000	INFINITY
JB	7.000000	2.000000	INFINITY

(continued)

JC	8.000000	2.000000	INFINITY
JD	9.000000	4.000000	INFINITY
JE	5.000000	4.000000	INFINITY
JF	4.000000	5.000000	INFINITY
JG	3.000000	7.000000	INFINITY
JH	5.000000	4.000000	INFINITY
JI	7.000000	3.000000	INFINITY
JJ	5.000000	5.000000	INFINITY
JK	8.000000	1.000000	INFINITY
JL	9.000000	INFINITY	1.000000
JM	4.000000	5.000000	INFINITY
JN	3.000000	6.000000	INFINITY
JO	5.000000	4.000000	INFINITY
JP	7.000000	2.000000	INFINITY
JQ	8.000000	2.000000	INFINITY
JR	9.000000	1.000000	INFINITY
JS	6.000000	4.000000	INFINITY
JT	4.000000	5.000000	INFINITY
KA	5.000000	4.000000	INFINITY
KB	6.000000	3.000000	INFINITY
KC	4.000000	6.000000	INFINITY
KD	7.000000	6.000000	INFINITY
KE	8.000000	1.000000	INFINITY
KF	9.000000	1.000000	.000000
KG	5.000000	5.000000	INFINITY
KH	4.000000	5.000000	INFINITY
KI	8.000000	2.000000	INFINITY
KJ	6.000000	4.000000	INFINITY
KK	7.000000	2.000000	INFINITY
KL	7.000000	2.000000	INFINITY
KM	4.000000	5.000000	INFINITY
KN	9.000000	.000000	.000000
KO	9.000000	.000000	INFINITY
KP	3.000000	6.000000	INFINITY
KQ	3.000000	7.000000	INFINITY
KR	3.000000	7.000000	INFINITY
KS	10.000000	INFINITY	1.000000
KT	5.000000	4.000000	INFINITY
LA	6.000000	2.000000	INFINITY
LB	7.000000	1.000000	INFINITY
LC	9.000000	3.000000	.000000
LD	12.000000	INFINITY	3.000000
LE	3.000000	5.000000	INFINITY
LF	4.000000	4.000000	INFINITY
LG	8.000000	1.000000	INFINITY
LH	7.000000	1.000000	INFINITY
LI	9.000000	.000000	INFINITY

LJ	6.000000	3.000000	INFINITY
LK	7.000000	1.000000	INFINITY
LL	4.000000	4.000000	INFINITY
LM	3.000000	5.000000	INFINITY
LN	6.000000	2.000000	INFINITY
LO	7.000000	1.000000	INFINITY
LP	3.000000	5.000000	INFINITY
LQ	6.000000	3.000000	INFINITY
LR	4.000000	5.000000	INFINITY
LS	2.000000	7.000000	INFINITY
LT	1.000000	7.000000	INFINITY
MA	5.000000	4.000000	INFINITY
MB	6.000000	3.000000	INFINITY
MC	7.000000	3.000000	INFINITY
MD	4.000000	9.000000	INFINITY
ME	3.000000	6.000000	INFINITY
MF	7.000000	2.000000	INFINITY
MG	8.000000	2.000000	INFINITY
MH	9.000000	INFINITY	1.000000
MI	4.000000	6.000000	INFINITY
MJ	5.000000	5.000000	INFINITY
MK	3.000000	6.000000	INFINITY
ML	5.000000	4.000000	INFINITY
MM	2.000000	7.000000	INFINITY
MN	1.000000	8.000000	INFINITY
MO	2.000000	7.000000	INFINITY
MP	8.000000	1.000000	INFINITY
MQ	9.000000	1.000000	INFINITY
MR	5.000000	5.000000	INFINITY
MS	4.000000	6.000000	INFINITY
MT	5.000000	4.000000	INFINITY
NA	4.000000	8.000000	INFINITY
NB	5.000000	7.000000	INFINITY
NC	6.000000	7.000000	INFINITY
ND	3.000000	13.000000	INFINITY
NE	7.000000	5.000000	INFINITY
NF	3.000000	9.000000	INFINITY
NG	4.000000	9.000000	INFINITY
NH	7.000000	5.000000	INFINITY
NI	8.000000	5.000000	INFINITY
NJ	9.000000	4.000000	INFINITY
NK	3.000000	9.000000	INFINITY
NL	4.000000	8.000000	INFINITY
NM	2.000000	10.000000	INFINITY
NN	3.000000	9.000000	INFINITY
NO	12.000000	INFINITY	4.000000

(continued)

NP	5.000000	7.000000	INFINITY
NQ	6.000000	7.000000	INFINITY
NR	7.000000	6.000000	INFINITY
NS	8.000000	5.000000	INFINITY
NT	3.000000	9.000000	INFINITY
OA	6.000000	1.000000	INFINITY
OB	7.000000	.000000	.000000
OC	4.000000	4.000000	INFINITY
OD	3.000000	8.000000	INFINITY
OE	5.000000	2.000000	INFINITY
OF	6.000000	1.000000	INFINITY
OG	4.000000	4.000000	INFINITY
OH	7.000000	.000000	.000000
OI	8.000000	.000000	INFINITY
OJ	5.000000	3.000000	INFINITY
OK	6.000000	1.000000	INFINITY
OL	7.000000	.000000	.000000
OM	3.000000	4.000000	INFINITY
ON	4.000000	3.000000	INFINITY
OO	5.000000	2.000000	INFINITY
OP	2.000000	5.000000	INFINITY
OQ	7.000000	1.000000	INFINITY
OR	8.000000	.000000	.000000
OS	4.000000	4.000000	INFINITY
OT	5.000000	2.000000	INFINITY
PA	2.000000	5.000000	INFINITY
PB	3.000000	4.000000	INFINITY
PC	4.000000	4.000000	INFINITY
PD	2.000000	9.000000	INFINITY
PE	5.000000	2.000000	INFINITY
PF	7.000000	.000000	INFINITY
PG	8.000000	INFINITY	1.000000
PH	4.000000	3.000000	INFINITY
PI	3.000000	5.000000	INFINITY
PJ	2.000000	6.000000	INFINITY
PK	4.000000	3.000000	INFINITY
PL	3.000000	4.000000	INFINITY
PM	5.000000	2.000000	INFINITY
PN	6.000000	1.000000	INFINITY
PO	6.000000	1.000000	INFINITY
PP	7.000000	.000000	INFINITY
PQ	8.000000	1.000000	.000000
PR	4.000000	4.000000	INFINITY
PS	5.000000	3.000000	INFINITY
PT	3.000000	4.000000	INFINITY
QA	5.000000	4.000000	INFINITY
QB	6.000000	3.000000	INFINITY

QC	7.000000	3.000000	INFINITY
QD	4.000000	9.000000	INFINITY
QE	5.000000	4.000000	INFINITY
QF	3.000000	6.000000	INFINITY
QG	4.000000	6.000000	INFINITY
QH	7.000000	2.000000	INFINITY
QI	2.000000	8.000000	INFINITY
QJ	8.000000	2.000000	INFINITY
QK	9.000000	INFINITY	2.000000
QL	4.000000	5.000000	INFINITY
QM	2.000000	7.000000	INFINITY
QN	3.000000	6.000000	INFINITY
QO	1.000000	8.000000	INFINITY
QP	4.000000	5.000000	INFINITY
QQ	6.000000	4.000000	INFINITY
QR	7.000000	3.000000	INFINITY
QS	4.000000	6.000000	INFINITY
QT	7.000000	2.000000	INFINITY
RA	5.000000	3.000000	INFINITY
RB	6.000000	2.000000	INFINITY
RC	8.000000	1.000000	INFINITY
RD	9.000000	3.000000	INFINITY
RE	4.000000	4.000000	INFINITY
RF	3.000000	5.000000	INFINITY
RG	2.000000	7.000000	INFINITY
RH	5.000000	3.000000	INFINITY
RI	6.000000	3.000000	INFINITY
RJ	7.000000	2.000000	INFINITY
RK	8.000000	INFINITY	1.000000
RL	5.000000	3.000000	INFINITY
RM	6.000000	2.000000	INFINITY
RN	3.000000	5.000000	INFINITY
RO	2.000000	6.000000	INFINITY
RP	3.000000	5.000000	INFINITY
RQ	2.000000	7.000000	INFINITY
RR	1.000000	8.000000	INFINITY
RS	4.000000	5.000000	INFINITY
RT	5.000000	3.000000	INFINITY
SA	4.000000	3.000000	INFINITY
SB	5.000000	2.000000	INFINITY
SC	6.000000	2.000000	INFINITY
SD	7.000000	4.000000	INFINITY
SE	4.000000	3.000000	INFINITY
SF	5.000000	2.000000	INFINITY
SG	6.000000	2.000000	INFINITY
SH	3.000000	4.000000	INFINITY

(continued)

SI	4.000000	4.000000	INFINITY
SJ	8.000000	INFINITY	1.000000
SK	6.000000	1.000000	INFINITY
SL	5.000000	2.000000	INFINITY
SM	4.000000	3.000000	INFINITY
SN	3.000000	4.000000	INFINITY
SO	1.000000	6.000000	INFINITY
SP	3.000000	4.000000	INFINITY
SQ	4.000000	4.000000	INFINITY
SR	2.000000	6.000000	INFINITY
SS	5.000000	3.000000	INFINITY
ST	6.000000	1.000000	INFINITY
TA	5.000000	4.000000	INFINITY
TB	6.000000	3.000000	INFINITY
TC	3.000000	7.000000	INFINITY
TD	7.000000	6.000000	INFINITY
TE	8.000000	1.000000	INFINITY
TF	9.000000	.000000	1.000000
TG	5.000000	5.000000	INFINITY
TH	7.000000	2.000000	INFINITY
TI	4.000000	6.000000	INFINITY
TJ	8.000000	2.000000	INFINITY
TK	9.000000	1.000000	.000000
TL	3.000000	6.000000	INFINITY
TM	2.000000	7.000000	INFINITY
TN	1.000000	8.000000	INFINITY
TO	5.000000	4.000000	INFINITY
TP	4.000000	5.000000	INFINITY
TQ	5.000000	5.000000	INFINITY
TR	2.000000	8.000000	INFINITY
TS	4.000000	6.000000	INFINITY
TT	6.000000	3.000000	INFINITY
UA	5.000000	4.000000	INFINITY
UB	3.000000	6.000000	INFINITY
UC	6.000000	4.000000	INFINITY
UD	7.000000	6.000000	INFINITY
UE	8.000000	1.000000	INFINITY
UF	9.000000	INFINITY	1.000000
UG	4.000000	6.000000	INFINITY
UH	3.000000	6.000000	INFINITY
UI	2.000000	8.000000	INFINITY
UJ	5.000000	5.000000	INFINITY
UK	6.000000	3.000000	INFINITY
UL	3.000000	6.000000	INFINITY
UM	5.000000	4.000000	INFINITY
UN	2.000000	7.000000	INFINITY
UO	7.000000	2.000000	INFINITY

UP	8.000000	1.000000	INFINITY
UQ	9.000000	1.000000	INFINITY
UR	3.000000	7.000000	INFINITY
US	2.000000	8.000000	INFINITY
UT	1.000000	8.000000	INFINITY
VA	5.000000	3.000000	INFINITY
VB	6.000000	2.000000	INFINITY
VC	7.000000	2.000000	INFINITY
VD	4.000000	8.000000	INFINITY
VE	3.000000	5.000000	INFINITY
VF	2.000000	6.000000	INFINITY
VG	5.000000	4.000000	INFINITY
VH	6.000000	2.000000	INFINITY
VI	7.000000	2.000000	INFINITY
VJ	5.000000	4.000000	INFINITY
VK	4.000000	4.000000	INFINITY
VL	3.000000	5.000000	INFINITY
VM	2.000000	6.000000	INFINITY
VN	6.000000	2.000000	INFINITY
VO	7.000000	1.000000	INFINITY
VP	8.000000	1.000000	.000000
VQ	9.000000	.000000	1.000000
VR	4.000000	5.000000	INFINITY
VS	5.000000	4.000000	INFINITY
VT	3.000000	5.000000	INFINITY
WA	4.000000	3.000000	INFINITY
WB	5.000000	2.000000	INFINITY
WC	7.000000	1.000000	INFINITY
WD	8.000000	3.000000	INFINITY
WE	5.000000	2.000000	INFINITY
WF	4.000000	3.000000	INFINITY
WG	6.000000	2.000000	INFINITY
WH	3.000000	4.000000	INFINITY
WI	5.000000	3.000000	INFINITY
WJ	4.000000	4.000000	INFINITY
WK	2.000000	5.000000	INFINITY
WL	1.000000	6.000000	INFINITY
WM	4.000000	3.000000	INFINITY
WN	3.000000	4.000000	INFINITY
WO	2.000000	5.000000	INFINITY
WP	6.000000	1.000000	INFINITY
WQ	7.000000	1.000000	INFINITY
WR	8.000000	INFINITY	1.000000
WS	5.000000	3.000000	INFINITY
WT	4.000000	3.000000	INFINITY
XA	4.000000	4.000000	INFINITY

(continued)

XB	3.000000	5.000000	INFINITY
XC	6.000000	3.000000	INFINITY
XD	7.000000	5.000000	INFINITY
XE	3.000000	5.000000	INFINITY
XF	5.000000	3.000000	INFINITY
XG	7.000000	2.000000	INFINITY
XH	8.000000	INFINITY	.000000
XI	9.000000	.000000	INFINITY
XJ	4.000000	5.000000	INFINITY
XK	3.000000	5.000000	INFINITY
XL	2.000000	6.000000	INFINITY
XM	1.000000	7.000000	INFINITY
XN	5.000000	3.000000	INFINITY
XO	4.000000	4.000000	INFINITY
XP	7.000000	1.000000	INFINITY
XQ	8.000000	1.000000	INFINITY
XR	6.000000	3.000000	INFINITY
XS	5.000000	4.000000	INFINITY
XT	3.000000	5.000000	INFINITY

RIGHTHAND SIDE RANGES

ROW	CURRENT RHS	ALLOWABLE INCREASE	ALLOWABLE DECREASE
2	10.000000	7.000000	10.000000
3	5.000000	20.000000	3.000000
4	12.000000	22.000000	4.000000
5	8.000000	7.000000	7.000000
6	10.000000	13.000000	10.000000
7	30.000000	22.000000	4.000000
8	13.000000	7.000000	7.000000
9	25.000000	.000000	25.000000
10	14.000000	7.000000	1.000000
11	30.000000	7.000000	7.000000
12	25.000000	20.000000	7.000000
13	53.000000	2.000000	3.000000
14	9.000000	7.000000	7.000000
15	10.000000	22.000000	10.000000
16	34.000000	7.000000	7.000000
17	30.000000	19.000000	1.000000
18	10.000000	5.000000	2.000000
19	6.000000	5.000000	2.000000
20	12.000000	22.000000	4.000000
21	11.000000	5.000000	2.000000
22	13.000000	5.000000	7.000000
23	20.000000	31.000000	1.000000
24	5.000000	7.000000	5.000000
25	10.000000	7.000000	7.000000

26	12.000000	INFINITY	12.000000
27	24.000000	INFINITY	7.000000
28	20.000000	3.000000	2.000000
29	35.000000	3.000000	2.000000
30	25.000000	INFINITY	.000000
31	20.000000	7.000000	5.000000
32	24.000000	1.000000	19.000000
33	28.000000	7.000000	7.000000
34	13.000000	1.000000	7.000000
35	50.000000	4.000000	22.000000
36	25.000000	2.000000	5.000000
37	60.000000	7.000000	7.000000
38	23.000000	INFINITY	13.000000
39	30.000000	INFINITY	20.000000
40	36.000000	INFINITY	22.000000
41	32.000000	INFINITY	31.000000
42	25.000000	1.000000	19.000000
43	15.000000	7.000000	7.000000
44	13.000000	7.000000	13.000000
45	28.000000	INFINITY	28.000000

Appendix 2

Standard Normal Curve Areas

Z	0	1	2	3	4	5	6	7	8	9
0.0	.0000	.0040	.0080	.0120	.0160	.0199	.0239	.0279	.0319	.0359
0.1	.0398	.0438	.0478	.0517	.0557	.0596	.0636	.0675	.0714	.0754
0.2	.0793	.0832	.0871	.0910	.0948	.0987	.1206	.1064	.1103	.1141
0.3	.1179	.1217	.1255	.1293	.1331	.1368	.1406	.1443	.1480	.1517
0.4	.1554	.1591	.1628	.1664	.1700	.1736	.1772	.1808	.1844	.1879
0.5	.1915	.1950	.1985	.2019	.2054	.2088	.2123	.2157	.2190	.2224
0.6	.2258	.2291	.2324	.2357	.2389	.2422	.2454	.2486	.2518	.2549
0.7	.2580	.2612	.2642	.2673	.2704	.2734	.2764	.2794	.2823	.2852
0.8	.2881	.2910	.2939	.2967	.2996	.3023	.3051	.3078	.3106	.3133
0.9	.3159	.3186	.3212	.3238	.3264	.3289	.3315	.3340	.3365	.3389
1.0	.3413	.3438	.3461	.3485	.3508	.3531	.3554	.3577	.3599	.3621
1.1	.3643	.3665	.3686	.3708	.3729	.3749	.3770	.3790	.3810	.3830
1.2	.3849	.3869	.3888	.3907	.3925	.3944	.3962	.3980	.3997	.4015
1.3	.4032	.4049	.4066	.4082	.4099	.4115	.4131	.4147	.4162	.4177
1.4	.4192	.4207	.4222	.4236	.4251	.4265	.4279	.4292	.4306	.4319

1.5	.4332	.4345	.4357	.4370	.4382	.4394	.4406	.4418	.4429	.4441
1.6	.4452	.4463	.4474	.4484	.4495	.4505	.4515	.4525	.4535	.4545
1.7	.4554	.4564	.4573	.4582	.4591	.4599	.4608	.4616	.4625	.4633
1.8	.4641	.4649	.4656	.4664	.4671	.4678	.4686	.4693	.4699	.4706
1.9	.4713	.4719	.4726	.4732	.4738	.4744	.4750	.4756	.4761	.4767
2.0	.4772	.4778	.4783	.4788	.4793	.4798	.4803	.4808	.4812	.4817
2.1	.4821	.4826	.4830	.4834	.4838	.4842	.4846	.4850	.4854	.4857
2.2	.4861	.4864	.4868	.4871	.4875	.4878	.4881	.4884	.4887	.4890
2.3	.4893	.4896	.4898	.4901	.4904	.4906	.4909	.4911	.4913	.4916
2.4	.4918	.4920	.4922	.4925	.4927	.4929	.4931	.4932	.4934	.4936
2.5	.4938	.4940	.4941	.4943	.4945	.4946	.4948	.4949	.4951	.4952
2.6	.4953	.4955	.4956	.4957	.4959	.4960	.4961	.4962	.4963	.4964
2.7	.4965	.4966	.4967	.4968	.4969	.4970	.4971	.4972	.4973	.4974
2.8	.4974	.4975	.4976	.4977	.4977	.4978	.4979	.4979	.4980	.4981
2.9	.4981	.4982	.4982	.4983	.4984	.4984	.4985	.4985	.4986	.4986
3.0	.4987	.4987	.4987	.4988	.4988	.4989	.4989	.4989	.4990	.4990
3.1	.4990	.4991	.4991	.4991	.4992	.4992	.4992	.4992	.4993	.4993
3.2	.4993	.4993	.4994	.4994	.4994	.4994	.4994	.4995	.4995	.4995
3.3	.4995	.4995	.4995	.4996	.4996	.4996	.4996	.4996	.4996	.4997
3.4	.4997	.4997	.4997	.4997	.4997	.4997	.4997	.4997	.4997	.4998
3.5	.4998	.4998	.4998	.4998	.4998	.4998	.4998	.4998	.4998	.4998
3.6	.4998	.4998	.4999	.4999	.4999	.4999	.4999	.4999	.4999	.4999
3.7	.4999	.4999	.4999	.4999	.4999	.4999	.4999	.4999	.4999	.4999
3.8	.4999	.4999	.4999	.4999	.4999	.4999	.4999	.4999	.4999	.4999
3.9	.5000	.5000	.5000	.5000	.5000	.5000	.5000	.5000	.5000	.5000

Appendix 3

Mainframe Packages for Mathematical Programming

Many programs other than LINDO may be used to solve the problems discussed in this text. Some are available as microcomputer packages, but any microcomputer can be used to access any mainframe package available to dial-in service. The performance of these packages is superior to the microcomputer routines because of the mainframes' greater memory and speed. An example of those available to my students in Southern California is given by accessing the information files held on the California State University Central Cyber in Los Angeles. This results in the following information:

```
                              MINOS
        ENTER NAME OF FILE TO BE LISTED OR
          D TO SEE ANOTHER DIRECTORY OR
          T TO TERMINATE INFO.


     ? minos


     DO YOU WANT THIS FILE LISTED AT THE TERMINAL (YES OR NO) ? y


     MINOS      08/24/84


     MINOS (MODULAR IN-CORE NONLINEAR OPTIMIZING SYSTEM) IS A PROGRAM FOR
     SOLVING LARGE SCALE LINEAR AND NONLINEAR PROGRAMMING PROBLEMS
     INVOLVING SPARSE LINEAR CONSTRAINTS. IT WAS WRITTEN AT THE SYSTEMS
     OPTIMIZATION LABORATORY AT STANFORD UNIVERSITY. MINOS IS A LARGE
     PROGRAM, REQUIRING 140500 OCTAL WORDS OF MEMORY TO EXECUTE.


     BOTH VERSION 4 AND VERSION 5 OF MINOS HAVE BEEN INSTALLED ON THE
     CENTRAL CYBER 730/760.


     TO FIND OUT HOW TO ACCESS VERSION 4 ENTER:


             FIND,MINOS,HELP.


     THE DOCUMENTATION FOR MINOS 4 IS CALLED:


                      MINOS USER'S GUIDE
                             BY
              BRUCE A MURTAGH AND MICHAEL SAUNDERS


                    TECHNICAL REPORT 77-9
                      FEBRUARY 1977


     TO FIND OUT HOW TO ACCESS VERSION 5 ENTER:


             FIND,MINOS5,HELP.


     THE DOCUMENTATION FOR MINOS 5 IS CALLED:


                    MINOS 5.0 USER'S GUIDE
                             BY
              BRUCE A MURTAGH AND MICHAEL SAUNDERS


                    TECHNICAL REPORT 83-20
                      DECEMBER 1983
```

THIS CAN BE PURCHASED FROM:

        OFFICE OF TECHNOLOGY LICENSING
        105 ENCINA HALL
        STANFORD UNIVERSITY
        STANFORD CA 94305

AT A COST OF $15-00.

                        MPOS

ENTER NAME OF FILE TO BE LISTED OR
    D TO SEE ANOTHER DIRECTORY OR
    T TO TERMINATE INFO.
? MPOS

DO YOU WANT THIS FILE LISTED AT THE TERMINAL (YES OR NO) ? Y

MPOS   12/14/82
        MPOS (MULTI PURPOSE OPTIMIZATION SYSTEM) IS AN INTEGRATED
SYSTEM OF COMPUTER PROGRAMS TO SOLVE OPTIMIZATION PROBLEMS. THE
SYSTEM WAS DEVELOPED TO PERMIT THE USER TO STATE HIS MP PROBLEM IN
ENGLISH AND ALGEBRAIC NOTATION AND ACCESS ONE OF MANY WELL KNOWN
LINEAR PROGRAMMING, INTEGER PROGRAMMING OR QUADRATIC PROGRAMMING
ALGORITHMS.
        IN GENERAL, MPOS IS EXECUTED WITH A CONTROL STATEMENT OF EITHER
OF THE FOLLOWING FORMS:
            FIND(MPOS)
        OR
            FIND,MPOS,I=FN1,L=FN2,S=FN3,
            D=FN4,A=FN5,NR,HELP.

THE FIRST FORM SELECTS THE DEFAULT FILE NAMES FOR ALL FILES USED BY
MPOS. THE SECOND FORM MAY BE USED TO CHANGE ONE OR MORE FILE NAMES
FROM THE DEFAULTS NORMALLY SELECTED.
        THE PARAMETERS IN THE SECOND FORM ARE ANY COMBINATION OF THE
        FOLLOWING ITEMS IN ANY ORDER;

PARM	DEFAULT	DESCRIPTION
I=FN1	INPUT	FILE CONTAINING MODEL DESCRIPTIONS.
L=FN2	OUTPUT	FILE CONTAINING ALL RESULTING OUTPUT OF EXECUTION.
S=FN3	SAVFIL	FILE MPOS CREATES CONTAINING TRANSLATED INPUTS. ALSO THIS IS ACCESSED WHEN 'GETFILE' COMMAND IS ISSUED.

(continued)

D=FN4   DATFIL   THE FILE MPOS WILL READ DATA FROM ALTERNATE INPUT
                 UNIT.

A=FN5   APXFIL   RESULTING APEX DATA FILE WHEN 'APEX1' OR 'APEX2' IS
                 ISSUED AS AN ALGORITHM.

NR   REWIND   DO NOT REWIND FILES I,L,D,A,F AND S. IF OMITTED, ALL
                 OF THESE FILES WILL BE POINTED TO THE TOP.

HELP   NO   IF SPECIFIED, SHORT HELP MESSAGE WILL BE LISTED TO
                 HELP USER OF ACCESS METHOD.

     FOR EXAMPLE, TO USE THE FILE 'MPOSIN' WHICH CONTAINS THE MPOS
INPUT AND HAVE THE OUTPUT TO A FILE NAMED 'MPOSPR', THE FOLLOWING
CONTROL STATEMENT WOULD BE USED TO RUN MPOS:

          FIND,MPOS,I=MPOSIN,L=MPOSPR.

THERE IS AN ONLINE EXAMPLE 'MPOSEX2' ACCESSIBLE FROM THE ACCOUNT
NUMBER '7ETPDOC'.

     SINCE MPOS IS A BATCH ORIENTED SYSTEM, IT IS ADVISABLE TO USE
BOTH FILE INPUT AND OUTPUT. INPUT FILES MUST NOT CONTAIN LINE
NUMBER.

IF THIS FILE IS NOT LOCAL, AN ATTEMPT WILL BE MADE TO "GET" OR
"ATTACH".

IN TIMESHARING ENVIRONMENT, THESE FILES WILL BE REWOUND BEFORE
EXECUTION UNLESS THE "NR" PARAMETER IS SPECIFIED.

MPOS RUNS ON CENTRAL MEMORY (CM) OF 65000B AND THUS DOES NOT NEED
ANY SPECIAL VALIDATION TO USE. BUT VERY LARGE MODEL/DATA CAN BE
EXECUTED BY INCREASING CENTRAL MEMORY VALIDATION.
     THE MPOS VERSION 4.0 USER'S GUIDE MAY BE PURCHASED FROM:

          NORTHWESTERN UNIVERSITY
          VOGELBACK COMPUTING CENTER
          2129 SHERIDAN ROAD
          EVANSTON, ILLINOIS 60201

THE COST IS $4.25 PER MANUAL, PLUS SHIPPING. REFER TO MANUAL NO. 320
WHEN ORDERING.

Index